4th International Conference on Design, Mechanical and Material Engineering (D2ME 2019)

IOP Conference Series: Materials Science and Engineering
Volume 686

Busan, South Korea
26 - 28 September 2019

ISBN: 978-1-7138-0938-8
ISSN: 1757-8981

Printed from e-media with permission by:

Curran Associates, Inc.
57 Morehouse Lane
Red Hook, NY 12571

CURRAN ASSOCIATES INC.
proceedings.com

Some format issues inherent in the e-media version may also appear in this print version.

Additional copies of this publication are available from:

Curran Associates, Inc.
57 Morehouse Lane
Red Hook, NY 12571 USA
Phone: 845-758-0400
Fax: 845-758-2633
Email: curran@proceedings.com
Web: www.proceedings.com

TABLE OF CONTENTS

DESIGN ENGINEERING

MECHANICAL AND MATERIAL ENGINEERING

IOP Conf. Series: Materials Science and Engineering **686** (2019) 011001 doi:10.1088/1757-899X/686/1/011001

PREFACE

4th International Conference on Design, Mechanical and Material Engineering (D2ME 2019) was held in Busan, South Korea from Sept.26 to 28, 2019. D2ME 2019 was organized by Hong Kong Society of Mechanical Engineers (HKSME), supported by Royal Melbourne Institute of Technology University (RMIT University), Swinburne University of Technology and The Australian National University.. The conference provides a useful and wide platform both for display the latest research and for exchange of research results and thoughts in Design, Mechanical, Material Engineering, Surface Engineering Applications and other topics. The participants of the conference were from almost every part of the world, with background of either academia or industry, even well-known enterprise. The success and prosperity of the conference is reflected high level of the papers received.

The proceedings are a compilation of the accepted papers and represent an interesting outcome of the conference. This book covers 2 chapters: Design Engineering and Mechanical and Material Engineering.

We would like to acknowledge all of those who supported D2ME 2019. Each individual and institutional help were very important for the success of this conference. Especially we would like to thank the organizing committee for their valuable advices in the organization and helpful peer review of the papers.

We sincerely hope that D2ME 2019 will be a forum for excellent discussions that will put forward new ideas and promote collaborative researches. We are sure that the proceedings will serve as an important research source of references and the knowledge, which will lead to not only scientific and engineering progress but also other new products and processes.

Prof. Raj Das, RMIT University, Australia

IOP Conf. Series: Materials Science and Engineering **686** (2019) 011002 doi:10.1088/1757-899X/686/1/011002

COMMITTEES

General Chair

Dr. Raj Das, RMIT University, Australia

Conference Co-Chairs

Prof. Zhan Chen, AUT University, Auckland, New Zealand

Prof. LAU Kin Tak Alan, Swinburne University of Technology, Australia

Prof. Qinghua Qin, The Australian National University, Australia

Program Committee Chair

Prof. Gong Hao, National University of Singapore, Singapore

Academic Committee

Prof. Amaresh Chakrabarti, Indian Institute of Science (IISc), Bangalore, India

Prof. Bi Zhang, Southern University of Science and Technology, China

Advisory Chair

Prof. Guy Littlefair, Deakin University, Australia

International Technical Committee

Prof. Wong Shaw Voon, Universiti Putra Malaysia, Malaysia

Prof. Li Chang, The University of Sydney, Australia

Prof. Alan C. Lin, National Taiwan University of Science and Technology

Assoc. Prof. Chun, Hyunjin, Nanjing University of Aeronautics and Astronautics, China

Dr. Abhishek Kundu, Cardiff University, UK

Dr. Khalid Arif, Massey University, New Zealand

Dr. Pierre Kerfriden, Cardiff University, UK

IOP Conf. Series: Materials Science and Engineering **686** (2019) 011002 doi:10.1088/1757-899X/686/1/011002

IOP Conf. Series: Materials Science and Engineering **686** (2019) 011003 doi:10.1088/1757-899X/686/1/011003

Peer review statement

All papers published in this volume of *IOP Conference Series: Materials Science and Engineering* have been peer reviewed through processes administered by the proceedings Editors. Reviews were conducted by expert referees to the professional and scientific standards expected of a proceedings journal published by IOP Publishing.

IOP Conf. Series: Materials Science and Engineering **686** (2019) 012001 doi:10.1088/1757-899X/686/1/012001

Numerical Prediction of Sand Erosion in Elbows

Nawaf H. Saeid* and Roslynna Rosli

Mechanical Engineering Programme, Universiti Teknologi Brunei, Jalan Tungku Link, Gadong BE 1410, Brunei Darussalam

* E-mail: nawaf.saeid@utb.edu.bn

Abstract. In the present article, turbulent flow of crude oil and sand particles in piping system with a 90° elbow is analysed using computational fluid dynamics simulations. The flow of the crude oil is modelled as continuous fluid which is govern by the continuity and Reynold-averaged Navier–Stokes equations. The Discrete phase model is used for sand particle flow. The governing parameters considered in the present study are: the inlet velocity, spherical sand particle size and sand mass flow rate. The maximum erosion rate is found on the outer surface of the elbow where the sand and fluid impinging the surface before changing the flow direction. It is found that the erosion rate is increasing with the increase of the inlet velocity of the crude oil. The numerical results show that the large sand particles with constant sand flow rate causing less erosion rate than those of smaller diameter. The results show also that the increase of the sand mass flow rate leads to the increase of the erosion rate.

1. Introduction

In the oil and gas industry sand is a source of several flow assurance problems. If sand erosion is not monitored, predicted and controlled properly, entire production processes can be impeded and even shut down for long periods of time. Sand erosion is a mechanical process where it is occurring when a sand particle repeatedly impinges on the inner surface of the pipe system gradually removed its material. In order to protect the pipeline and equipment from potential hazards or to design the pipeline and equipment, it is necessary to understand the wearing of piping system caused by sand impingements. The estimation of sand erosion in a multiphase flow is a complicated process, depends on many parameters, such as: sand properties and particle tracks, fluid flow, surface conditioning and multi-phase effects among others.

Calculating impact speed is one of the challenging parts of erosion prediction. Low particle velocity has low particle kinetic energy. Hence, low impact to the pipe wall create minor erosion. Higher particle velocity has higher kinetic energy resulting in higher material erosion rate of the pipes. Okonkwo et al. [1] studied various combinations of impact angles and particle speeds and demonstrated that the predominant erosion mechanism happening under lower impact angles and higher speeds.

The literature review shows a vast use of computational fluid dynamics (CFD) techniques in predicting sand erosion in the oil and gas applications. CFD modeling can predict the detailed information on the location and the rate of erosion. The results of the CFD modeling have the potential to optimize the design prior to fabrication and testing. The literature shows many erosion models using either single phase or multi-phase modeling. The simple and easy to employ sand erosion model is using the Discrete Phase Model (DPM), which applicable for sand volume fractions less than 10 % [2].

Zhang et al. [3] used CFD methodology to study the erosion wear of fracturing pipeline under the action of multiphase flow in oil & gas industry and found that the significant factors influencing the erosion rate is the flow velocity of fluid and the impact angle of the solid particles. Karimi et al. [4] utilized CFD to calculate erosion caused by fine particles in different geometries. The geometries include submerged jet impingement and elbows. Simulation results are compared with experimental data available in the literature for elbows and conducted during this investigation for fine particles submerged in an impinging slurry jet. Droubi et al. [5] analysed numerically the sand erosion in piping system with 90° bend. The Erosion rate was found to decrease with the increase of either the pipe diameter or/and the bend radii. The capability of CFD for multiphase flow simulation using Volume of Fluid (VOF) method has been tested by Zahedi et al. [6] for air-water flow with emphasis on erosion prediction. They concluded that a larger experimental database is necessary to study the behaviour and relation between superficial gas and liquid velocities with phase velocities in a bend and their effects on particle impact speeds and angles. Huey and Saeid [7] considered numerical simulations to study the performance analysis and sand erosion in a centrifugal pump. The CFD simulation results show that the erosion rate increases with increasing rotational speed and particle size. Erosion mostly occurred at the leading edge of the blade near the shroud of the impeller at different particle diameters and impeller rotational speed. Recently, Saeid [8] considered the discrete phase model to numerically predict the sand erosion in a check valve used in oil production system.

The literature review show that most of the CFD-based erosion models focus on the pipe bend with gas–solid flow or water–sand flows. Therefore, the aims of the current study are to simulate the crude oil flow in a pipe elbow. The study will include identifying the governing parameters effecting the sand erosion in the piping system in order to minimize the damage in the system. The piping system used in the present investigations is shown in Figure 1. The diameter of the piping system is D = 100 mm and the bend radius R = 1.5D. The inlet pipe length is extended to be 30D in order to eliminate the effects of the boundary layers entrance region and ensure the fully developed flow at the elbow. Similarly, the outlet pipe is extended to be 15D to eliminate the effects of the reversed flow to the solution domain.

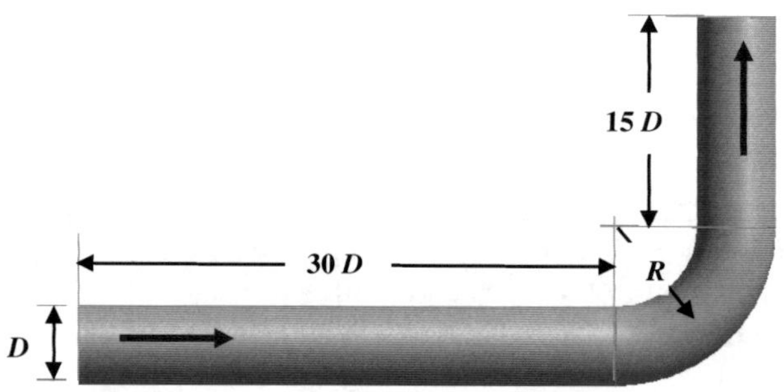

Figure 1. Flow geometry (Not to scale).

2. Mathematical Model

The crude oil is treated as continuous phase and the governing equations are based on the mass and momentum balance. The sand particles are treated as the dispersed phase, which can be modelled using the Discrete Phase Model (DPM). The flow of the continuous phase (crude oil) is selected in the turbulent regime. The continuity and the Reynolds-averaged Navier-Stokes (RANS) equations are:

$$\frac{\partial}{\partial x_i}(\rho u_i) = 0 \tag{1}$$

$$\frac{\partial}{\partial x_j}(\rho u_i u_j) = -\frac{\partial p}{\partial x_i} + \frac{\partial}{\partial x_j}\left\{\mu\left(\frac{\partial u_i}{\partial x_j} + \frac{\partial u_j}{\partial x_i} - \frac{2}{3}\delta_{ij}\frac{\partial u_l}{\partial x_l}\right)\right\} + \frac{\partial}{\partial x_j}\left(-\rho\overline{\acute{u}_i\acute{u}_j}\right) + S \tag{2}$$

where u_i and $\acute{u}_i$ are the average and fluctuating components of the velocity vector and the overbar represents the average value of the fluid velocity. S is the momentum source term. The Boussinesq approximation is used to relate the Reynolds stresses to the average velocity gradients as:

$$-\rho\overline{\acute{u}_i\acute{u}_j} = \mu_t\left(\frac{\partial u_i}{\partial x_j} + \frac{\partial u_j}{\partial x_i}\right) - \frac{2}{3}\left(\rho k + \mu_t\frac{\partial u_k}{\partial x_k}\right)\delta_{ij} \tag{3}$$

where the eddy viscosity (μ_t) calculation depends on the turbulence model and k is the turbulent kinetic energy of the flow. In the present study, two turbulence models are used. They are the realizable $k-\varepsilon$ with scalable wall functions and the SST $k-\omega$ with near surface wall corrections model [2].

The discrete phase model is employed, in which the particle trajectories are calculated through the solution domain. The calculation of trajectories is done by solving the force balance for a single particle using Newton's second law as follows [2]:

$$\frac{du_{pi}}{dt} = F_D(u_i - u_{pi}) + \frac{g_i(\rho_p-\rho)}{\rho_p} + F_i \tag{4}$$

where u_{pi} is the particle velocity vector in the i direction and ρ_p is the density of the particle. The buoyancy force term is neglected in the present study as the sand particles and crude oil have insignificant difference in densities. F_i is an additional acceleration of the fluid surrounding the particle, and F_D is the drag force/ particle mass and F_i is the force needed to accelerate/decelerate the fluid around the particle. These terms are defined by [2]:

$$F_D = \frac{18\mu}{\rho_p d_p^2}\frac{C_D Re_d}{24} \quad \text{and} \quad F_i = \frac{1}{2}\frac{\rho}{\rho_p}\frac{d(u_i-u_{pi})}{dt} \tag{5}$$

where μ is the dynamic viscosity of the continuous phase (fluid), dp is the diameter of the particle, and Re_d is the Reynolds number, which is given by: $Re_d = \rho d_p|u_{pi} - u_i|/\mu$. The drag coefficient C_D in equation (5) for smooth spherical particles applicable for wide ranges of Re given by [2]:

$$C_D = a_1 + \frac{a_2}{Re_d} + \frac{a_3}{Re_d^2} \tag{6}$$

The default values of the model constants (a_1, a_2 and a_3) are used in the present investigations [2]. The rate of erosion is defined as the velocity of the mass removal of material from a surface per unit area (kg/m^{2}s). The aim of the current study is to monitor the particle erosion rate at wall boundaries, which is given in the following correlation [2]:

$$R_{erosion} = \sum_{p=1}^{N_p}\frac{\dot{m}_p C(d_p) f(\alpha) v^{b(v)}}{A_{face}} \tag{7}$$

where $\dot{m}_p$ = mass flow rate of a particle (kg/s), $C(d_p)$ = particle diameter function, $f(\alpha)$ = impact angle function, v = relative solid particle velocity, $b(v)$ = function of particle velocity and A_{face} = area of the mesh face at the surface. The default values of $C(d_p) = 1.8\times10^{-9}$, $f(\alpha) = 1$, and $b(v) = 0$ are used in the present study.

3. Numerical Solution Method

The geometry was created in ANSYS 18.1 [2]. The geometry is then meshed by dividing it into smaller tetrahedral cells. Five inflation layers are generated near walls and the cells are clustered near the elbow area as shown in Figure 2. The domain is divided into 398421 tetrahedral elements after performing the suitable mesh sensitivity analyses. The mesh is then transferred to the solver, which is FLUENT software [2]. Medium density crude oil is used as the working fluid and it is considered as incompressible Newtonian fluid with density ρ = 850 kg/m^3 and dynamic viscosity μ = 0.009 kg/m/s [9]. The sand material is considered as inert anthracite spherical particles with uniform diameter and density of 1550 kg/m^3 injected from the inlet face with different mass flow rates. The no-slip boundary condition is selected at all the solid surfaces of the system. The boundary condition at the inlet is selected to be constant velocity inlet in the range 1m/s to 100 m/s and the pressure at the outlet is set to atmospheric (zero gage). The boundary conditions for particles are set as reflected by the walls and escape through the inlet and outlet boundaries. The inlet sand particles velocity is set to zero for all cases. The numerical solution of the governing equations is based on the second order upwind scheme. SIMPLE algorithm [2] is employed to solve the discretized equations. This is an iterative solution and needs under relaxation to control the changes in the values during iteration. The values of the under relaxation factors of 0.3, 0.7 and 0.8 for pressure, momentum and turbulence model equations respectively are used. The under relaxation factors of 0.5 is used for the interphase exchange of momentum during the calculation. The maximum residual in continuity equation and all variables were lower than 10^{-3} at the end of the iterations. The numerical solution of the Discrete Phase Model is computed by integrating equation (4) using the implicit Euler method.

4. Results and Discussion

The velocity profile of crude oil flow in the inlet pipe (at 0.2 m from the elbow) with inlet velocity of 30 m/s is used for validation. The power-law velocity profile express as [10]:

$$\frac{u}{u_{max}} = \left(1 - \frac{r}{R}\right)^{1/n} \tag{8}$$

where r is the radial distance from the centreline of the pipe and R is the total radius of the pipe. The results presented in Figure 3 show that prediction using the realizable $k - \varepsilon$ with scalable wall functions is almost similar to that obtained using SST $k - \omega$ model. However, the solution is converged faster when the realizable $k - \varepsilon$ with scalable wall functions is used. Therefore the realizable $k - \varepsilon$ with scalable wall functions it is used for results generation in the present study. Different mesh sizes are used for grid independent study and the accuracy of the results is verified by checking the velocity profiles and mass balance of the flow using the realizable $k - \varepsilon$ with scalable wall functions. The total number of cells (control volumes) in the fine mesh is 581 921 cell. The moderated mesh contains 398 421 cell and it is 252 384 cell for the coarse mesh for the same geometry. The results presented in Figure 3 for the velocity profile in the same location show negligible difference using moderated and fine mesh sizes with same trend as the power law profile. Therefore, the moderated size of the mesh is used to generate the results in the present study. The global mass balance is monitored and satisfied in the converged solution within ± 0.01% in all cases.

Figure 2. Mesh geometry (moderate mesh)

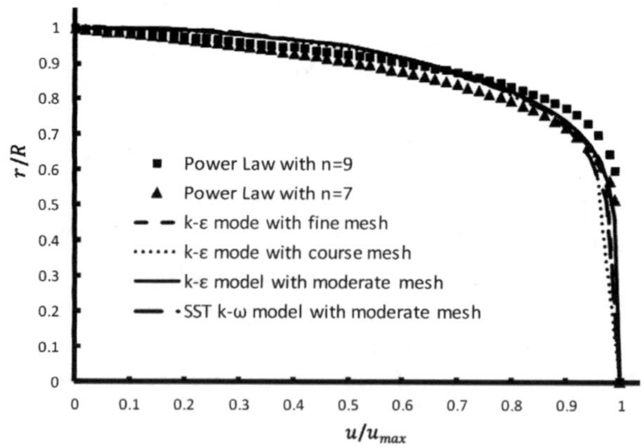

Figure 3. Velocity profiles with inlet velocity 30 m/s.

To understand the flow structure and the particles-fluid interaction and its effect on the surface erosion, the governing equations are solved and the erosion rate is calculated for different conditions. Figure 4 shows the crude oil and particle flow details near the elbow at the med-section where the flow direction changed suddenly with inlet velocity of 30 m/s. The spherical sand particles are considered with a diameter of 1 μm and sand flow rate of 10^{-20} kg/s. The fluid pressure at the elbow is maximum at the outer surface of the elbow where the fluid imping the elbow and on the opposite side the pressure is minimum due to the fluid acceleration as shown in Figure 4(a). The discrete phase model results show the particle tracking with a velocity almost similar to the fluid velocity as shown in Figures 4(a) and 4(b). The maximum erosion rate happening on the outer surface of the elbow where the sand and fluid impinging the wall as shown in Figure 4 (c). After impinging the outer surface of the elbow, the sand particles are directed to the inner surface of the outlet pipe and create another region of high erosion rate as shown in Figures 4(b) and 4(c).

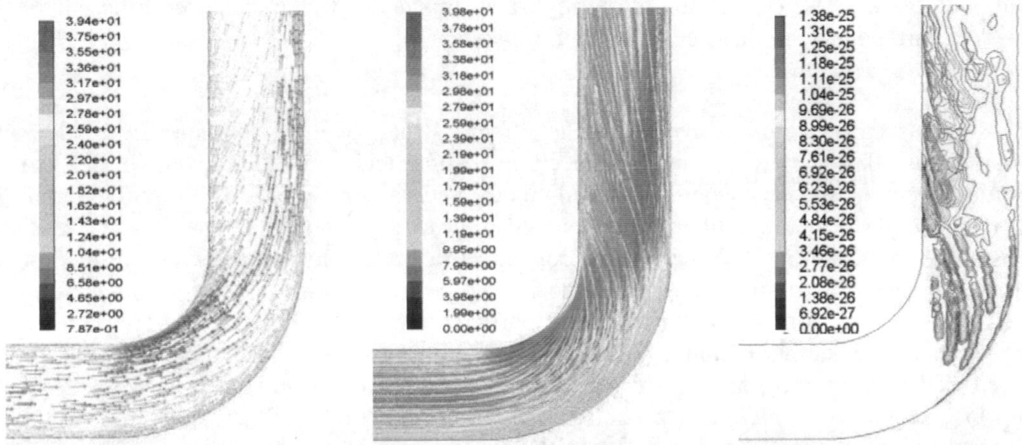

(a) Velocity vectors (m/s) (b) Particle tracking velocity (m/s) (c) Erosion rate (kg/m^{2}s)

Figure 4. Fluid and particle flow details near the elbow with inlet velocity of 30 m/s.

The simulation results show that the inlet velocity of the crude oil, the size of the sand particles and its mass flow rate are important parameters on the rate of sand erosion in the system. In practice, the sand production rate during the crude oil exaction is a complex process with different sand shape and size especially in the new oil wells. To simplify the problem, it is assumed that the sand particles are of spherical shape with a diameter in the range of 1 μm to 50 μm and flow rate in the range of 1×10^{-20} kg/s to 3×10^{-20} kg/s with crude oil inlet velocity in the range 1 m/s to 100 m/s. Figure 5(a) show that the erosion rate are increasing with the increase of the inlet fluid velocity while keeping the inlet sand particles velocity at zero. The sand particles are carried by the fluid and therefore will have more kinetic energy with high inlet velocity of the fluid which leads to increase the erosion rate. The location of the maximum erosion rate is always same as that presented in Figure 4(c).

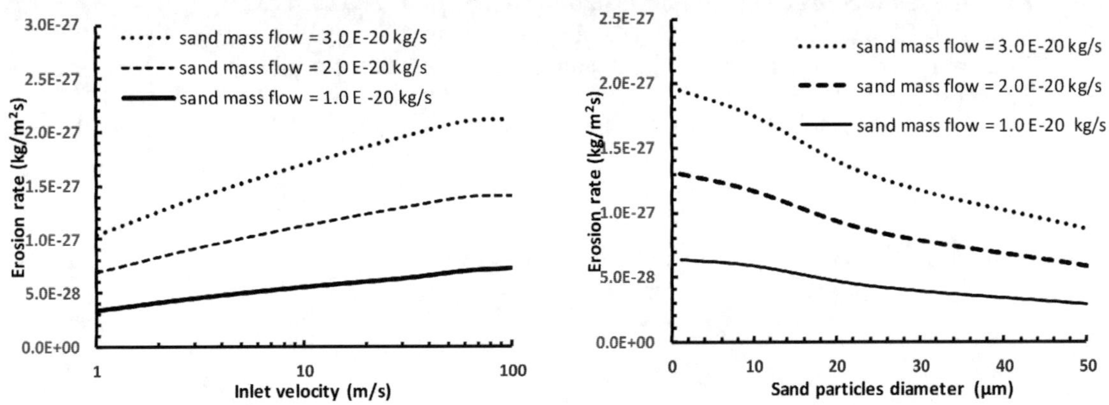

(a) Effect of inlet velocity (sand diameter of 1μm) (b) effect of sand diameter (inlet velocity 30 m/s)

Figure 5. Variation of the area-weighted average erosion rate with inlet velocity and sand diameter.

It is observed that the large sand particles with constant sand flow rate causing less erosion rate than those of smaller diameter as shown in Figure 5 (b). This is due to less number of large size particles in the same sand flow rate compare to the smaller particles. The fluid-sand flow interaction leads to different particle impact angles and velocities at different particle diameter. The particles with small diameter can be carried out by the fluid and impact on the elbow surfaces. However the flow of the big size particles need more fluid kinetic energy and will have less impact force on the elbow surfaces and therefore less erosion rate. The results presented in Figure 5 also show the increase of the erosion rate with the increase of the sand mass flow rate. This is due to the increase of the number of impacts of sand particles on the surface of the piping system.

5. Conclusions

Parametric study is carried out to investigate the effect of the governing parameters on the sand erosion in a crude oil flow through a 90° pipe elbow. Numerical simulations were performed for different crude oil inlet velocity, spherical sand diameter and sand flow rate. The simulations in the present study were performed by solving the fluid dynamics field followed by particle tracking and finally erosion rate calculation. The maximum erosion rate is calculated on the outer surface of the elbow where the sand and fluid coming from the inlet pipe impinging the surface facing the flow before changing the direction. After that the sand particles are directed to the inner surface of the outlet pipe which create another region of high erosion rate. It is found that the maximum erosion and the area-weighted average erosion rate are increasing with the increase of the inlet velocity of the crude oil while keeping the inlet sand particles velocity at zero due to fluid-sand interaction. The numerical results show the large sand particles with constant sand flow rate causing less erosion rate than those of smaller diameter. This is because the flow of the big size particles need high fluid kinetic energy and therefore have less impact and less erosion rate compared with small size particles. Due to the increase of the number of impacts on the surface of the piping system the more sand flow rate generates more erosion in the system. Finally the current findings suggest to lower production velocity and avoid sudden change in the flow direction if possible to reduce the sand erosion.

References

[1] Okonkwo PC, Sliem MH, Sk MH, Shakoor RA, Mohamed AMA, Abdullah AM and Kahraman R 2018 *Coatings* **8** 343-1-14
[2] ANSYS Inc. ANSYS FLUENT User's Guide, Canonsburg, PA 15317. 2011
[3] Zhang J, Kang J, Fan J and Gao J 2016 *Journal of Natural Gas Science and Engineering* **32** 334
[4] Karimi S, Shirazi SA and McLaury BS 2017 *Wear* **376-377** 1130–1137
[5] Droubi MG, Tebowei R, Islam SZ, Hossain M and Mitchell E 2016 Ninth International Conference on Computational Fluid Dynamics, Istanbul, Turkey, July 11-15, 2016.
[6] Zahedi P, Zhang J, Arabnejad H, McLaury B, Shirazi S 2017 *Wear* **376-377** 1260
[7] Huey YS and Saeid NH, 7th Brunei International Conference on Engineering and Technology 2018, Bandar Seri Begawan; Brunei Darussalam; 12 - 14 November 2018; Code 147647.
[8] Saeid NH 2018 *Journal of Mechanical Engineering and Sciences* **12** 3988
[9] McKnight D and Friedman D 2016 Washington, DC: The National Academic Press
[10] Saeid, N 2013 *Noise Control Engineering Journal* **61** 355

IOP Conf. Series: Materials Science and Engineering **686** (2019) 012002 doi:10.1088/1757-899X/686/1/012002

Optimal Plastic Analysis of Structures under Uncertain Conditions

Majid Movahedi Rad

Széchenyi István University, Department of Structural and Geotechnical Engineering
Egyetem square 1, Győr, Hungary

E-mail: majidmr@sze.hu

Abstract. In this study optimum plastic shakedown analysis of framed steel structures with semi-rigid connections (SRC) between beam and columns subjected to multi-parameter static loading presented. Since shakedown analysis does not provide information concerned with the accumulated residual displacements, complementary strain energy of the residual force (CSERF) considered as constrain to evaluate the post yield behavior of framed steel structures. The constraints on the CSERF are modelled using probabilistic and deterministic methods. To evaluate the maximum shakedown load-multipliers a numerical example is introduced, shakedown load-multipliers are calculated and safe loading domains of the framed steel structure are illustrated. The numerical results show that the constraints on CSERF, SRC and probabilistic given constraints can influence on the value of the shakedown load-multipliers.

1. Introduction

Generally two methods have been used to calculate the uncertainties in structural engineering. The first approach is the deterministic method, where a safety factor is applied. The second method is probabilistic analysis and design, in which the design information are known within a certain bound and probability distributions are given. The deterministic method despite being favorably, the accurate safety is not clear for assumed safety factor. In the probabilistic method, the uncertainties are given randomly, where the most frequent value of the random variable are assumed to the highest number in the probability density function. In mechanical engineering the uncertainties have an extensive performance and requires precise computations. Reliability based design has a long history with important books and papers. Several international courses and meetings provide a proper tool about the practical and theoretical problems in the field of mechanical engineering like stochastic load modelling and stochastic computational mechanics, reliability-based optimization response surface methodology and reliability-based design (Melchers and Beck [1]; Wang and Katafygiotis [2]; Lógó [3]; Marti [4] and Marti and Stöckl [5]) proposed reliability programming for the solution method of different stochastic problems.

In structural optimization plastic analysis and design methods have a significant role since they provide information in respect to the post yield behaviour of structures which lead to have an economical design. However, as a consequence of this interest, large accumulated residual deformations might occur, which lead to the collapse of the structures. To avoid this undesirable situation in the literature several methods proposed which provide appropriate tools to control the residual deformations of the structures (see e.g. Atkociunas et al. [6]; Liepa et al. [7]; Simon and Weichert [8]; Levy et al. [9]; Weichert and Maier [10]; Tin-Loi [11]; König [12]; Polizzotto [13]; Kaneko and Maier [14]; Corradi [15]; Ponter [16] and Maier [17]). Along with others Kaliszky and Lógó [18-19] proposed that the post yield behavior of the structures can be limited by introducing a bound on CSERF. In this study, a parametric investigation is introduced to evaluate the influence of SRC and constraints on the shakedown load-multiplier. The SRC of beam-column are modelled using appropriate internal springs at the connections. Furthermore, the constraints on CSERF are modelled using probabilistic and deterministic methods.

2. Bounded plastic deformation

2.1. Deterministic problem

Consider an elastic-plastic solid body subjected to external force P_0. Consequently, plastic force Q^p occurs inside the body. When the external force is reduced elastic deformations appear and consequently elastic force $-Q^e$ occurs inside the body. Accordingly, after full unloading the residual force Q^r remains inside the body.

$$Q^r = Q^p - Q^e \tag{1}$$

where Q^e is:

$$Q^e = T^{-1}GK^{-1}P_0 \tag{2}$$

here K: stiffness matrix; G: geometrical matrix and T: flexibility matrix. Using positive definite function, CSERF can be calculated as follows:

$$C_r = \frac{1}{2}\sum_{i=1}^{n}\frac{1}{S_i}\int_0^{l_i}\left(Q_i^p(s) - Q_i^e(s)\right)^2 ds \geq 0 \tag{3}$$

applying equation (1):

$$C_r = \frac{1}{2}\sum_{i=1}^{n}\frac{1}{S_i}\int_0^{l_i}\left(Q_i^r(s)\right)^2 ds \geq 0 \tag{4}$$

here S_i defines flexural and tensile stiffnesses of beam and trusses elements. The residual deformations constrained by considering a limit value on the amount of CSERF:

$$\frac{1}{2}\sum_{i=1}^{n}\frac{1}{S_i}\int_0^{l_i}\left(Q_i^r(s)\right)^2 ds - C_{r0} \leq 0 \tag{5}$$

where C_{r0} is a proper bound for CSERF, C_r.

In the case of truss elements the CSERF can be introduced as follows:

$$C_r = \frac{1}{2E}\sum_{i=1}^{n}\frac{l_i}{A_i}(N_i^r)^2 \tag{6}$$

where N_i^r: residual normal force of the truss elements; A_i: cross sectional area of the truss elements; l_i: the length of the elements; E: the elastic modulus. $(i = 1, 2, \ldots, n)$: number of elements.

For beam members the CSERF can be defined as follows:

$$C_r = \frac{1}{2E}\sum_{i=1}^{n}\frac{1}{I_i}\int_0^{l_i}(M_i^r(s))^2 ds \tag{7}$$

here M_i^r: residual moment of the beam elements; I_i:the moment of inertia for beam elements. Supposing residual moments M_{i1}^r and M_{i2}^r acting at the two ends of a beam element the integral part in equation (7) can be defined as:

$$\int_0^{l_i}(M_i^r(s))^2 = \frac{1}{3}[(M_1^r)^2 + (M_1^r)(M_2^r) + (M_2^r)^2]. \tag{8}$$

Then the plastic deformations are constrained by introducing an appropriate limit value for C_{r0}.

$$\frac{1}{6E}\sum_{i=1}^{n}\frac{l_i}{I_i}\left[(M_1^r)^2 + (M_1^r)(M_2^r) + (M_2^r)^2\right] - C_{r0} \leq 0 \tag{9}$$

The limit functions for truss elements can be introduced by equation (10):

$$C_r = \frac{1}{2E}\sum_{i=1}^{n}\frac{l_i}{i}(N_i^r)^2 - C_{r0} \leq 0 \tag{10}$$

2.2. Probabilistic problem

Introducing the basic concepts of the static theorem, the structural failure can be expressed by $S_R \leq S_S$. Here S_R expresses the bound for statically admissible forces S_S and $f_R(S_R)$ and $f_S(S_S)$ are corresponding distribution function.

The failure function can be defined by the following equation:

$$P_f = P[S_R \leq S_S] = \iint_{S_R \leq S_S} f_R(S_R)\, f_S(S_S) dS_R dS_S. \tag{11}$$

The above problem alternatively can be expressed in term of the limit state function (LSF):

$$G(S_R, S_S) = S_R - S_S \leq 0. \tag{12}$$

Assuming that $G \leq 0$ shows the failure event, the probability that a failure occurs can be defined by

$$P_f = F_G(0) \tag{13}$$

here $F_G(0)$ is the cumulative distribution function (CDF) of the LSF. Assuming both $S_R$and S_S have the normal distribution $N(\mu_{S_R}, \sigma_{X_R}^2)$ and $N(\mu_{S_S}, \sigma_{X_S}^2)$ respectively. In this case, G also has normal distribution with the mean value equal to $\mu_G = \mu_{S_R} - \mu_{S_S}$ and standard deviation $\sigma_G^2 = \sigma_{S_R}^2 + \sigma_{S_S}^2$, hence,

$$P_f = \Phi\left(-\frac{\mu_G}{\sigma_G}\right) = \Phi(-\beta) \tag{14}$$

here $\beta = \mu_g/\sigma_g$ is safety index and $\Phi(.)$: the normal cumulative density function.

The failure probability of the structural system is computed by the following integral:

$$P_f = P[X \in D_f] = P[G(x) \leq 0] = \int_{D_f} f_X(x) dx. \tag{15}$$

D_f: domain of failure. Assuming the bound on the CSERF is given randomly and $\bar{C}_{r0}$considered as mean value and standard deviation: σ_w. Then equation (15) can be expressed from the following equation:

$$P_f = P[G(S) \leq 0] = \int_{D_f} f(\bar{C}_{r0}, \sigma_w) dS. \tag{16}$$

3. Determination of the loading domain

In most practical cases structures are subjected to general loading which consists of a set of proportional loads which can vary independently of each other. In this general case, the load carrying capacity of a structure cannot be characterized by a single collapse load multiplier but all the combinations of the independent loads under which the structure will collapse have to be defined.

To illustrate the problem of general loading, consider a structure which is subjected to forces $P_0^{(j)}$ at points $j = 1, 2, 3, \dots, s$ with given locations and directions. Assume that the forces can be devide into groups which form independent proportional loads such that:

$$P_j = m_1 P_0^{(1)} + m_2 P_0^{(2)} + \dots + m_s P_0^{(s)} = \sum_{j=1}^{s} m_j P_0^{(j)} \tag{17}$$

Here $P_0^{(j)}, j = 1, 2, 3, \dots, s$ denotes the independent group of forces and m_j are the associated load multipliers. These multipliers can vary independently but are limited by relationship

$$L(m_j) \leq 0 \tag{18}$$

Which defines all the possible combination of the external loads which can occur during the loading process. In an s-dimensional load space with the coordinate axes $m_1, m_2, \dots, m_s$ the function $L(m_j) =$

0 can be defined by a closed hyper-surface bounding, the domain of possible load, as shown for the case s=2 in Figure 1 (by the conditions $m_1 \geq 0$ and $m_2 \geq 0$).

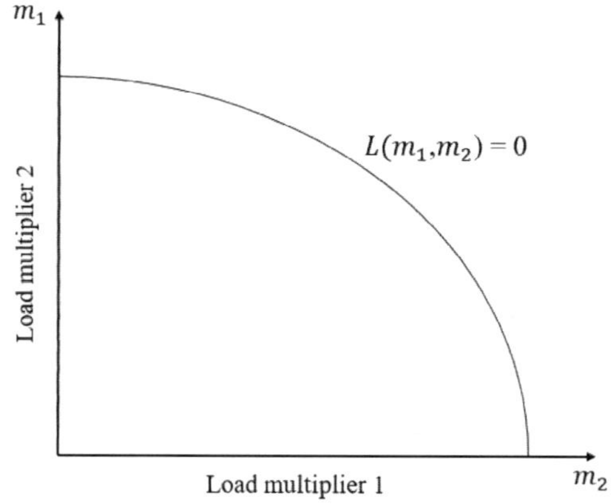

Figure 1. Safe domain in the case of 2 load-multipliers

4. Shakedown analysis

During the shakedown analysis all the possible distributions of the residual forces should be considered. These residual internal forces form a self-equilibrating system. The solution mechanism can be expressed by introducing the static principal of shakedown theorem, hence equilibrium equation will be defined from equation (19):

$$GQ^r = 0 \tag{19}$$

where G is the equilibrium matrix, Q^r denotes the residual bar forces or the residual bending moments.

The yields condition can be defined as follow:

$$-\boldsymbol{Q^p} \leq \boldsymbol{Q^r} + \boldsymbol{maxQ^e} \leq \boldsymbol{Q^p} \tag{20}$$

here Q^p is plastic limit force. Equation (21) defines elastic internal force:

$$Q^e = T^{-1}GK^{-1}m_{sh}P_0 \tag{21}$$

where P_0: constant load; m_{sh}: shakedown multiplier; K: stiffness matrix and T: flexibility matrix.

In the case of deterministic problem shakedown load-multiplier will be maximized when accumulated residual displacements are controlled with applying C_{ro}:

$$\left.\begin{array}{c} m_{sh} = max \\ GQ^r = 0 \\ -Q^p \leq Q^r + maxQ^e \leq Q^p \\ Q^e = T^{-1}GK^{-1}m_{sh}P_0 \\ \frac{1}{2}\sum_{i=1}^{n}\frac{1}{S_i}\int_0^{l_i}\left(Q_i^r(s)\right)^2 ds - C_{ro} \leq 0 \end{array}\right\} \tag{22}$$

Assuming residual moments M_1^r and M_2^r acting at the ends of a beam element equation (22) can be written as follow:

$$\left.\begin{array}{c} m_{sh} = max \\ GM^r = 0 \\ -M^p \leq M^r + maxM^e \leq M^p \\ M^e = T^{-1}GK^{-1}m_{sh}P_0 \\ \frac{1}{6E}\sum_{i=1}^{n}\frac{l_i}{I_i}\left[(M_1^r)^2 + (M_1^r)(M_2^r) + (M_2^r)^2\right] - C_{ro} \leq 0 \end{array}\right\} \tag{23}$$

Supposing probabilistic problem the bound on the CSERF is given randomly as explained in section 2.2.

5. Numerical example

To evaluate the method proposed above a simple frame with SRC is considered (Figure 2). The constant load values are $P_0^{(1)} = 10$KN and $P_0^{(2)} = 15$KN. At the beam to column connections joints the steel frame has semi-rigid connection. $EI_{column} = 8172.4 \cdot 10^6$ KNcm^2, $EI_{beam} = 4080.3 \cdot 10^6$KNcm^2. First moment of area of columns and beam are $S_{co} = 215$ cm^{3}and $S_{bo} = 130$cm^3, respectively. The influence of SRC are considered by using spring coefficient in the stiffness matrix for beam elements. The aim is to find maximum shakedown load multiplier for given SRC and bounds on the CSERF.

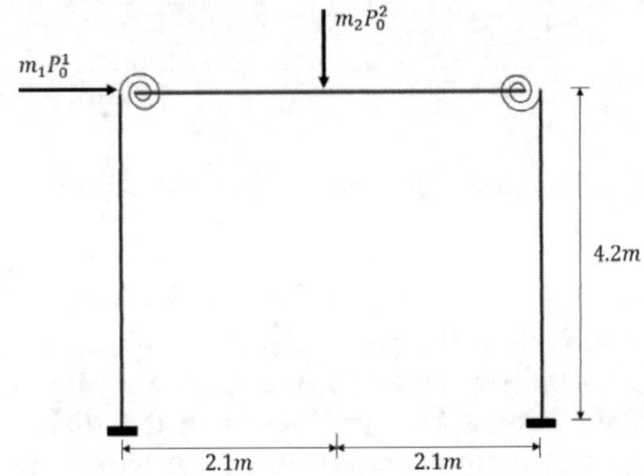

Figure 2. Frame structure with SRC between beam and columns

The numerical results using deterministic and probabilistic problems, are presented in Figures 3 and 4, respectively. In Figure 3 the variation of the shakedown load-multiplier is presented in function of the connection rigidities and limit values on the CSERF. One can see that the limit values on the CSERF and the stiffnesses of the SRC influences significantly the plastic behavior of the frame structure.

In Figure 4 safe domain of the structure for five different load combinations is illustrated. For each loading case $[m_1P_1,\ m_2P_2]$ a shakedown load-multiplier m_{sh} can be determined. Using these load-multipliers a safe domain, for five load combinations, can be generated in $m_1$and m_2 plane. Mean value is $\bar{C}_{r0} = 1.5$, standard deviation $\sigma_C = 0.1$ and failure probability of the structure (P_f =0.00069; 0.000013; 10×10^{-7}). The numerical results are in agreement with the expectation that the increase of the probability of failure results in bigger load-multipliers.

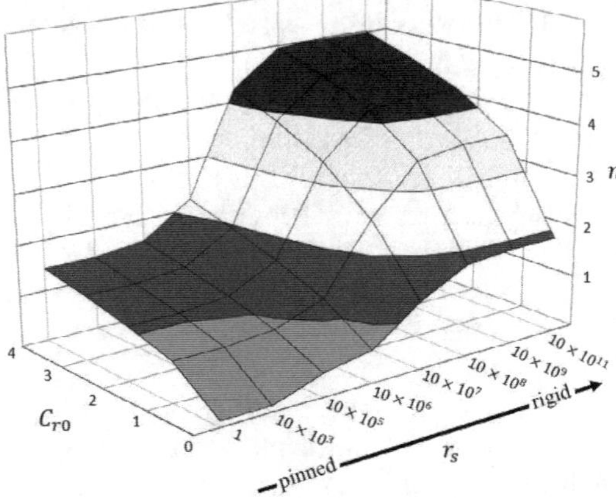

Figure 3. Variation of the shakedown load-multiplier in function of the connection rigidities and limit values on the CSERF.

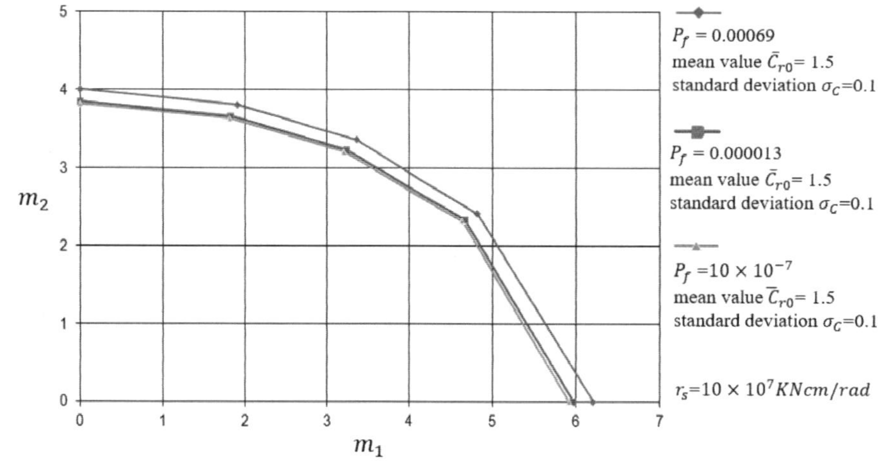

Figure 4. Safe domain of the structure

6. Conclusions

In this research optimal plastic shakedown analysis introduced to evaluate the maximum shakedown load-multipliers. Furthermore to control the post yield behaviour and residual displacements of the structures deterministic and probabilistic constraints are applied on the CSERF. The SRC between the beam and columns is described using spring coefficient in the stiffness matrix. Shakedown load-multipliers are calculated and safe domain of the structure for deterministic and probabilistic conditions under different load combinations is illustrated, reliable bound on the given probability of failures are presented. The numerical results show that the constraints on CSERF, stiffness of the SRC and probabilistic given conditions can influence on the magnitude of the shakedown load-multipliers.

Acknowledgment

The paper was written with the support of the project titled "Internationalisation, initiatives to establish a new source of researchers and graduates and development of knowledge and technological transfer as instruments of intelligent specialisations at Széchenyi István University" (project number: EFOP-3.6.1-16-2016-00017).

References

[1] Melchers R E and Beck A T 2017 Structural Reliability Analysis and Prediction *John Wiley & Sons Ltd* Third Edition.
[2] Wang J and Katafygiotis L S 2014 Reliability-based optimal design of linear structures subjected to stochastic excitations *Structural Safety* 47 29-38
[3] Lógó J 2007 New Type of Optimality Criteria Method in Case of Probabilistic LoadingConditions *Mechanics Based Design of Structures and Machines* **35**(2) 147-162.
[4] Marti K 2005 *Stochastic Optimization Methods* Springer-Verlag,Berlin-Heidelberg.
[5] Marti K and Stöckl G 2004 Stochastic Linear Programming Methods in Limit Load Analysis and Optimal Plastic Design under Stochastic Uncertainty *ZAMM* **84**(10-11) 666-677.
[6] Atkociunas J, Liepa L, Gervyte A, Merkeviciute D 2018 The scanning method for analysing the residual displacements of the framed structures at shakedown, *Computers & Structures* 207 148-154.
[7] Liepa L, Blaževičius G, Merkevičiūtė D and Atkočiūnas J 2016 Structural shakedown: a new methodology for estimating the residual displacements *Journal of Civil Engineering and Management* **22**(8) 1055-1065.
[8] Simon J W and Weichert D 2012 Shakedown analysis of engineering structures with limited kinematical hardening *International Journal of Solids and Structures* **49** 2177–2186.

[9] Levy N H, Einav I and Hull T 2009 Cyclic shakedown of piles subjected to two-dimensional lateral loading *International Journal for Numerical and Analytical Methods in Geomechanic* **33**(10) 1339-1361.
[10] Weichert D and Maier G 2002 Inelastic behaviour of structures under variable repeated loads *Springer* Vienna.
[11] Tin-Loi F 2000 Optimum shakedown design under residual displacement constraints, *Structural and Multidisciplinary Optimization* **19**(2) 130–139.
[12] König J A 1987 Shakedown of elastic-plastic structures (Fundamental studies in engineering) *Elsevier Publishing Company* Warsaw Poland.
[13] Polizzotto C 1982 A unified treatment of shakedown theory and related bounding techniques *Solid Mechanics Archives* 7 19-75.
[14] Kaneko I and Maier G 1981 Optimum design of plastic structures under displacements constraints *Computer Methods in Applied Mechanics and Engineering* **27**(3) 369–392.
[15] Corradi L 1977 Deformation bounding techniques in elastoplastic analysis, *In: Cohn, M.Z.; Maier, G. (eds.) Engineering plasticity by mathematical programming* New York: Pergamon Press 495-516.
[16] Ponter A R S 1972 An upper bound on the small displacements of elastic, perfectly plastic structures *Journal of Applied Mechanics* **39**(4) 959–963.
[17] Maier G 1969 Complementary plastic work theorems in piece wise-linear elastoplasticity *International Journal of Solids and Structures* 5 261–270.
[18] Kaliszky S and Lógó J 2002 Plastic Behaviour and Stability Constraints in the Shakedown Analysis and Optimal Design of Trusses *Journal of Structural and Multidisciplinary Optimization* **24**(2), pp. 118-124.
[19] Kaliszky S and Lógó J 1997 Optimal plastic limit and shakedown design of bar structures with constraints on plastic deformation *Engineering Structures* **19**(1) 19-27.

IOP Conf. Series: Materials Science and Engineering **686** (2019) 012003 doi:10.1088/1757-899X/686/1/012003

Evaluation of Visual Preference for Seasonal Streetscape Design Using 3D Simulation Technology

Chun Hyunjin[1, 2, a] and Wang Yue[3]

[1] School of Arts, Nanjing University of Aeronautics and Astronautics, NJ 211106, China
[2] School of Architecture, Southeast University, NJ 210096, China
[3] School of Industrial Design, Nanjing University of the Arts, NJ 210013, China No. 29 JIangjung Road, Jiangning District, Nanjing 211106, Jiangsu, China

[a]tough4324@naver.com

Abstract. As 3D technology advances, many designers are using 3D technology in their design research. Environment designers, in particular, conduct public design research using 3D simulation technology. In the field of environmental design, green space is an important factor in determining the image of public design. Green space, however, has different visual preferences as the seasons change. Thus, this research aimed to analyze the visual preferences of streetscape according to the seasons change using 3D simulation technology. The result of the study shows that the autumn image of maple background and ginkgo background are more preferred than the winter image of maple background and ginkgo background. In conclusion, the change of seasons is a crucial factor in determining the visual preference of the landscape. Designers, therefore, should consider the variation of seasons as well as the shape of streetscape to improve the visual preference of a streetscape. These findings could also be applied as theoretical data to conduct the public design.

1. Introduction

1.1. Research Background and Purpose
As 3D (3 Dimensions) technology advancing, it is being utilized in various fields. In the past, 2D technology were widely adopted in design fields. With the development of 3D technology, tremendous changes have been gone through in artistic design. 3D technology can visualize the hypothetical work in advance. Environment designers conduct public design research using 3D technology. In the meanwhile, environment designers use 3D simulation to create a suitable public space. In particular, green space is an important factor in determining the image of public design. However, green space has different visual preferences as the seasons change. In addition, trees have been planted without any design criteria according to seasonal changes. Thus, this research aimed to analyze the visual preferences of streetscape according to the seasons change using 3D simulation technology. These findings could be used as theoretical basis data to create the design scheme.

1.2. Literature Review
This study explored seasonal images and landscape preferences in public spaces. Previous studies carried out on streetscape were as follows. Shin Jae Yoon etc. (2014) analyzed preferences and visual

characteristics of each type of streetscape using 3D simulation [1]. Kim Ji Hye etc. (2009) studied various factors affecting the evaluation of the streetscape through the structural equation model [2]. Sung Hyun Chan (2003) analyzed if the environmental function and effect of trees along road was actually realized in city [3]. Jung Jin Hyung (2005) analyzed methods to improve the green space [4]. These studies had revealed that green space was the key factor in determining the visual preference of streetscape. However, most of these previous studies conducted analysis according to functional aspects of green space [1-4]. Since green space is made up of plants, visual preferences of streetscape may vary depending on the seasons change [5]. Thus, this study, unprecedentedly from the perspective of seasons change, analyzes the landscape preference of streetscape. These findings can also be employed as basic data for establishing streetscape plan.

2. Theoretical Background

2.1. Design Using 3D Technology
Designers use 3D technology in various fields. Previously, 3D technology was mainly used in product design, but nowadays 3D technology is used in most areas of design. Designers use various 3D programs such as Sketch up, 3D max, Auto cad, and Vector works. In particular, designers conduct design research using 3D simulation. In public design research, 3D simulation techniques are being used in a variety of ways. Recently, 3D technology becomes more important due to the development of 3D printing. Using 3D printers has the advantage of reducing the period and cost of design. So 3D printing technology will be used not only in product design but also in architectural design and environmental design in the future.

2.2. Characteristics of Public Design
Public design is to design not only public space but also public facilities. This public design shows the beauty of the region. Aesthetic, economic, and functional factors are of equal importance in public design. Aesthetically pleasing public space are landmarks, and their design is produced from a creative cognizance which designers pursue in order to match a streetscape to its surroundings [6].

3. Research Method

3.1. Selection of Background and Landscape Adjectives
In this study, background images, selected before 3D simulation was performed, were 40 photos taken with clear seasonal streetscape. The photo shoot was conducted on a streetscape of Korea, where there were many ginkgo and maple trees. Preliminary survey was conducted based on 40 photos. The subjects who participated in the survey were 30 graduates of the department of landscape architecture from Seoul National University. First, the subjects were given 40 photos and asked to select two of them, respectively representing the maple streetscape and the ginkgo streetscape. Among these selected photos, the highest number of times selected was chosen as background image. The selected pictures were as follows (figure 1, figure 2).

Figure 1 Maple Streetscape

Figure 2 Ginkgo Streetscape

Representative landscape adjectives were chosen to describe the landscape simulation to analyze the aesthetic of streetscape. The process of selection included several steps, the first, two of which were to understand the meaning of landscape adjectives through reviewing related literature [7-8] and then to select fifty pairs of landscape adjectives appropriate for analyzing streetscape. The third step was to conduct a primary survey for selecting pairs of adjectives that were most appropriate for evaluating the streetscape [7-8]. This research surveyed 25 graduate students of Konkuk University and 25 graduate students of Seoul National University; the total number of respondents equaling 50. Each respondent was given the pool of fifty adjective pairs and directed to select nine pairs which seemed most appropriate for describing streetscape. From the results of this survey, a total of nine pairs of landscape adjectives that were most frequently chosen by the respondents were taken for use in this experiment (Table. 1).

Table 1. Landscape Adjectives

Landscape Adjectives		
Artificial - Natural	Unstable - Stable	Ugly - Beautiful
Flimsy - Sophisticated	Confusing - Orderly	Unfamiliar-Familiar
Charmless - Attractive	Boring - Interesting	Simple- Splendid

3.2. 3D Simulation

3D simulation image model selected eight streetscape types, which were spring (figure 3), summer (figure 4), autumn (figure 5), winter (figure 6) of maple background and spring (figure 7), summer (figure 8), autumn (figure 9), winter (figure 10) of ginkgo background. The elements of 3D simulation such as camera, light's location, shape were in equal conditions; season was the only variable factor (Table 2). Landscape image for 3D simulation used computer programs such as Auto-CAD (Auto desk, 2005), Google Sketch-UP (Google Inc., 2007), Adobe Photoshop CS (Adobe System. 2003), 3D MAX. This research formed eight images, including the original maple streetscape and original ginkgo streetscape.

Table 2. Landscape Simulation

Figure 3 Spring of Maple Backgroun

Figure 4 Summer of Maple Background

Figure 5 Autumn of Maple Background

Figure 6 Winter of Maple Background

Figure 7 Spring of Ginkgo Background

Figure 8 Summer of Ginkgo Background

Figure 9 Autumn of Ginkgo Background

Figure 10 Winter of Ginkgo Background

3.3. Selection of Survey Group and Method of Analysis

Data for this research was collected via three surveys from June 10th, 2018 to September 10th, 2018. All subjects saw the face of an image of standardized dimensions (210×297mm) for a length of 8 seconds per image [7-8]. Twenty-five graduate students of Konkuk University in Korea and twenty-five graduate students of Seoul National University in Korea were surveyed, for a total of fifty participants. The participants were majored in civil engineering and landscape architecture. Students of these majors were chosen on the basis of their technical experience in the field; participants of civil engineering understand streetscape design concepts with a degree of professional expertise. Participants of landscape architecture were chosen for their understanding of streetscape designs and estimation of the visual quality of such landscapes. After the participants fully understood the evaluation method and purpose of the study, they were then asked to evaluate the streetscape images according to image and preference. The visual image evaluation was a 7-point semantic differential scale; evaluation about visual preference used the 7-point Likert scale [7-8]. Statistical analysis utilized Excel and KESS (Korean Educational Statistics Software) to abstract descriptive statistical values about the streetscape images. This research conducted a two-way analysis of variance and Duncan's test (post hoc tests) to understand visual preference differences [7-8] according to background and season.

4. Results and Discussion

4.1. Assessment and Variation of Images According to Background and Season

The results of the survey of background and season assessment are summarized as follows. The images portraying spring and summer from maple background were both rated highly as 'Beautiful' and 'Natural', with the autumn image additionally rating high as 'Beautiful' and 'Splendid'. The maple landscape of autumn gave a beautiful image to the surroundings. The winter image from maple background rated low as 'Attractive'. Low 'Attractive' ratings were thought to occur because the bare branches and dull color of the winter image had a negative impact on the natural surroundings of the image (Table 3).

Table 3. Average Value of Image Assessment in Maple Background

Landscape Adjectives	Maple Background			
	Spring image	Summer image	Autumn image	Winter image
Artificial - Natural	4.14	4.16	4.08	4.04
Unstable - Stable	3.72	3.64	3.82	3.78
Ugly - Beautiful	4.06	4.04	4.22	3.74
Flimsy - Sophisticated	3.88	3.65	3.86	3.72
Confusing - Orderly	3.43	3.62	3.52	3.44
Unfamiliar-Familiar	3.72	3.34	3.66	3.56
Charmless - Attractive	3.88	3.96	4.02	3.22
Boring - Interesting	3.84	3.54	3.88	3.68
Simple- Splendid	3.88	3.74	4.04	3.54

The images portraying spring and summer from ginkgo background were both rated highly as 'Beautiful' and 'Natural', with the autumn image additionally rating high as 'Attractive' and 'Splendid'. The green landscapes of spring and summer from similar images. And the yellow leaves of autumn brought an attractive image. The winter image rated low as 'Beautiful'. The summer from ginkgo background also rated highly as 'Splendid', and the autumn from ginkgo background rating also rated highly as both 'Beautiful' and 'Splendid'. All bridges received a highly 'Natural' rating,

IOP Conf. Series: Materials Science and Engineering **686** (2019) 012003 doi:10.1088/1757-899X/686/1/012003

with the winter image from ginkgo background also rated low as 'Beautiful'. As such, the landscape of autumn gives a more positive image than the landscape of winter. Because autumn had a positive effect on space by the leaf color of trees (Table 4).

Table 4. Average Value of Image Assessment in Ginkgo Background

Landscape Adjectives	Ginkgo Background			
	Spring image	Summer image	Autumn image	Winter image
Artificial - Natural	4.12	4.12	4.08	4.14
Unstable - Stable	3.82	3.86	3.76	3.84
Ugly - Beautiful	3.94	3.82	4.22	3.64
Flimsy - Sophisticated	3.88	3.65	3.74	3.72
Confusing - Orderly	3.54	3.62	3.64	3.44
Unfamiliar-Familiar	3.72	3.64	3.72	3.56
Charmless - Attractive	3.88	3.96	4.12	3.46
Boring - Interesting	3.82	3.74	3.76	3.88
Simple- Splendid	3.84	3.76	4.02	3.52

4.2. Assessment and Variation of Images According to Background and Season

This study analyzed visual preferences according to different backgrounds and seasons. The results were taken at a significance of less than 0.05, and showed that, above all, autumn image of maple background and ginkgo background were more highly preferred than other images. The summer image in maple background received the highest preference rating, and the winter image in maple background had the lowest preference. In maple background, the autumn image had the highest preference rating, and the preference of winter image had the lowest rating. In ginkgo background, the autumn image had the highest preference rating, and the winter image was, again, the lowest. In both of maple and ginkgo backgrounds, the autumn image was highly evaluated for visual preference because the color and fallen leaves of these image seem to have a good correspondence with the surroundings. These results suggested that different seasons and backgrounds would give different visual preferences for streetscape, even if the shape of landscape was the same (Table 5).

Table 5. Statistical Analysis of Visual Preferences

Landscape		Average	Standard deviation	F-Value	Significant probability	post hoc tests a=0.05
Maple Background	Spring image	4.02	1.18	15.120	0.002	b
	Summer image	3.88	1.14			c
	Autumn image	4.36	1.06			a
	Winter image	3.32	1.02			d
Ginkgo Background	Spring image	4.10	1.12	7.224	0.001	b
	Summer image	3.96	1.32			c
	Autumn image	4.32	1.14			a
	Winter image	3.88	1.18			c

This study also conducted a Duncan test (post hoc tests) to analyze the variation in landscape according to background and season. The level of significance was less than 0.05, and it was apparent that the different visual preferences according to background and season were significant. The spring images in maple background and ginkgo background also had a similar evaluation in visual preference. The autumn images in maple background and ginkgo background received a similar evaluation in visual preference. The visual preference for the summer image in maple background, ginkgo background and winter image in ginkgo background were also similar across the board. These results proved that visual preferences changed according to background and season.

5. Conclusion

Streetscape is an important factor in regional landscape. However, when designing a street, designers often don't carefully consider the aesthetic function of street, so some inappropriately designed streetscapes frequently destroy the beauty of the regional landscape. Preferences of streetscape may vary depending on the season. Thus, the landscape preference of streetscape was analyzed according to the season in this study. The result of the study on visual preference according to season and background showed that the autumn image of maple background and ginkgo background were more highly preferred than the winter image of maple background and ginkgo background. The statistical analysis on visual preferences according to background and season showed that the autumn image of maple background was preferred with the highest overall score. The visual preference of the autumn image of ginkgo background received the second highest score, and the winter image of maple background was the lowest. To conclude, variation of seasons is an important element in visual preference. Therefore, to improve the visual preference of a streetscape, designers should consider the variation of season as well as the shapes. Further research is needed to determine the visual preference of many different components such as light and shape. Therefore, further researches will be done to thoroughly analyze other factors determining the visual preference of streetscape.

References

[1] Shin, Jae Yoon etc. 2014 *Evaluation of Streetscape by Street Planting Types using Dynamic Simulation - In the case of Ginkgo planted on the Dongdaegu-ro in Daegu -*. Journal of Korea Institute of traditional landscape architecture. **42** Vol 1 pp 89-103

[2] Kim Ji Hye etc. 2009 *A Study on Analysis of Evaluation Structure on Streetscape Image - Focused on Samcheongdong-gil in Seoul -*. Journal of Korea Planners Association. **44** Vol 1 pp 273-286

[3] Sung Hyun Chan 2003 *An Empirical Study on the Function and Effect of Urban Openspace Focusing on Urban Roadside Trees -*. Journal of Korea Institute of landscape architecture. **31** Vol 2 pp 48-57

[4] Jung Jin Hyung 2005 *A Study of the Planting Characteristics of Street Trees and Herbaceous Plants in Gangwon-do.* Journal of Korea Institute of landscape architecture. **33** Vol 5 pp 57-68

[5] Kim Soo Yeon etc. 2010 *Importance-Performance Evaluation of Streetscape on Insa-dong road.* Journal of Korea Planners Association. **45** Vol 7 pp 179-191

[6] Kevin Lynch 1960 *The image of the city*. Cambridge, MA: MIT press.

[7] Feimer N R 1979 *Personality and Environment Perception: Alternative Predictive System and Implications for Evaluative Judgement.* Ph.D. Dissertation. University of California. Berkely.

[8] Im Seung-Bin 1991 *Landscape analytics*. Seoul: Seoul National University Press.

Performance Improvement and Parameter Influence of Thread Shaft by Axial Self-infeed Rolling Process

Shuowen Zhang[1,*] Shuqin Fan[1,2] Shengdun Zhao[1] and Yongfei Wang[1]

[1] School of Mechanical Engineering, Xi'an Jiaotong University, No.28 Xian Ning West Road, Xi'an, People's Republic of China

[2] State Key Laboratory of Materials Processing and Die & Mould Technology, Huazhong University of Science and Technology, No.1037 Luoyu Road, Wuhan, People's Republic of China

[*]E-mail: zhangsw_edu@163.com

Abstract. The rolling process is an important processing technology for improving the performances of thread shaft. In this paper, the axial self-infeed rolling process is investigated. The principle of the axial self-infeed rolling process is introduced and the performances of the formed thread shaft with copper T2 are evaluated. In addition, the major diameter of the formed thread shaft and the rolling torque during the axial self-infeed rolling process are studied. The hardness of the formed thread shaft with copper T2 is improved 74.2% and the microstructure is fibrous tissue in the bottom of the tooth groove. The size of the blank has greater impact on the major diameter and the forming torque than the rotating speed of the blank.

1. Introduction

The thread shaft is widely used in the industry which has important effects on the reliability and security of machine. The rolling process for thread shaft manufacturing is an advanced manufacturing technology which has the advantages: high fatigue strength and high reliability [1-3]. The rolling process of thread shaft is similar to forming the spline [4-6] and the gear [7,8].

Domblesky et al. modeled two-dimensional and three-dimensional rolling process with the help of finite element analysis to study the effect of material and technological parameters on the rolling process [9,10]. Peter et al. proposed a new method for the thread shaft manufacturing with special flat dies and conducted the finite element analysis and the laboratory research [11]. Lee et al. studied the forming process using a rigid-plastic finite element method and analyzed the thread rolling of fasteners and the results of the finite element analysis were compared with the experiments [12]. Qi et al. studied the forming hollow threads through three dies cold rolling and the stress state of the hollow blank was also analyzed. The minimum thickness equations were derived [13]. Peter Groche et al. analyzed the rolling process with flat dies and the influence of friction was investigated by three-dimensional simulation [14]. Philipp Kramer et al. studied the defect detection in thread rolling processes with flat die. The effects of process parameters on the quality of formed threads were systematically studied [15]. The radial feed rolling process has some disadvantages: huge forming force and torque, harsh demands for the forming equipment and the length of the formed thread shaft limits by the rolling dies. In order to solve the existing problems of the radial feed rolling process, the axial self-infeed rolling process of thread shaft is applied to form the long screw [16]. But the axial rolling process was rarely studied. In this paper, the principle of the axial self-infeed rolling process is described. The axial self-infeed rolling process experiments are carried out on a self-developed equipment and the performances of the formed thread shaft with copper T2 are studied. In addition, the major diameter and the forming torque with various materials and different parameters are analyzed in detail.

2. The Principle of The Axial Self-infeed Rolling Process

The rolling die system mainly consists by three rolling dies (rolling die-A, rolling die-B and rolling die-C) which is shown in Figure 1(a). The structures of rolling die are shown in Figure 1(b), which includes three parts: the exit part, the correction part and the pre-rolling part. The pre-rolling part has the angle α_p which is convenient for entrancing and pre-rolling the blank. The correction part calibrates the pre-forming tooth and ensures that the formed tooth profile is standard. The exit part has the angle α_e which benefits the formed thread shaft to return smoothly.

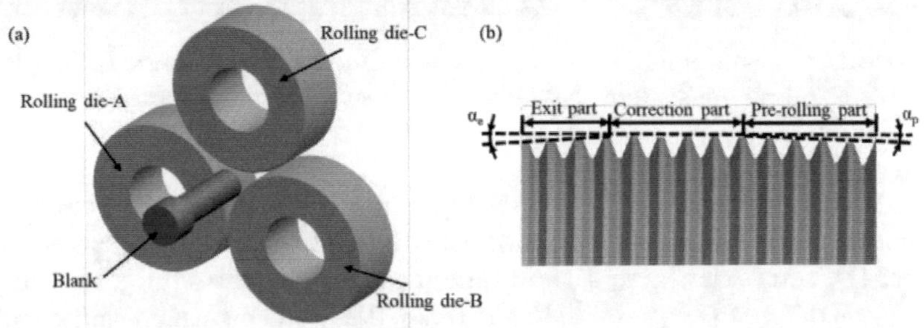

Figure 1. The schematic diagrams of rolling die: (a) the rolling die system (b) the structure of rolling die.

Figure 2 shows the procedures of the axial self-infeed rolling process of thread shaft which include before rolling, during rolling and after rolling. The working procedures are listed as follows: (1) Before the rolling process, the blank rotates with the speed n_b and the rolling dies remain stationary. (2) During the rolling process, the rolling dies are pushed to contact with the rotating blank. Due to the role of the meshing, with the rotation of the blank the rolling die also begins to rotate and the rotating speed is n_d. Meanwhile, the rolling dies feed along axial direction with the speed v. The metal deformation occurs on the blank under the action of the pre-rolling part and the correction part. (3) After the rolling process, the blank reversal rotating and the rolling dies return to their initial position, the rolling process end.

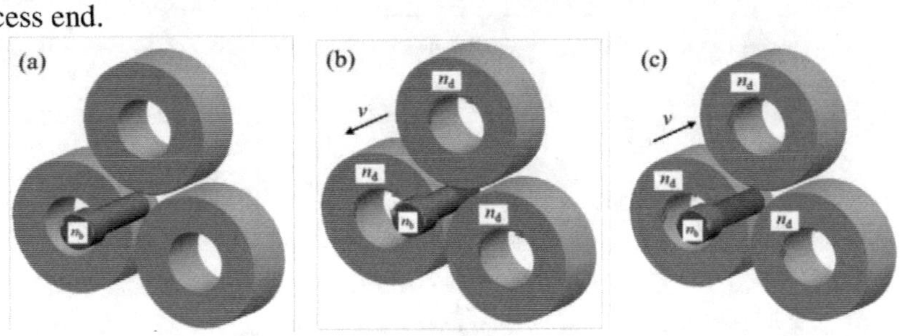

Figure 2. The working process of the axial self-infeed process: (a) before rolling (b) during rolling (c) after rolling

Figure 3(a) shows the local structure of rolling die system and Figure 3(b) shows the structure of the self-developed experimental equipment. The experimental equipment mainly includes four parts: the driving part, the deceleration part, the clamping part and the rolling die system. The driving part is composed of a servo motor which is used to rotate the blank. The deceleration part mainly includes the planetary reducer which is used to improve the torque and ensure the blank rotation with desired speed. The clamping part is a three-jaw chuck which is used to fix the blank. The rolling die system consists of three rolling dies and various accessories which is the vital component for the forming equipment. The rolling die system is the forming part for the thread shaft.

In these experiments, the blank materials are AISI 1045, AISI 1035 and copper T2. The major diameter and pitch of the formed thread shaft respectively are 24mm and 3mm.The rotating speed of the blank is from 10r/min to 60r/min. And without lubricant was applied in the rolling process.

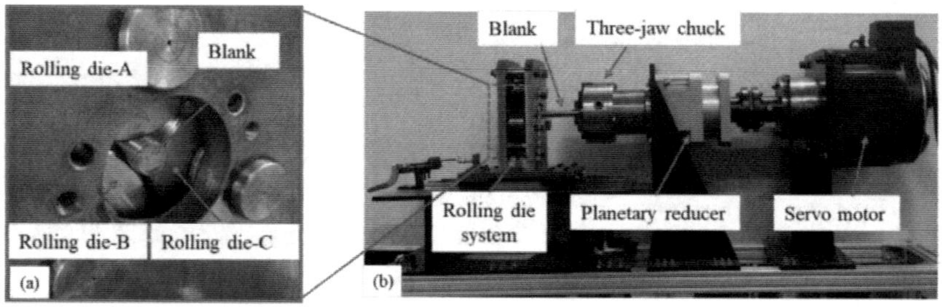

Figure 3. The experimental equipment of the axial self-infeed rolling process: (a) the local structure of rolling die system (b) the whole experimental equipment

3. The Experimental Results

The formed thread shaft with copper T2 was cut the cross-section for the microhardness tests and metallographic analysis. The hardness values of formed thread shaft were measured through a micro-sclerometer (Type: HVT-100A). Figure 4 shows the measuring device and points. In order to ensure the accuracy, three teeth (tooth Ⅰ, tooth Ⅱ and tooth Ⅲ) were measured and each tooth includes seven points.

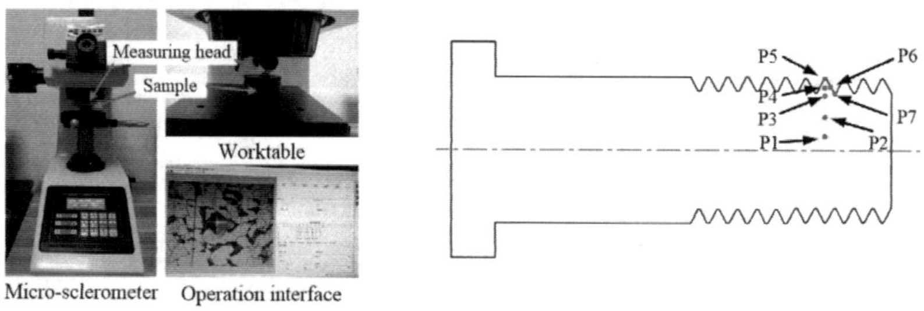

Figure 4. The hardness measuring device and points

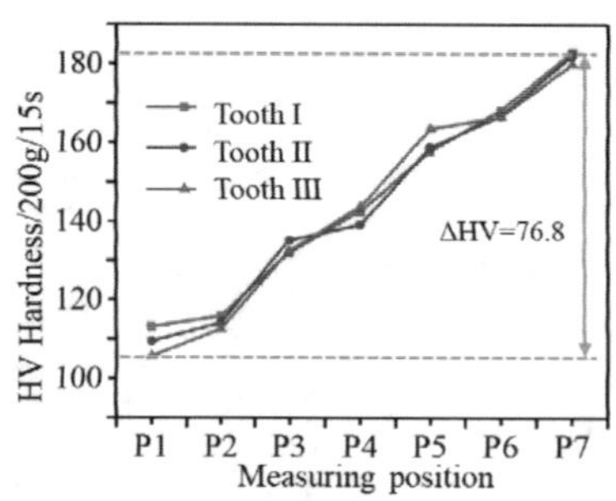

Figure 5. The measuring hardness values of the formed thread shaft with copper T2

Figure 5 shows the hardness values of formed thread shaft with copper T2 in different teeth and positions. The hardness values gradually increase from HV 103.5 to HV 182.1 corresponding to the points from P1 to P7. P1 represents the hardness value of the core of the formed thread shaft and P7 represents the hardness value of the bottom of tooth space of the formed thread shaft. The differential value of the hardness values between the P1 and P7 is 76.8 (ΔHV=76.8). In the core region of the formed thread shaft, the hardness value is equal to that of the blank. The maximum hardness value of the formed thread shaft is improved 74.2% than that of the core of the formed thread shaft.

The microstructures of the formed thread shaft with copper T2 show in Figure 6. The microstructure in the bottom of tooth groove region is obviously fibrous tissue and the grain elongated (Figure 6(a)). The streamlines are also obvious which are extended from the bottom of tooth groove to the two flanks of tooth profile, the microstructure in the flank of tooth profile is also dense and streamlined, but the degree of fibrosis is less than that in the bottom of tooth groove (Figure 6(b)). In the top of tooth profile, the grain is slightly refined and the grain size slightly decreases (Figure 6(c)). With the increase of depth inward from the tooth surface to the center of the formed thread shaft, the fibrous tissue gradually disappears and the grain size increases. The influent degree in different region through axial self-infeed rolling process for the formed thread shaft in descending order is the bottom of tooth groove, the flank of tooth profile and the top of tooth profile. The characteristics of the microstructure are corresponding to the hardness distribution of the formed thread shaft.

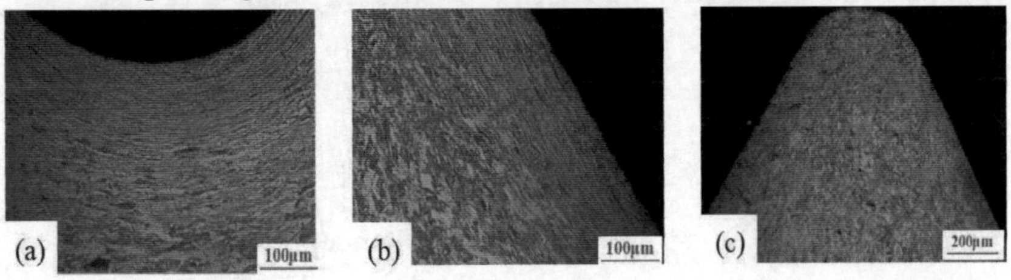

Figure 6. The microstructure of the formed thread shaft with copper T2: (a) the bottom (b) the flank (c) the top

3.1. The Major Diameter Analysis

The major diameter of thread shaft is an important indicator for forming precision. In this section, the major diameter is selected as representation to evaluate the influences of the various materials, the blank diameter and the rotating speeds. In theses experiments, the blank diameter increase from 22.2mm to 22.8mm and the length is 200mm.

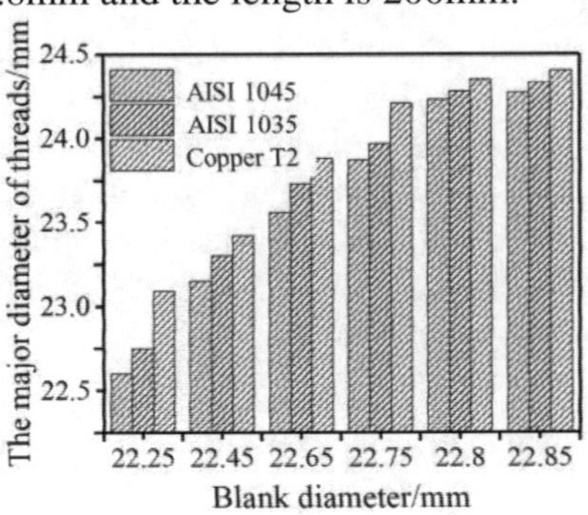

Figure 7. The major diameter of the formed thread shaft with various materials under different blank diameters

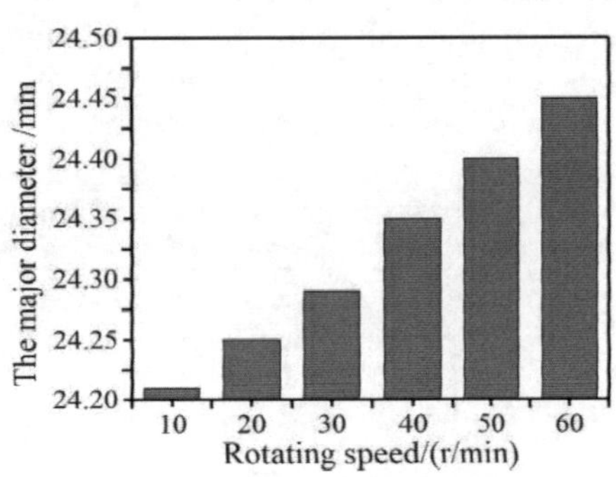

Figure 8. The major diameter of the formed thread shaft under different rotating speeds

Figure 7 shows the major diameter of the formed thread shaft with various materials under the different sizes of the blank when the rotating speed of the blank is 10r/min. When the blank diameter is less than 22.75mm, the blank with copper T2 forms the thread shaft with the largest major diameter and the blank with AISI 1045 forms the thread shaft with the smallest major diameter. With the size of the blank is gradually increasing, when the blank diameter is larger than 22.75mm the major diameters of the thread shafts with different materials are almost equal.

Figure 8 shows the major diameter of the formed thread shaft under various rotating speeds when the blank diameter is 22.75mm and the material of the blank is copper T2. When the rotating speed of the blank increases from 10r/min to 60r/min, the major diameter of thread shaft increases from 24.21mm to 24.45mm.

When the blank diameters are less than 22.75mm, the addendum of the formed thread shaft is no constraint by the dedendum of the roller which is shown in Figure 9(a). The height of tooth is determined by the material hardness and the material flow velocity. Under the same blank diameter, the blank with AISI 1045 steel has higher hardness therefore the material flow is difficult which leads to the size of the formed thread shaft is less. And the rotating speed of the blank determines the deformation velocity on the surface of the blank. With the increasing rotating speed of the blank, the material flow velocity is large. The size of the formed thread shaft increases slightly as well.

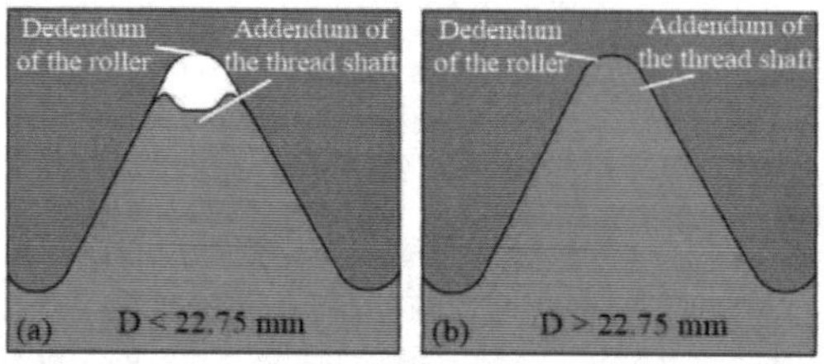

Figure 9. The profile of the formed thread shaft under different blank diameters

As shown in Figure 9(b), when the blank diameters are larger than 22.75mm, the dedendum of the roller contacts with the addendum of the formed thread shaft and the tooth profile of the roller is filled to the full. Therefore, the major diameters with different materials are almost the same.

3.2. The Rolling Torque Analysis

The rolling torque is an important parameter for the forming equipment which limits the development of forming process. The rolling torque under different sizes and rotating speeds of the blank with various materials are studied. The torque value was calculated by the output current of the servo motor and the maximum rolling torque values are selected to analyze in this section.

Figure 10 is the rolling torque under different sizes of the blank and the rotating speed of the blank is 10r/min. For the same blank diameter, the rolling torque with kinds of materials is different. The rolling torque of forming copper T2 blank is smallest, but which of forming AISI 1045 steel blank is largest. And when the size of the blank is less than 22.75mm, the rolling torque increases slowly. The diameter of the blank increases from 22.75mm to 22.85mm the rolling torque rises sharply.

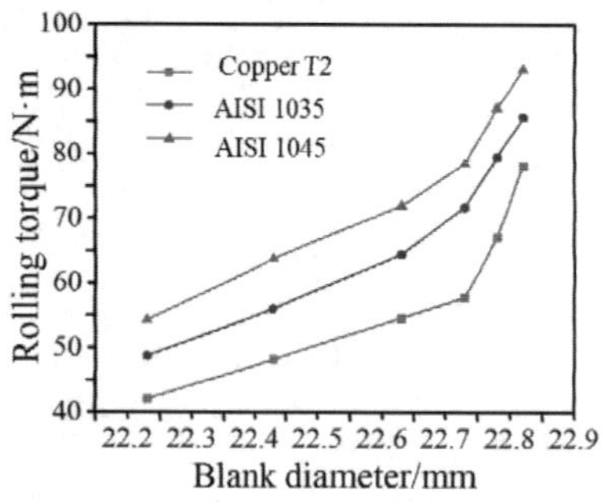

Figure 10. The rolling torque under size of the blank

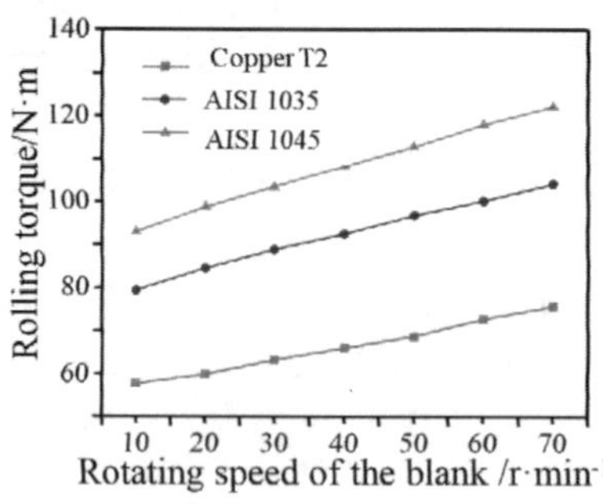

Figure 11. The rolling torque under different rotating speed of the blank

Figure 11 shows the effect of the rotating speed of the blank on the rolling torque when the blank diameter is 22.75mm. With the increasing of rotating speed of the blank, the rolling torque increases but the amplitude of increasing is small.

Under the same rotating speed of the blank, the rolling torque of the blank with copper T2 is smallest and which of the blank with AISI 1045 steel is largest.

The material with high hardness is difficult to deform, therefore the blank with AISI 1045 steel needs largest torque to be formed. When the blank diameter is larger than 22.75mm, the radial force between the blank and the dies increases dramatically and the addendum of the formed thread contacts with the dedendum of the rolling die, as shown in figure 9. The frictional force between the blank and the rolling die causes a significantly increase of the torque during the forming process. With the increasing of the rotating speed of the blank, the material strain rate increases and which causes an increase of the rolling torque.

4. Conclusion

From the above, the experiments of the thread shaft were carried out. Following is the results:

(1) The maximum hardness value of the formed thread shaft with copper T2 distributes in the bottom of the tooth groove and which is improved 74.2%. Through axial self-infeed rolling process, the microstructure in the bottom of tooth groove and the flank of tooth profile is dense and streamlined.

(2) When the blank diameter is larger than 22.75mm, the major diameter of the formed thread shaft remains consistent but the rolling torque increases sharply. With the rotational speed of the blank increase the major diameter and the torque are slightly increased.

(3) The size of the blank has greater impact on the major diameter of the formed thread shaft and the rolling torque than the rotating speed of the blank.

ACKNOWLEDGEMENTS

The authors gratefully acknowledge the contribution of subjects: National Natural Science Foundation of China for key Program (Grant No. 51335009) State Key Laboratory of Materials Processing and Die & Mould Technology, Huazhong University of Science and Technology (Grant No. P2019-028).

References

[1] Lange K 1985 *Handbook of Metal Forming* (New York:McGraw-Hill)

[2] Tschätsch H 2006 *Metal Forming Practice: Processes -Machines-Tools* (Berlin: Springer)

[3] Song J L, Liu Z Q and Li Y T 2017 *Cold Rolling Precision Forming of Shaft Parts* (Beijing: Springer)

[4] Cui M Cao, Zhao S D, Chen C, Zhang D W and Li Y Y 2017 Process parameter determination of the axial-pushed incremental rolling process of spline shaft. *Int J Adv Manuf Technol* **90** 1-11.

[5] Cui M C, Zhao S D, Zhang, D W, Chen C, Fan S Q and Li Y Y. Deformation mechanism and performance improvement of spline shaft with 42CrMo steel by axial-infeed incremental rolling process. *Int J Adv Manuf Technol* **88** 1-10.

[6] Zhang D W, Li Y Y, Fu J H and Zheng Q G 2007 Mechanics analysis on precise forming process of external spline cold rolling. *Chin Journal Mech Eng* **20** 54–58.

[7] Amir A K, Jun N, David S and Richard V 2007 Investigation of work hardening of flat-rolled helical-involute gears through grain flow analysis, FE-modeling, and strain signature. *Int J Mach Tool Manu* **47** 1285–1291.

[8] Amir A K, Jun N, David S, Richard V and Garrold D 2007 Diagnosis of involutometric issues in flat rolling of external helical gears through the use of finite-element models. *Int J Mach Tool Manu* **47** 1257-1262.

[9] Domblesky J P, and Feng F 2002 A parametric study of process parameters in external thread rolling. *J Mater Process Technol* **121** 341–349.

[10] Domblesky J P and Feng F 2002 Two-dimensional and three-dimensional finite element models of external thread rolling. *Proc Instn Mech Engrs Part B, Journal of Engineering Manufacture* **216** 507–517.

[11] Pater Z, Gontarz A, and Weroñski W 2004 New method of thread rolling. *J Mater Process Technol* **153–154** 722–728.

[12] Lee M C, Jang S J, Han S S, Yoon D K and Joun M S. New finite-element model of thread rolling. *Steel Res. Int.* **81** 214–217.
[13] Qi H P, Li Y T, Fu J H and Liu Z Q 2008 Minimum wall thickness of hollow threaded parts in three-die cold thread rolling. *Int J Mod Phys B*, **22** 6112–6117.
[14] Peter G and Philipp K 2018 Numerical investigation of the influence of frictional conditions in thread rolling operations with flat dies. *Int J Mater Form*, **11** 687-703.
[15] Philipp K and Peter G 2018 Defect detection in thread rolling processes-experimental study and numerical investigation of driving parameters *Int J Mach Tool Manu*, **129** 27-36.
[16] Ivanov, V 1998 Rolling of long screws. *J Mater Process Technol*, **82** 1-12.

Research on UX Evaluation Model of Computer Input Devices (Keyboard and Mouse) Based on AHP

Zhaohui Huang and Haoda Li*

417, Dongyi, Huazhong University of Science and Technology, Wuhan, Hubei Province, China

* 543471520@qq.com

Abstract. With the rapid development of the Internet, there are more and more computer input devices for different usage scenarios, and a lot of factors need to be considered comprehensively when users select computer input devices. At present, there are still some problems need to be solved in this field: how to evaluate the user experience of computer input devices in order to give reasonable suggestions to consumers when making purchasing decisions; how can design and development personnel find the shortcomings of design solutions in the design decision process? Aiming at the above problems, a user experience evaluation system for computer input devices is proposed.

Firstly, based on the user experience 5E model proposed by Whitney Quesenbury, the user experience evaluation system of computer input device was constructed by using the questionnaire method, interview method, and expert opinion method. Then, the user experience evaluation model of computer input device was constructed by AHP.

This paper takes computer input equipment as the research object and studies the user experience evaluation method of computer input device through the combination of qualitative and quantitative research methods. Established a more scientific user experience evaluation system to guide the design and development personnel in the design decision process. And provides reasonable suggestions for consumers to purchase and has certain reference value in the field of user experience evaluation research.

1. Introduction

1.1 Research object

The main research object of this paper is the computer input device (keyboard and mouse). The computer input device inputs data and information to the computer and is the main device for information exchange between the user and the computer system. Keyboard and mouse are indispensable computer input devices, and user usage frequency are the highest in all computer input devices, with certain representativeness and research value. This paper will mainly establish user experience evaluation model for keyboard and mouse.

1.2 Research problems

In the field of academic research, the research on user experience evaluation model is mainly for internet products, such as mobile APP, website, system, software etc. But relatively few researches on hardware devices. This paper focuses on the equipment for information exchange between hardware and software. The evaluation criteria include both the hardware experience and the experience of software. The research object has certain representativeness and research value. The existing user experience evaluations on keyboards and mice are mostly evaluations and purchase evaluations on the web. These evaluations cover different content and no uniform standard. It is difficult to bring a systematic reference to design developers or potential consumers.

2. Establishment of UX evaluation system

2.1 Establishment of criteria level evaluation index

User experience professor Whitney Quesenbery proposed the user experience 5E evaluation model, which includes five aspects: Engaging, Easy to learn, Effective, Error tolerant and Efficient.
Since the original 5E model is more concerned with the user experience of Internet products such as software and websites, some of the indicators may not be suitable for hardware products. Therefore, according to the original model content, considering the characteristics of the hardware products, the description of the indicators of the 5E model has been modified to some extent[Table 1].[1]

Table 1. Descriptions of model for computer input devices (keyboard and mouse) based on 5E model

5E Model	Description	Computer input device (keyboard and mouse) user experience 5E model	Contents
Engaging	Good-looking and easy to use, giving users a pleasant experience	Satisfaction	Five senses experience of keyboard and mouse
Error tolerant	Prevent errors	Error tolerant	Error tolerant design of keyboard and mouse
Easy to learn	There is no obstacle to initial use and deeper use	Easy to learn	Usage guidance of keyboard and mouse
Efficient	Basic features available	Efficient	Basic features of keyboard and mouse
Effective	Additional features help increasing efficiency	Effective	Additional features of keyboard and mouse

2.2 Establishment of alternatives level evaluation indicators

When selecting the user interview sample, the consumers with the purchase intention are mainly selected. In order to avoid the subjective bias caused by the understanding of the keyboard and mouse, and to improve the applicability and persuasiveness of the research results. 4 industrial designers, 1 self-media writer, 2 computer game players are selected as interview sample.

Due to the diversification of keyboards and mice on the market, consumers have different motivations when purchasing products. In order to make the research results more scientific and objective, according to different motivations, three different representative products are selected [Table 2]. Reference equipment for interviews. First, through the completion of specific tasks[2], the user can form a subjective cognition of the keyboard and mouse experience, and then conduct in-depth interviews [Figure 1]. The interview form is a preliminary conclusion to evaluate the evaluation indexes of the keyboard and mouse through direct inquiry.

Table 2. Representative products

Motivation	Using scenarios	Usage time	Keyboard	Mouse	Connection method
Gaming	Playing games	2-3h	Dareu First	Razer	Cable
Working	Writing	3-5h	Logitech K380	Logitech M590	Bluetooth
Personalized	Personalized experiences	1-2h	Hangshi HB066	Microsoft Scuplt Ergonomic	Bluetooth

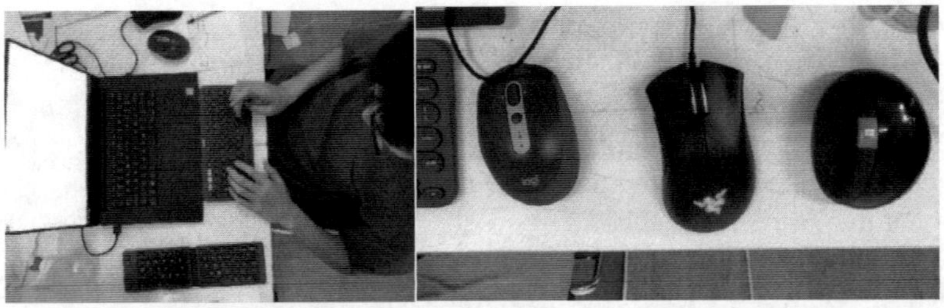

Figure 1. Interview based on task scenarios

Through classification and integration, a total of 41 indicators were obtained [Table 3].

Table 3. Descriptions of 5E model for computer input devices (keyboard and mouse)

Satisfaction	Color, Material, Shape, Desktop appearance uniform, Surface curvature, Wrist placement, Arm opening width, Keyboard width, Keyboard placement, Product size, Palm, Roller noise, Case noise, The comfort of the feedback sound, The length of the keystroke, Product weight, The moderate force required for the button, The friction of the material, The softness of the material.
Error tolerant	Keyboard button shape, Keyboard button size, Mouse movement sensitivity, Keyboard button layout.
Easy to learn	Unique interaction, Easy to understand interactive way, Fast connection to the computer, Easy to learn additional features.
Efficient	Length of cable, Material of cable, Stability of bluetooth, USB transmission stability, Basic function available, Surface wear resistance, Connection stability, Button usage life.
Effecive	Multi-device adaptation, Multiple additional functions, Adjustable sensitivity, Smooth use, Product portability, High-efficiency combination keys, Wheel flexibility, Connection speed.

2.3 First round indicator screening

There are many user experience evaluation indicators obtained through interviews, but indicators including repeated meanings, such as Color and Material can be combined into appearances, which also include indicators that may occur for specific scenarios, such as different connection methods. With a mouse, using a wired connection may not require consideration of the ease of connection. Using a wireless connection (such as Bluetooth) requires learning how to connect the product. The basis of the screening is mainly to merge, eliminate and correct the indicators with repeated meanings, superfluous or unclear levels [3], [4], [5]. According to the analysis and screening, the results of the first indicator screening resulted in a total of 19 indicator indicators.

2.4 Second round indicator screening

In order to further improve the objectivity and effectiveness of the indicators, this study organized an expert opinion group to score and screen the indicators. In order to achieve better accuracy, the study identified the number of experts as 10 people, including industrial designers, master students in user experience, and professors in industrial design. The expert opinion method uses the form of scores and in-depth interviews to screen the above indicators.

The expert opinion review in this paper uses the Likert five-point scale to measure the indicators, mainly for the reasonable setting of the indicators. The rating is: completely unreasonable, unreasonable, neutral, reasonable, and completely reasonable.

According to the results of the expert opinion method, the indicators were screened for the second time. The principles of screening were as follows: 1. User experience characteristics of computer input devices (keyboard and mouse); 2. Indicators need to be related to user experience theory; The indicator description language is easy to understand; 4. The indicator meaning is clearly expressed; 5. The indicators with the same meaning are merged; 6. The missing indicators should be added. Based on the above principles, after discussion and research with experts. Five criteria-level indicators were retained, and 17 alternatives-level indicators were revised and derived.

The satisfaction level has covered the visual experience, the auditory experience, the tactile experience and the human-machine experience. It can fully summarize the user experience evaluation of computer input devices (keyboard and mouse).

At the Error tolerant level, since the size and shape of the buttons may affect the occurrence of wrong clicks, the indicator is corrected to be “Button shape and size are easy to click”. The “Button layout” is corrected to “Button layout conforms to the usage habits”.

At the level of Easy to learn, users usually have a deeper understanding of the keyboard and mouse. Therefore, the easy-to-learn is usually reflected in the additional functions provided by the product. The indicator of “Additional functions are easy to learn” has already been representative.

At the Effective level, the meanings of the two indicators on the connection method are duplicated. The “Stable and effective connection method” is deleted, and the “Connection stability” is retained. Correct the "Lifetime" to "Long product life".

At the Efficient level, because the additional functions of the keyboard and mouse tend to bring efficiency, it is very important for users to use these attachment functions more simply, and the additional feature diversity may only make the product function more complicated. So, correct the "Additional functional diversity" to "Function buttons that simplify operation".

2.5 Questionnaires

In order to improve the practicability and scientific of the evaluation indicators, this paper will reorganize the evaluation system after the expert review and revision into a questionnaire to quantify the form of scoring, and finalize the screening of the indicators to finally determine the computer input device (keyboard and Mouse) user experience evaluation system. The method used in the sample data is the questionnaire survey method, which collects the data needed for the research by widely distributing the questionnaire.

The survey period was 3 days, a total of 241 questionnaires were issued, and 176 valid questionnaires met the data sample size requirements of the Analytic Hierarchy Process. After descriptive statistical analysis of the sample data [Table 4], the mean and variance of the criterion layer are shown in the table, and the mean and variance of the alternative level are shown in the tables (the data is accurate to two decimal places).

Table 4. Average and standard deviation data of criteria level summary table

Criteria	Average deviation	Standard deviation
Satisfaction	4.42	0.71
Easy to learn	4.06	0.71
Effective	4.15	0.69
Error tolerant	4.17	0.72
Efficient	4.24	0.73

Table 5. Average and standard deviation data of alternatives level summary table

Criteria	Alternatives	Average deviation	Standard deviation
Satisfaction	Dimensions and shapes fit the human factors	4.51	0.69
	Good appearance	4.25	0.81
	Comfortable surface	4.01	0.90
	The button needs moderate strength	3.97	0.94
	Key feedback sounds comfort	3.94	0.77
Error tolerant	Button layout conforms to usage habits	4.29	0.76

	Button shape and size are easy to click	4.31	0.78
Easy to learn	Connect to your computer quickly and easily	4.31	0.82
	Intuitive interaction	4.24	0.75
	Additional functions are easy to learn	4.01	0.70
Effective	Button function available	4.56	0.61
	Long product life	4.60	0.65
	Stable connection	4.43	0.64
Efficient	Smooth use process	4.47	0.71
	Efficient response	4.13	0.80
	Moderate flexibility	4.02	0.84
	Function buttons that simplify operation	3.92	0.78

According to the results of the above survey data [Table 5], "The button needs moderate strength" in the satisfaction is eliminated, and the user experience evaluation system of the computer input device (keyboard and mouse) is finally determined [Table 6].

2.6 Final user experience evaluation system

Table 6. Descriptions of 5E model for computer input devices (keyboard and mouse)

Goal	Criteria	Alternatives
UX evaluation for computer input devices(keyboard and mouse)	Satisfaction	Dimensions and shapes fit the human factors
		Good appearance
		Comfortable surface
		Key feedback sounds comfort
	Error tolerant	Button layout conforms to usage habits
		Button shape and size are easy to click
	Easy to learn	Connect to your computer quickly and easily
		Intuitive interaction
		Additional functions are easy to learn
	Effective	Button function available
		Long product life
		Stable connection
	Efficient	Smooth use process
		Efficient response
		Moderate flexibility
		Function buttons that simplify operation

3. User experience evaluation model based on AHP

To calculate the weights by using AHP, giving the judgment matrix between each indicator of each level. The AHP combines qualitative analysis with quantitative analysis. The AHP is to decompose complex problems into several levels and several elements, and at the same level. Simply compare, judge and calculate, and obtain the importance of different alternatives, to provide decision-making basis for selecting the optimal solution. Constructing a judgment matrix: giving a judgment on the relative importance of each factor at each level, and using numerical values, generally using a scale of 1 to 9. [Table 7]

Table 7. UX evaluation index system weight

Goal	Criteria	Alternatives	Single weight	Comprehensive weight
User experience evaluation index system for computer input devices (keyboard and mouse)	Satisfaction 0.3982	Dimensions and shapes fit the human factors	0.3407	0.1357
		Good appearance	0.2865	0.1141
		Comfortable surface	0.2026	0.0807
		Key feedback sounds comfort	0.1703	0.0678
	Error tolerant 0.2422	Button layout conforms to usage habits	0.6667	0.1615
		Button shape and size are easy to click	0.3333	0.0807
	Easy to learn 0.1693	Connect to your computer quickly and easily	0.4126	0.0699
		Intuitive interaction	0.3275	0.0554
		Additional functions are easy to learn	0.2599	0.0440
	Effective 0.1183	Button function available	0.4434	0.0525
		Long product life	0.3874	0.0458
		Stable connection	0.1692	0.0200
	Efficient 0.0720	Smooth use process	0.3431	0.0247
		Efficient response	0.2426	0.0175
		Moderate flexibility	0.2426	0.0175
		Function buttons that simplify operation	0.1716	0.0124

References

[1] Research on User Experience Evaluation Methods of Smartphone Based on Fuzzy Theory. Yan Zhou (Sch. of Mechatron. Eng., Harbin Inst. of Technol., Harbin, China); Shanshan Niu; Siyu Wang Source: 2015 7th International Conference on Intelligent Human-Machine Systems and Cybernetics (IHMSC). Proceedings, p 122-5, 2015

[2] Lim Tek Yong. User experience evaluation methods for mobile devices[P]. ,2013.

[3] Development of a user experience evaluation framework for wearable devices. Kim, Young Woo (Department of Information and Industrial Engineering, Yonsei University, Seoul, Korea, Republic of); Yoon, Sol Hee; Hwangbo, Hwan; Ji, Yong Gu Source: Lecture Notes in Computer Science (including subseries Lecture Notes in Artificial Intelligence and Lecture Notes in Bioinformatics), v 10298 10298 LNCS, p 53-67, 2017, Human Aspects of IT for the Aged Population: Applications, Services and Contexts- Third International Conference, ITAP 2017 Held as Part of HCI International 2017, Proceedings

[4] Eye tracking evaluation of user experience on large-scale displays. Schall, A. (Key Lime Interactive, Miami, FL, United States) Source: Universal Access in Human-Computer Interaction. Access to Today's Technologies. 9th International Conference, UAHCI 2015, held as part of HCI International 2015. Proceedings: LNCS 9175, p 98-108, 2015

[5] User Experience Evaluation Methods: Current State and Development Needs, Arnold P.O.S. Vermeeren1, Effie Lai-Chong Law2, Virpi Roto3, Marianna Obrist4, Jettie Hoonhout5, Kaisa Väänänen-Vainio-Mattila6, 31Delft

IOP Conf. Series: Materials Science and Engineering **686** (2019) 012006 doi:10.1088/1757-899X/686/1/012006

The Development of Shop Documentation Plug-In Based on NX Software

Jiaqi Zhao*, **Xiaodong Yang and Ligeng Wu**

Department of Precision Manufacturing Centre, Tianjin Institute of Aerospace Mechanical and Electrical Equipment, Tianjin, China

*royjuventus_29@126.com

Abstract. Shop documentation is a document filled with all relevant information of product processing, which is playing an important role in CNC machining. At present, although shop documentation could be generated from 3D modelling software, the contents in shop documentation are usually not up to operators' requirement. On this condition, the data structure of shop documentation should be rebuilt, and a new approach of generating shop documentation should be created as soon as possible. In this paper, the data structure of shop documentation is redesigned to make sure that the document contains the most suitable information, and a new kind of shop documentation plug-in is developed in Siemens NX software at the same time, which could generate PDF-format shop documentation very fluently. The plug-in is programmed by C++, using the function library of NX software which is called NX Open C. Finally, an example of generating shop documentation in NX software is given to prove that the shop documentation plug-in works properly.

1. Introduction

In the last decades, with the great development of computer technology, CAD and CAM have made great contributions to CNC machines. As the complication of workpiece shape is increasingly growing, multi-axis machines are being used more widely [1]. It has become particularly vital of rationality of using equipment and equipment utilization. Reasonable scheduling scheme makes the use of production resource more efficient, reduces the production cost of enterprises, and enhances the competitiveness of enterprises [2]. To this point, Bin Li has provided an insight into reasonable scheduling scheme, and he commits to analysing process parameters, reducing process time and waste of tools [3].

With the advance of NC technology, the majority of modern CNC machines are characteristic of high speed, high precision, high reliability, intelligence and compounding [4]. Complex machine structure makes CNC programming require software assistance [5]. For the purpose of maximizing the performance of CNC machines, the separation of CNC programmers and operators is a major trend.

At present, companies establish digital and paperless workshops [6], including models, database-based information management and equipment management and so on [7]. The workshop management is applied for CNC machine management and tool management. It is also convenient for transmitting CNC program [8], and they use the CNC program list to send message of process to operators [9]. However, it is difficult to transmit some key information to operators and operating-complex software elevates operators' learning burden and company training cost.

In order to convey information more effectively, shop documentation is extremely necessary. So far, shop documentation has been suitable for the products of similar shapes and the specific CNC machines [10]. However, most of shop documentation is accomplished manually, which is easier to contain human errors. Besides, the editable documents are likely to be modified by mistakes during delivery process. Shop documentation could be generated within NX software [11], but the shop documentation could not list important information, including clamping method, WCS, step description and some useful process parameters.

Compared to traditional research results, the shop documentation plug-in in this paper overcomes the shortcomings mentioned above. Firstly, the information, including model, tool parameters, WCS and other parameters of the manufacturing process, could be fully extracted by the plug-in. Secondly, the plug-in could generate read-only shop documentation in PDF-format. Thirdly, the shop documentation is suitable for duplication, storage and delivery.

2. System planning of shop documentation plug-in

The shop documentation plug-in could automatically extract views and manufacturing information of the model. The views include front view, side view, top view and axonometric view, while the manufacture information includes operation names, tool parameters, cutting parameters and non-cutting parameters. The data structure and workflow of the plug-in are illustrated as follows.

2.1. Data structure

The data structure of the plug-in is as shown in figure 1.

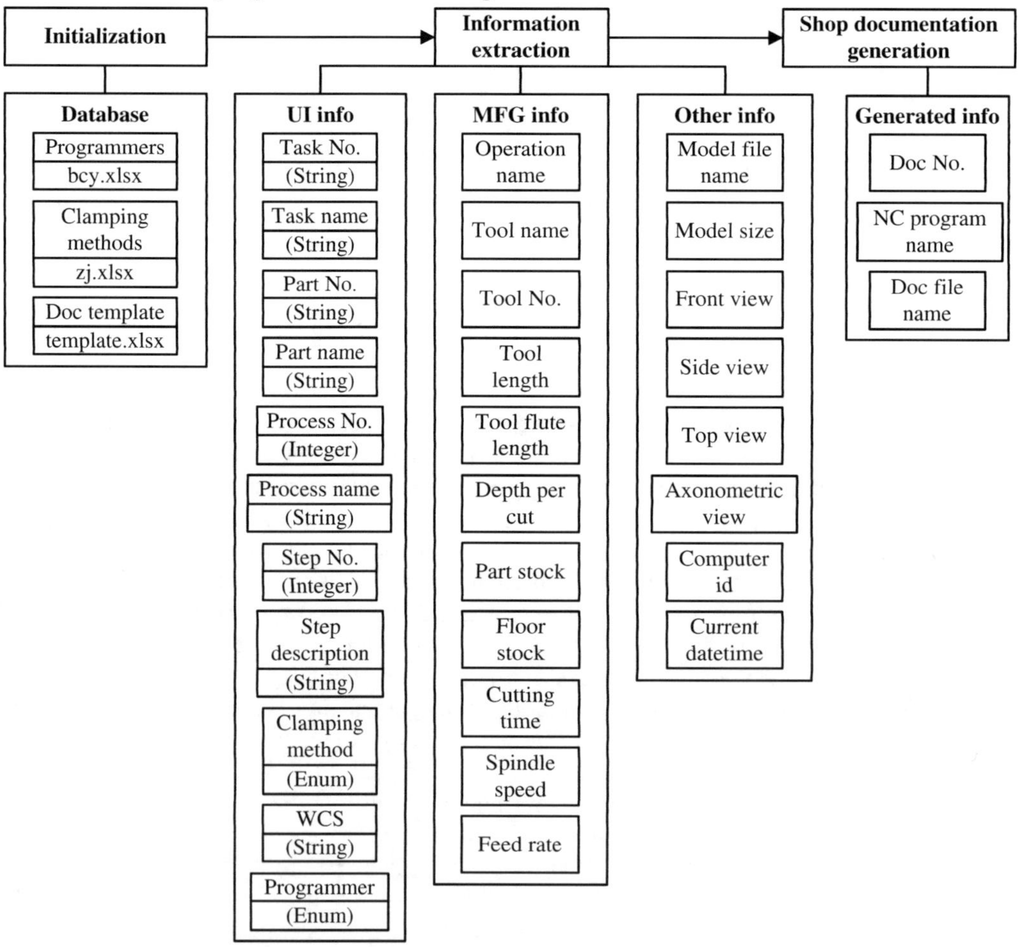

Figure 1. The data structure of shop documentation plug-in

IOP Conf. Series: Materials Science and Engineering **686** (2019) 012006 doi:10.1088/1757-899X/686/1/012006

2.2. Workflow
The workflow of the plug-in is as shown in figure 2.

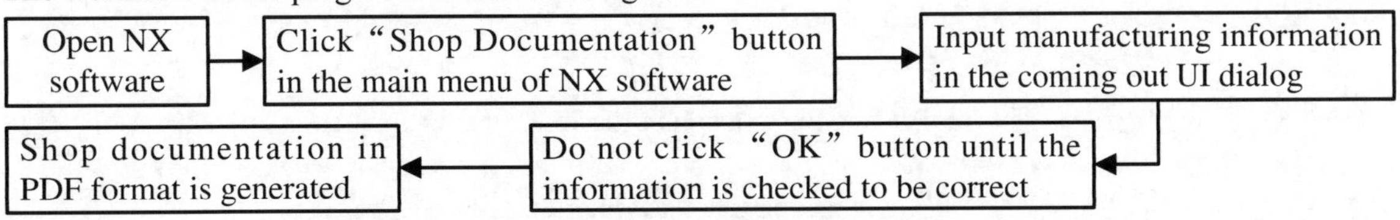

Figure 2. The workflow of shop documentation plug-in

3. The development of shop documentation plug-in

The plug-in is developed by C++ on the platform Microsoft Visual Studio version 2012, and it could be used in NX software version 10 or above. The key technologies of shop documentation plug-in development are as follows.

3.1. The design of UI dialog
The UI dialog is created under "Block UI Styler" module of NX software. In "Block UI Styler" module all the controls could be dragged straightforwardly to the UI dialog, and the UI dialog could be easily made as the designer's wish. The UI dialog interface is as shown in figure 3.

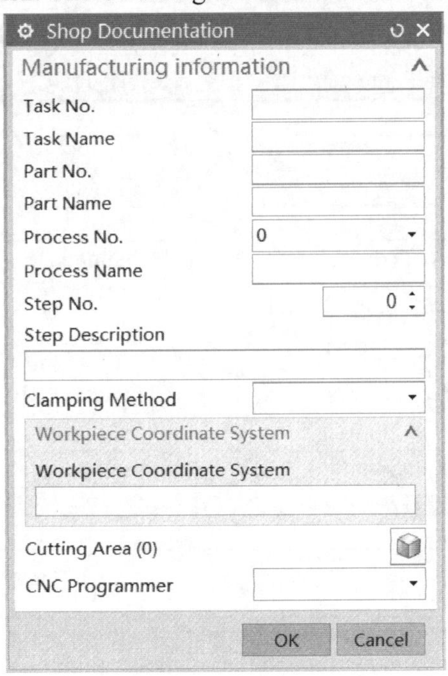

Figure 3. The UI dialog interface of shop documentation plug-in

3.2. The creation of model views
In NX software, the ACS (absolute coordinate system) is a conceptual location and orientation in model space. It is invisible and cannot be moved. All of the model's information is related to the ACS. The Machine Coordinate System (MCS) determines the orientation and origin of tool paths for all operations in the orient group. The MCS has an initial position which is matched to the ACS.

In actual processing, each face of the part may be machined. In order to express cutting area properly, the plug-in must set appropriate views to show the MCS and cutting faces. After this, the CNC programmer should set the top view which he wants to use, then the plug-in will calculate the remaining views. The calculation process is shown in figure 4.

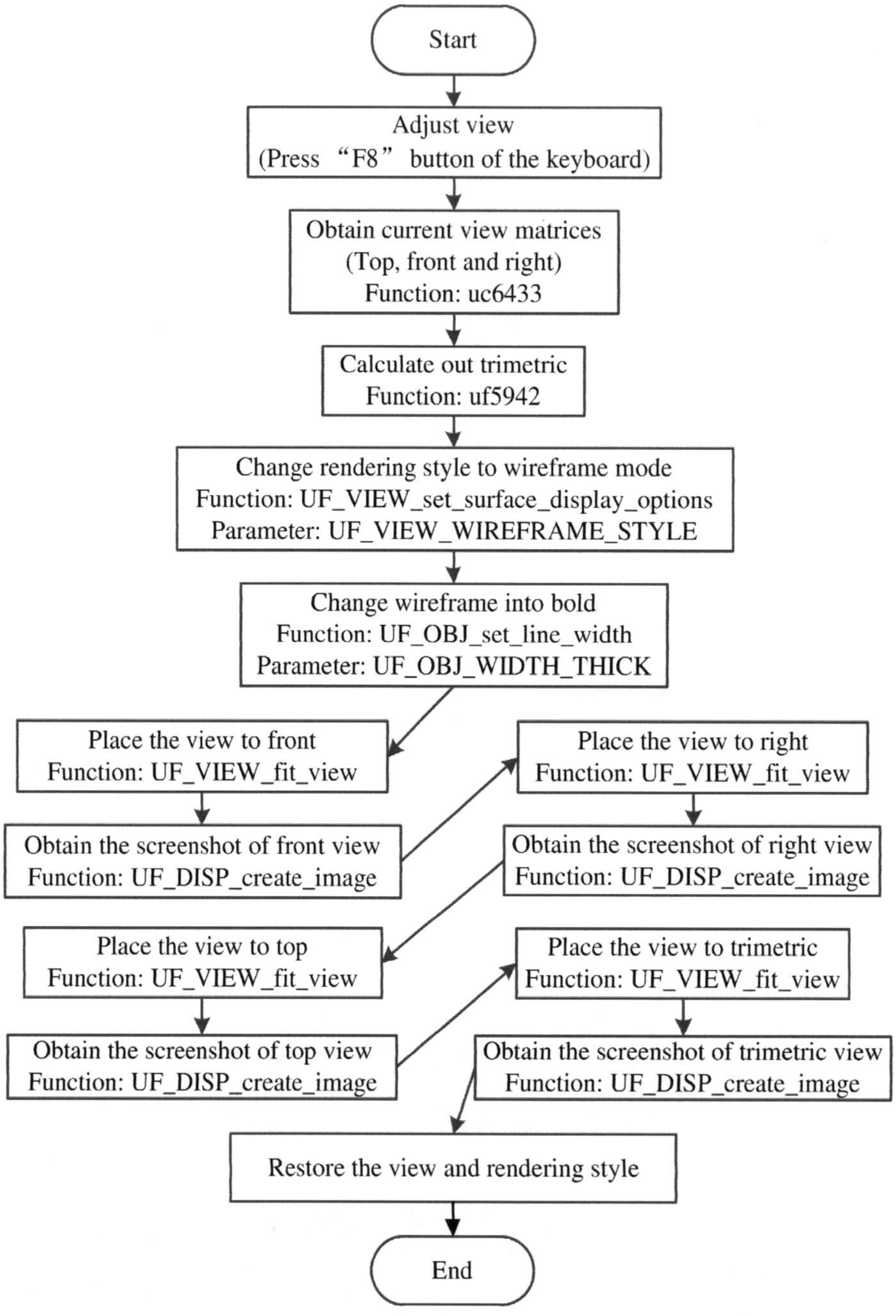

Figure 4. The flow chart of obtaining the screenshots of model views

a) Get the view matrix. The view matrix is a unit matrix, which is like $\begin{bmatrix} 1 & 0 & 0 \\ 0 & 1 & 0 \\ 0 & 0 & 1 \end{bmatrix}$. It is orientation of the current view relative to the ACS.
b) The plug-in analyses whether the axis of the view is parallel (or vertical) to ACS or not. If the view is eligible, the plug-in will keep going. Otherwise, it will remind the user to adjust views.
The view matrix is a unit matrix. In 3D space, it is a ball with a radius of 1, and the spherical centre coordinates is (0, 0, 0) (figure 5).When the axis of the view is parallel (or vertical) to ACS,

it intersects the unit ball at the quadrant point .The vector may be (1, 0, 0), (0, 1, 0), (0, 0, 1), (-1, 0, 0), (0, -1, 0) or (0, 0, -1). In order to determine the eligibility of the views conveniently, the function $|x|+|y|+|z|=1$ (figure 5) is established. In other words, when the vector meets the condition, the plug-in will proceed.

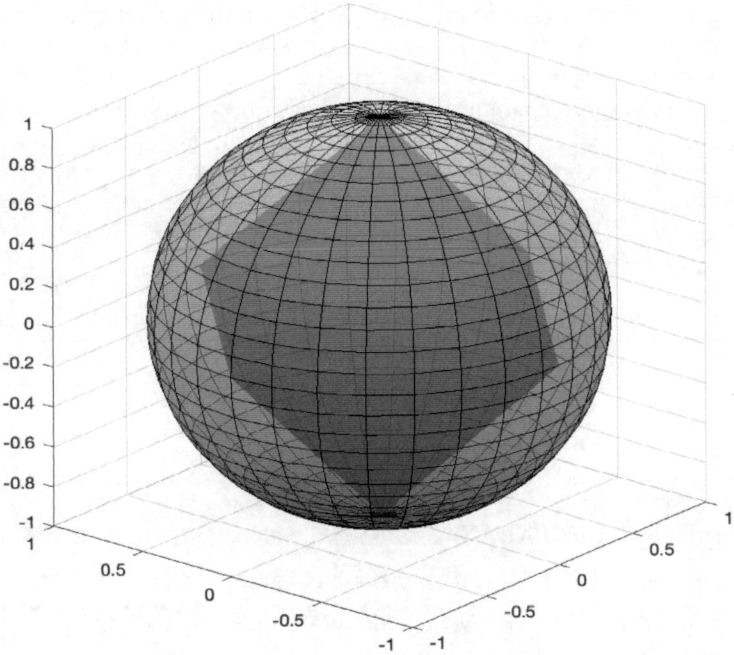

Figure 5. Unit ball and function $|x|+|y|+|z|=1$ graph

c) In the previous step, the plug-in obtains the view matrix $\begin{bmatrix} n_{11} & n_{12} & n_{13} \\ n_{21} & n_{22} & n_{23} \\ n_{31} & n_{32} & n_{33} \end{bmatrix}$, in which $|n_{11}|+|n_{12}|+|n_{13}|=1$, $|n_{21}|+|n_{22}|+|n_{23}|=1$, $|n_{31}|+|n_{32}|+|n_{33}|=1$. Then, for the purpose of displaying parts clearly, the plug-in will change rendering style to wireframe. In NX software, using the function by the parameters n_1 and n_2, it will return the view of X-Y plane, according to the user-defined view. In this way, the plug-in obtains the front view, side view and top view of the part.

d) The cutting area including several faces, the plug-in could present the cutting area in the trimetric view, which makes the processing information more intuitive.
The transformation method is $M_c \times M_v$. In this function, M_c is the view matrix. M_v is defined as $\begin{bmatrix} M_r & M_t \\ 0 & 1 \end{bmatrix}$. M_r is a rotate matrix. $M_r = R_X R_Y R_Z$.

The R_X is defined as: $\begin{bmatrix} 1 & 0 & 0 \\ 0 & \cos(roll) & \sin(roll) \\ 0 & -\sin(roll) & \cos(roll) \end{bmatrix}$

The R_Y is defined as: $\begin{bmatrix} \cos(pitch) & 0 & -\sin(pitch) \\ 0 & 1 & 0 \\ \sin(pitch) & 0 & \cos(pitch) \end{bmatrix}$

The R_Z is defined as: $\begin{bmatrix} \cos(yaw) & \sin(yaw) & 0 \\ -\sin(yaw) & \cos(yaw) & 0 \\ 0 & 0 & 1 \end{bmatrix}$

$(\text{roll} = 45°, \text{pitch} = 45°, \text{yaw} = 45°)$.

M_t is a translation matrix. As trimetric view does not require translation, $M_t = \begin{bmatrix}0\\0\\0\end{bmatrix}$. After that, the colour of cutting area in trimetric view changed into dark colour so that cutting area could be figured out.

e) Finally, the plug-in will direct view to the X-Y plane and change rendering style to shade.

3.3. The extraction of manufacturing parameters

In manufacturing process, a large amount of information is necessary to be passed to CNC machine operators, including the information of task, model, WCS, clamping, operation and tools. The plug-in requires proper CNC program, cutting area and essential basic data.

In the NX manufacturing application, most of operation data is set in the same way, including tools data, part stock, floor stock, spindle speed and feed rate. The functions and parameters needed by extracting data are as shown in table 1.

Table 1. Functions and parameters of the extraction of manufacturing data

MFG info	Data sort	Function	Parameter of function
Operation type	String	UF_OPER_ask_oper_type()	Tag of operation
Tool name	String	UF_OBJ_ask_name()	Tag of tool
Tool No.	Integer	UF_PARAM_ask_int_value()	UF_PARAM_TL_NUMBER
Tool length	Double	UF_PARAM_ask_double_value()	UF_PARAM_TL_HEIGHT
Tool flute length	Double	UF_PARAM_ask_double_value()	UF_PARAM_TL_FLUTE_LN
Part stock	Double	UF_PARAM_ask_double_value()	UF_PARAM_STOCK_PART
Floor stock	Double	UF_PARAM_ask_double_value()	UF_PARAM_STOCK_FLOOR
Spindle speed	Double	UF_PARAM_ask_double_value()	UF_PARAM_SPINDLE_RPM
Feed rate	Double	UF_PARAM_ask_subobj_ptr_value()	UF_PARAM_FEED_CUT
Cutting time	Double	UF_PARAM_ask_double_value()	142
Cut level	Double	UF_PARAM_ask_double_value()	Depending on operation type

The cut levels are double numbers, but the method of setting different. Planar/face milling use Depth per Cut. Fixed/variable contour milling uses the Stock Part Offset. Cavity/zlevel/zlevel_5axis uses Global Cut Depth. Hole milling uses Axis Pitch. Thus, when the plug-in extracts values of the above-mentioned operations, it will identify the types of the operations and use the matched parameters.

3.4. The generation of shop documentation

The shop documentation is generated from a template in the format of excel, and saved as a PDF format file. The program execution procedure of the generation of shop documentation is as shown in figure 6.

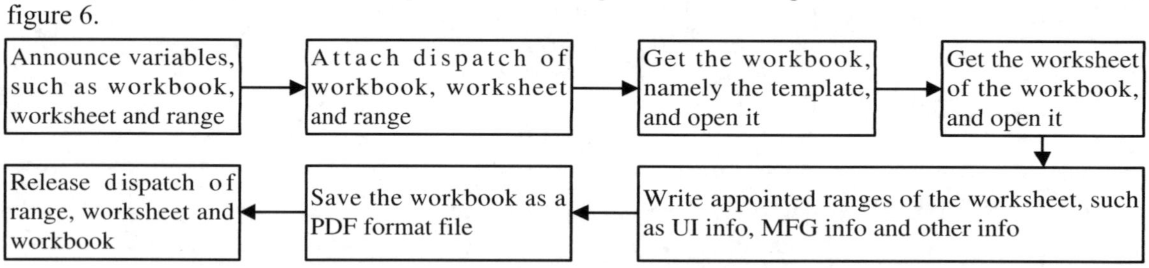

Figure 6. The program execution procedure of the generation of shop documentation

4. Experiment and verification

This paper will use a part model to test and verify the plug-in, which is as shown in figure 7.

4.1. The description of the experiment

In this workpiece, the MCS is set at center and all surfaces need milling-finishing. This processing work step as follows:

a) Use a D35R5 cutting tool to process four fillets and open rough.
b) Use a D17R0.8 cutting tool to process the rest of upper surface and semi-finished side.
c) Use a D6 cutting tool to finish the ground and side.
d) Use a R4 cutting tool to finish the slender groove on the top of the part.

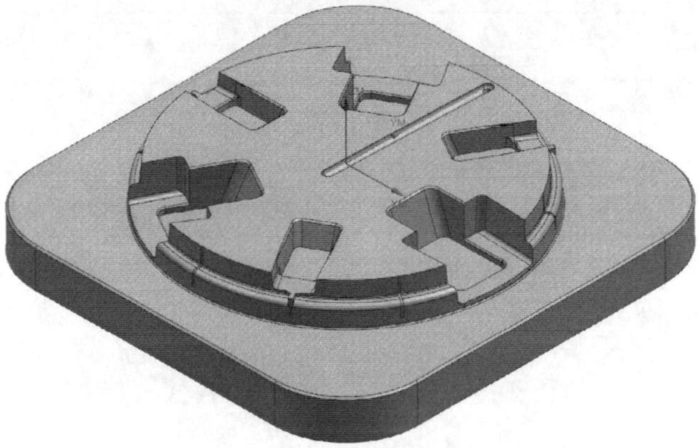

Figure 7. The part model used for the experiment

4.2. The verification of shop documentation plug-in

When users run the plug-in, they need to fill in a dialog, which is as shown in figure 8.

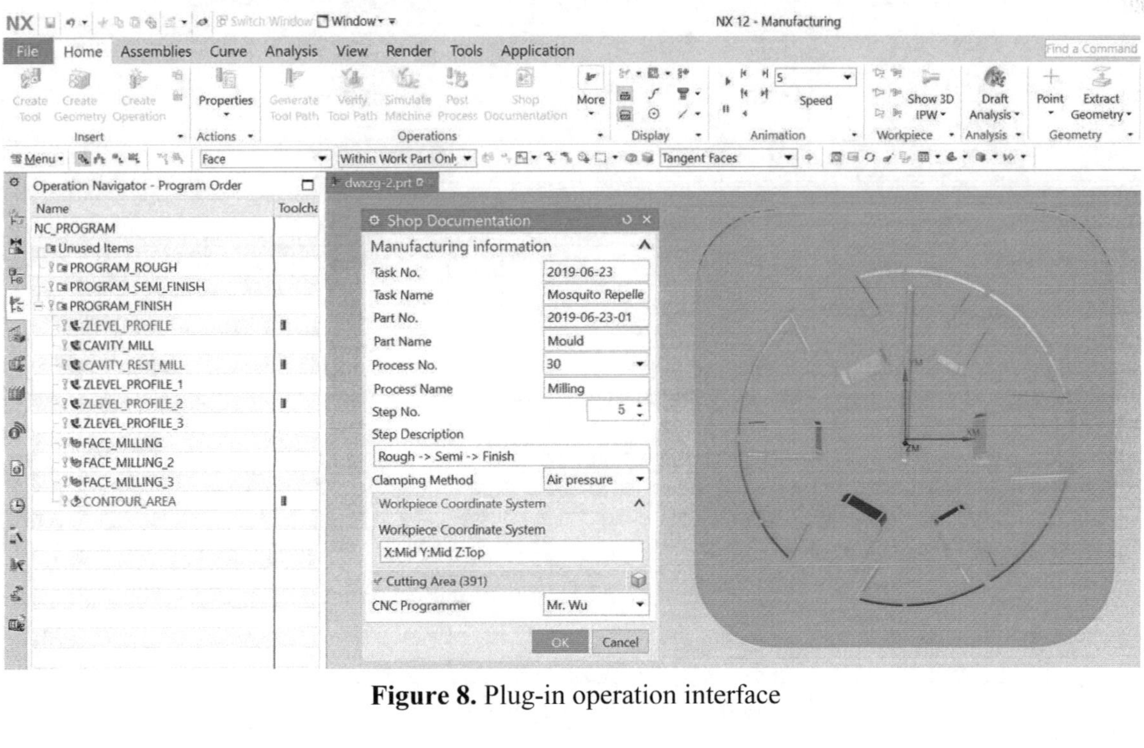

Figure 8. Plug-in operation interface

IOP Conf. Series: Materials Science and Engineering **686** (2019) 012006 doi:10.1088/1757-899X/686/1/012006

After filling in the UI dialog, selecting the cutting area, and setting the model views, shop documentation will be generated. The shop documentation generated from the plug-in is as shown in figure 9, which could verify the shop documentation plug-in is working properly and friendly to use.

Shop Documentation

File No.	JJCXD20190623233153		**Program No.**	dwxzg-2_30_5		**Cutting Time**	454.07min
Task No.	2019/6/23		**Task Name**	Mosquito Repelle		**Body Size**	330×340×64.7872
Part No.	2019-06-23-01		**Part Name**	Mould		**Modelfile Name**	dwxzg-2.prt
Process No.	30	**Process Name**	Mill	**Step No.**	5	**Step Description**	Rough->Semi->Finish
Programmer	Mr. Wu		**Computer**	yxd_s		**Date**	23/6/2019 23:31:53
Clamping	Air pressure		**WCS**	X:Mid.;YMid;Z:Top			

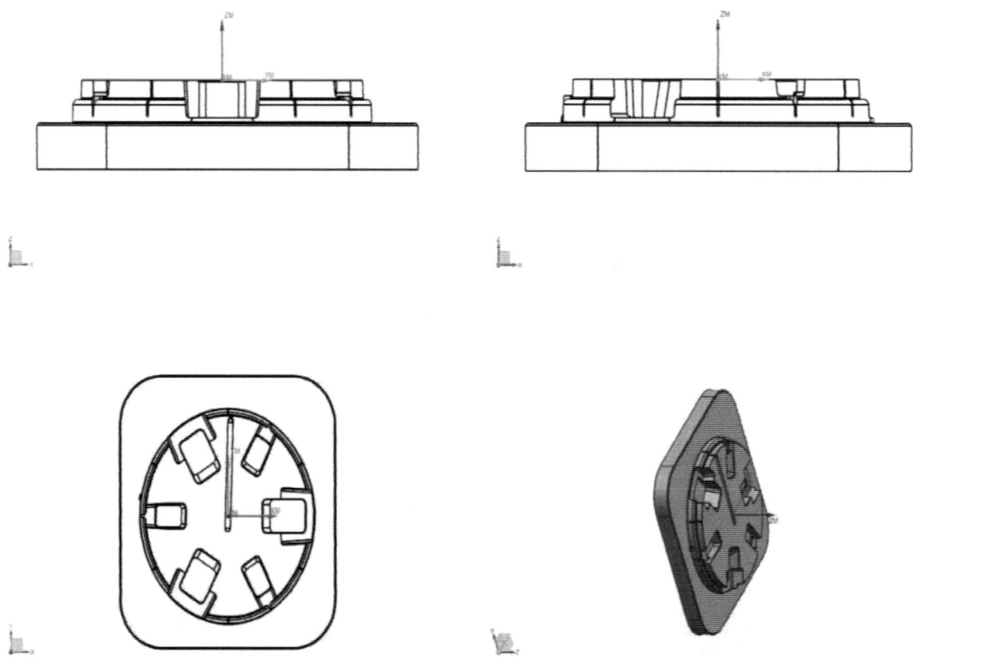

Operation Name	Tool Name	Tool No.	Tool/Flute Len.	Cut Levels	Part/Floor Stock	Time	Speed	Feed
ZLEVEL_PROFILE	D35R5	1	120/40	0.3	0/0	98.29min	2500	1500
CAVITY_MILL	D35R5	1	120/40	1	0.3/0.15	112.05min	2000	2500
CAVITY_REST_MILL	D17R0.8	2	80/40	0.25	0.35/0.15	57.88min	2000	2500
ZLEVEL_PROFILE_1	D17R0.8	2	80/40	0.35	0.15/0	86.46min	2500	1500
ZLEVEL_PROFILE_2	D6	6	60/30	1	0.12/0	33.23min	3500	1500
ZLEVEL_PROFILE_3	D6	6	60/30	1	0/0.05	33.26min	3500	1500
FACE_MILLING	D6	6	60/30	0	0/0	21.45min	2000	1000
FACE_MILLING_2	D6	6	60/30	0	0/0	2.08min	2000	1000
FACE_MILLING_3	D6	6	60/30	0	0/0	1.07min	2000	1000
CONTOUR_AREA	R4	4	75/40	4	0/0	8.30min	2000	1000

Figure 9. The shop documentation generated by the plug-in

IOP Conf. Series: Materials Science and Engineering **686** (2019) 012006 doi:10.1088/1757-899X/686/1/012006

5. Conclusion

In this paper, a kind of shop documentation plug-in has been successfully developed, which could extract all the relevant information of the manufacturing product from NX software, and combine all the information to generate shop documentation in PDF format. This development of the shop documentation plug-in makes all manufacturing information shown in a file, which not only reduces times of the communication between programmers and machine operators, but also makes NC programming mistakes easily to be found out and corrected. Overall, the development of this kind of shop documentation plug-in makes manufacturing process more effective and more efficient.

Acknowledgments

We wish to acknowledge the encouragement from our department leaders, Zhaohui Zhu and Huadong He. Besides, we also appreciate for the support from Xiaobao Li Skill Master Studio of China.

References

[1] Yibo Liu and Fangrong Liu 2013 Classification and trend of NC programs *Metal Processing* **7** 25

[2] Xin Zhou 2014 *Design and implementation of workshop scheduling system on distributed numerical control system* (China: University of Chinese Academy of Sciences) p 7

[3] Bin Li 2007 *Basic on Lean Production of CNC Machining* (China: The College of Automation Chongqing University) p 62

[4] Wei Liang and Xian Wang 2010 Development trend of CNC machine tools *Journal of Guilin college of Aerospace Technology* **58** 164

[5] Zhaoliang Liu and Zanbo Yang 2018 Deeply study the application of NX software in NC machining *World Nonferrous Metal* **2018(9)** 285-286

[6] Song Jin 2006 *Development of Paperless Production System Based on Embedded Terminal In Tool workshop* (China: the College of Automation Chongqing University) p 50

[7] Yu Zhang 2009 *Design and Implementation of CNC Workshop Management System* (China: Heilongjiang University of China) p 54

[8] Limao Mu 2003 The software development of long NC machining code transmission direct from PC machine to NC machine *Modular Machine Tool & Automatic Machining Technology* **2003(6)** 53-55

[9] Guoli Gao and Jiaqiang Huang 2010 Automatic generation of machining program list and network post-process for die & mould *Die & Mould Industry* **2010(4)** 16-17

[10] Haifeng Wang, Bin Wang and Xuehua Li *Research on CNC Machining Template and Program Automation* (China: High Level Equipment for Civil Aircraft Manufacturing Technology and Equipment) p 8-9

[11] Siemens NX10 Documentations (Germany: SIEMENS)

IOP Conf. Series: Materials Science and Engineering **686** (2019) 012007 doi:10.1088/1757-899X/686/1/012007

Development of a Parallel Hydraulic Hybrid Loader Through Modelling

Jixiang Yang[1], Yongming Bian[1], Guangjun Liu[1], Pengcheng Ji[1] and Qifan Chen[1]

[1]School of Mechanical Engineering, Tongji University, Shanghai 201804, China

*Corresponding Author's Email:ymbianmail@163.com

Abstract. With the aggravation of the energy crisis, people pay more and more attention to the rational utilization of energy. Thus the construction machinery field has started to take hybrid technology into consideration to save the energy. This paper analyses the working condition of a 5 tons loader and develops the hydraulic parallel hybrid scheme which has little research in the loader field. The system adopts a dynamic rule-based control strategy, which takes the required torque and the SOC value of the accumulator as the input to realize the output torque of the engine and the pump/motor. A power-train model is also developed in the paper. The results show the scheme can reduce the fuel consumption by 13.5% compared to the conventional loaders.

1. Introduction

With the aggravation of the global energy crisis [1, 2], hybrid, which is a new fuel-saving technology, has gained more attention in recent years [3-5]. Loader, as a typical construction machines, has the characteristics of frequent starts/stops, low energy utilization, high energy consumption and poor emissions. Numerous construction machinery manufacturers at home and abroad have started to study the hybrid power technology of construction machinery. The first hybrid loader was proposed by Hitachi Construction in 2003 [6], Volvo introduced L220F Hybrid wheel loader, an energy-efficient parallel hybrid electric drive system batteries for storage in 2008 and a LXI prototype hybrid wheel loader using a series hybrid system made a first appearance In Switzerland which showed that it could improve fuel efficiency by 50% in 2016 [7]. The CLG862-HYBIRD, China's First Hybrid Loader, proposed by Liugong in 2010, is a series-parallel hybrid wheel loader with super capacitor [8]. XCMG launched a parallel hydraulic hybrid system with engine, variable pump, hydraulic accumulator, hydraulic pump/motor devices which can reduce 15% fuel consumption by recovering of braking kinetic energy in the same year [9]. There is few literature on hybrid loader due to technical confidentiality and other reasons.

This paper reports the development of parallel hybrid versions of a loader through modelling. The models were made using two different software packages, LMS Imagine.Lab AMESim and AMESet based on a 5 tons loader. The energy-saving effect and performance of the system are studied which can prepare for parameter matching and subsequent test validation.

2. Parallel hydraulic hybrid loader

2.1. Working condition of loader

A typical working cycle of a 5 tons loader consists of six steps [10]. The first step is that the empty loader drives to the gravel bank from initial position, the second step is that the empty loader scoops and loads gravel which means the empty loader becomes to the full-load loader, the third step is that the full-load loader returns to the initial position, the fourth step is that the full-load loader drives to the unloading position, the fifth step is that the loader unloads the gravel and the full-load loader becomes to the empty loader, the final step is that the empty loader returns to the initial position. Generally speaking, for a loader with two forward gears and one backward gear, the first, second and fourth steps are carried out in the first forward gear, the third and final steps are carried out in the backward gear.

2.2. Structure of parallel hydraulic hybrid loader

As shown in figure 1, the structure of parallel hydraulic hybrid loader is illustrated. The brake energy is regenerated by pump/motor and stored in the accumulator and a coupler works as a transmission which can match the speed of pump/motor in order to prevent over-speed. The energy of the engine is divided into two parts. One is transmitted to drive the working device and control the steering of the loader and the other is converted by the torque converter, the gearbox and the transaxle to drive loader.

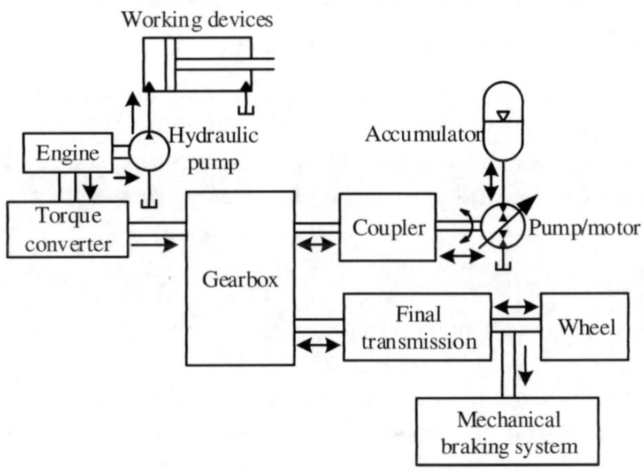

Figure 1. Structure of parallel hydraulic hybrid loader.

3. Hybrid propulsion system design

3.1. Modelling subsystems

Most key components of hybrid loader can be found in AMESim and the object-oriented modelling means more convenient modelling process. The characteristic map of engine and primary characteristics of a torque converter should be taken to ensure the correctness of the model. Part of the key components is similar to the ordinary loader and parameters can be reference to the ordinary loader, and the other components need continuous matching to find the best configuration. And the transmission ratio of coupler has a limit, it should guarantee that the pump/motor will not get over-speed [11], as shown in equation (1), among it n_{max} is the max speed of pump/motor, v_{max} is the max velocity of loader, the transmission ratio i_a is from coupler to the wheel, the transmission ratio of coupler is i_c.

$$i_c \le \frac{2\pi r_w n_{\max}}{60 v_{\max} i_a} \tag{1}$$

3.2. Control Strategy

Control strategy is the core issue of energy management allocation in hybrid power system [12-14]. In this paper, two control strategies are adopted and their fuel-saving effects are compared to each other.

One is the static logic threshold control strategy(SLTCS)—as shown in figure 2 and the other is dynamic rule-based control strategy(DRCS)—as shown in figure 3.

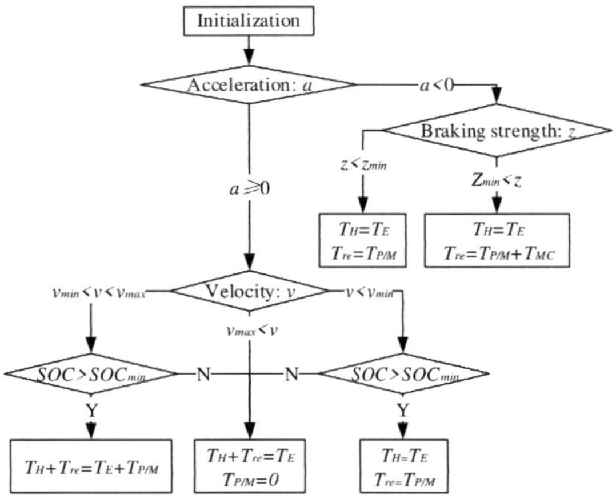

Figure 2. Static logic threshold control strategy.

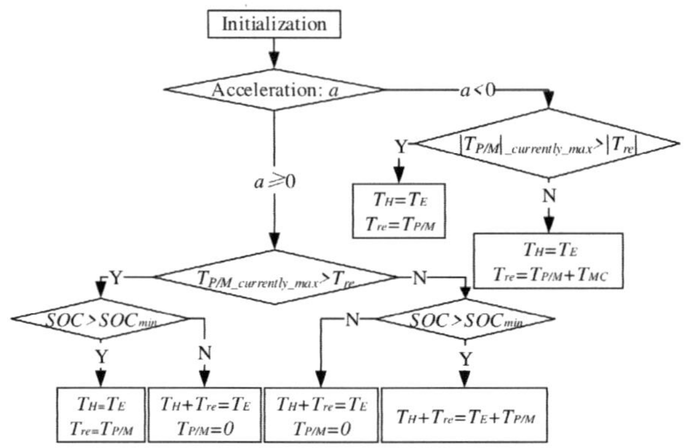

Figure 3. Dynamic rule-based control strategy.

The SLTCS first determines whether the loader is in braking state or not, braking state means the acceleration of loader $a<0$. In the braking state, if the braking strength $z<z_{min}$ then the torque of hydraulic system T_H will be provided by torque of engine T_E and the torque of drive system T_{re} will be provided by that of pump/motor $T_{P/M}$, otherwise the torque of drive system T_{re} shall be provided by the torque of pump/motor $T_{P/M}$ and the torque of mechanical brake system T_{MC}. When the loader is out of braking state, if the velocity $v<v_{min}$ and the accumulator charging state $SOC>SOC_{min}$, T_H will be supplied by T_E and T_{re} will be supplied by T_E and $T_{P/M}$; if the velocity $v_{min}<v<v_{max}$ and the $SOC>SOC_{min}$, T_H will be supplied by T_E and T_{re} will be supplied by $T_{P/M}$; if the velocity $v_{max}<v$ or the $SOC<SOC_{min}$, T_H and T_{re} will all be supplied by T_E.

The DRCS first determines whether the loader is in braking state or not, and the braking state means acceleration of loader $a<0$. In the braking state, if the $T_{P/M}$ is enough to stop the loader then T_H will be provided by T_E and T_{re} will be provided by $T_{P/M}$, otherwise the T_{re} shall be provided by $T_{P/M}$ and T_{MC}. When the loader is out of braking state, if the $T_{P/M}$ is enough to drive loader and the accumulator charging state $SOC>SOC_{min}$, T_H will be supplied by T_E and T_{re} will be supplied by $T_{P/M}$; if the $T_{P/M}$ cannot drive loader alone $SOC>SOC_{min}$,, T_H shall be supplied by T_E and T_{re} will be supplied by $T_{P/M}$ and T_E; if $SOC<SOC_{min}$ then T_H and T_{re} should all be supplied by T_E.

4. Simulation model

As shown in figure 4, the models were created by LMS Imagine.Lab AMESim and AMESET. The objective of the study was the preliminary sizing of the hybrid configurations for the loader. The hydraulic parallel hybrid loader is shown in figure 1, and some parameters are shown in table 1.

Table 1. Parameters of hydraulic parallel hybrid loader.

Specifications	Value	Specifications	Value
Displacement of pump/motor(ml r^{-1})	180	Maximum speed in first gear(km h^{-1})	11.8
Volume of accumulator(L)	30	Maximum speed in second gear(km h^{-1})	15.6
Ratio of coupler	2.73	Maximum reverse speed(km h^{-1})	37.4
Ratio of gearbox	2.547/0.683/1.864	Working pump(ml r^{-1})	100
Main drive ratio	4.928	Steering pump(ml r^{-1})	83.6
Wheel reducer ratio	4.625	Maximum speed in first gear(km h^{-1})	11.8

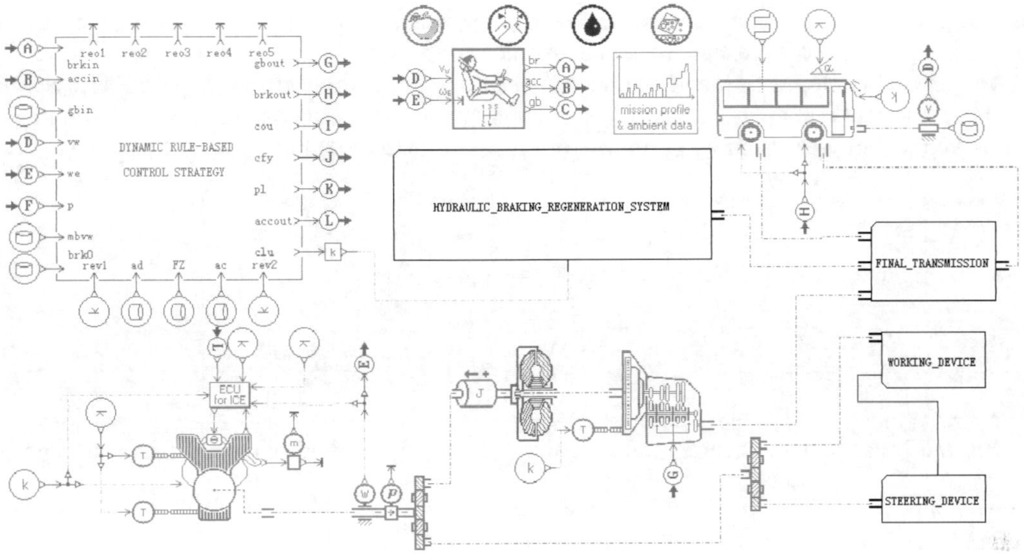

Figure 4. Simulation model in Amesim.

5. Results

5.1. Fuel Consumption

Parts of the simulation results shown in figure 5 demonstrate that the hybrid loader can reduce fuel consumption. After five typical working cycles, ordinary loader consumes 1.01kg diesel oil. And hybrid loader with DRCS can reduce fuel consumption by 13.5%, while the reduction rate of hybrid loader with SLTCS is 9.6%. The pump/motor works in the pump mode which can convert the kinetic energy of the loader into hydraulic energy and store it in the accumulator during the deboost phase (speed of loader drops and the loader is not at shovelling stage) of loader, then the hydraulic energy is applied to drive the loader when the loader is moving or shovelling and the pump/motor works in the motor mode at this time. The relationship between pressure of accumulator and speed of loader is shown in figure 6.

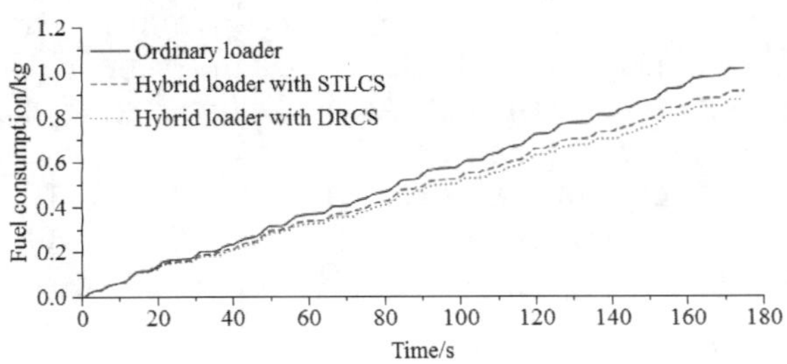

Figure 5. Fuel consumption of loader.

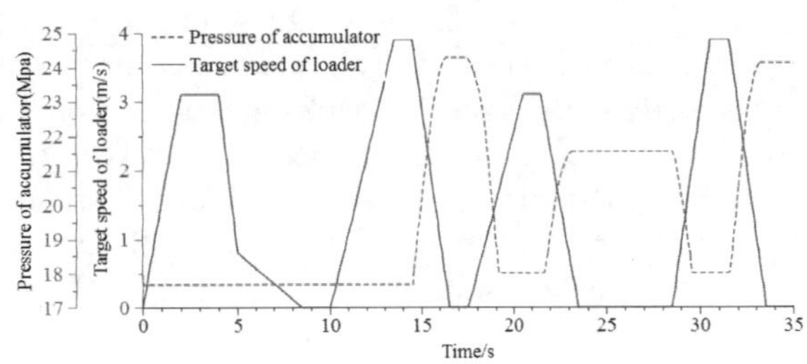

Figure 6. The relationship between pressure and speed.

However, hybrid loader with DRCS can recover more energy which means more fuel consumption decline as shown in figure 7. The pressure of accumulator for hybrid loader with DRCS increases from 18 Mpa to 24.3 Mpa in first recycling and the pressure of hybrid loader with STLCS raises from 18 Mpa to 23.5 Mpa. And high pressure means more hydraulic energy is stored when other factors are the same ones. The stored hydraulic energy can be calculated. In a typical working cycle, hybrid loader with DRCS can reclaim 231 KJ of energy, while that of hybrid loader with STLCS is 212.5 KJ. Because hybrid loader with DRCS can reduce the use of mechanical brake which means more kinetic energy is converted into hydraulic energy and stored instead of being consumed as thermal energy. As shown in figure 8, if the pressure of accumulator is increasing, the pump/motor will recover energy. At this time, lower mechanical brake strength means more energy will be recovered. This value in hybrid loader with DRCS is closed to 0 and is much lower than that of hybrid loader with STLCS which is around 0.01. The brake moment at the wheel is 2000 Nm when the mechanical brake strength is 0.01 and the brake moment generated by pump/motor ranges from 20000Nm to 28000Nm.

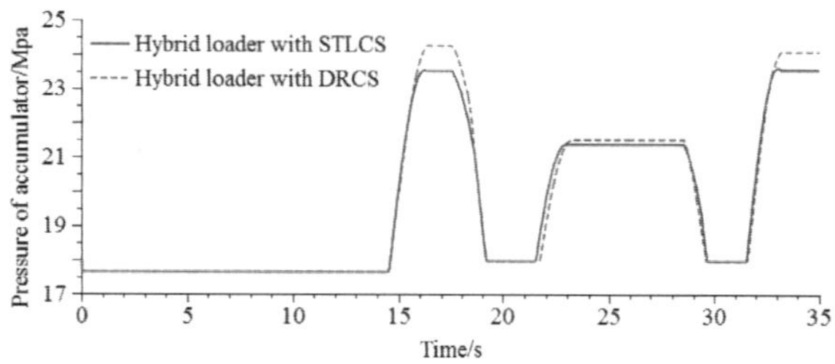

Figure 7. Pressure of accumulator.

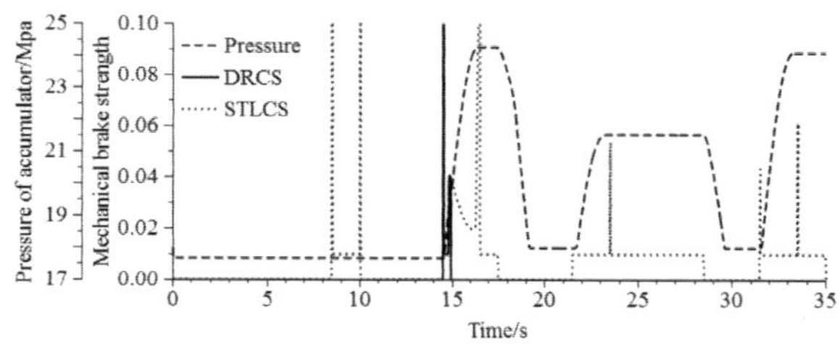

Figure 8. Mechanical brake strength.

5.2. Speed control

The general trend of the speed of hybrid loader is the same as that of ordinary loaders as shown in figure 9. But there is a problem that the hybrid loader has greater fluctuation than ordinary loader.

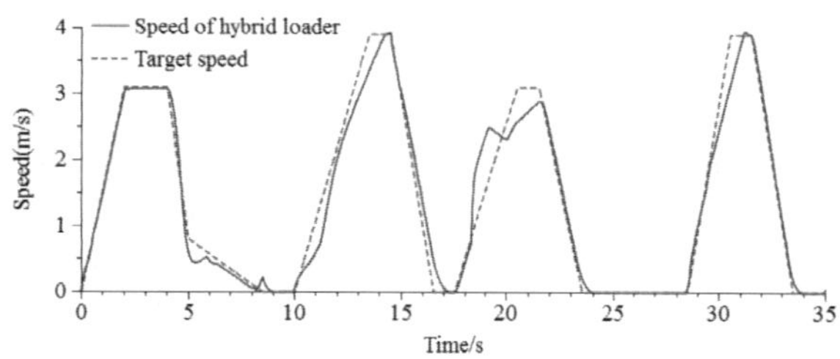

Figure 9. Speed of loader.

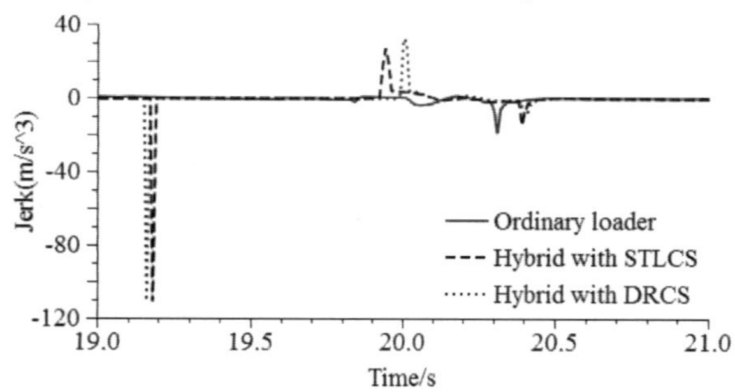

Figure 10. Jerk of loader.

This phenomenon is mainly reflected in the acceleration stage, especially when loader is full-load. As shown in figure 10, the jerk (derivative of acceleration) of the loaders are displayed. The jerk of the ordinary loader is close to 0 and is much lower than hybrid loader at about 19.2s, while that of the hybrid loader is -110.55 ms^{-3}.The reason for this abnormal jerk of hybrid loader is that the pump/motor stops driving the loader and the engine can't provide enough power in time. As shown in figure 11, the pressure of accumulator dropped to its lowest level which means pump/motor will stop. This problem can be solved by changing the control strategy. For example, to ensure the energy stored in accumulator is sufficient to drive the vehicle to the stage of uniform speed by changing the hydraulic energy release conditions or add a buffer stage before the accumulator runs out of stored energy. However, this is not within the scope of this paper.

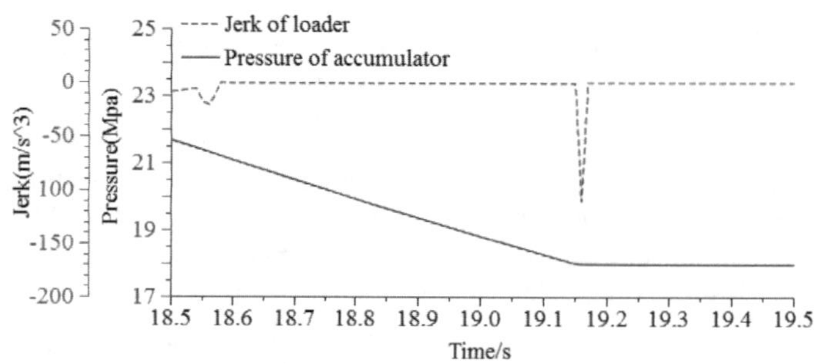

Figure 11. Reason for the abnormal jerk.

6. Conclusions

This paper discussed the process of modelling and sizing of a parallel hydraulic hybrid 5 tons loader in order to save energy. The simulation results indicate the purpose was practicable with a fuel consumption reduction for about 13.5% with DRCS, while that of hybrid loader with STLCS is 9.6%. On the other hand, the reason for the problem of abnormal jerk is illustrated and the solutions of the problem are briefly introduced. The optimization of system parameters and control strategies are absence and require further refinements and the development of the activity. However, the purpose of this paper is to provide a basic parameter matching and control strategy simulation which can show how the hybrid loader is working and avoid unreasonable component parameters and control strategies before formal design testing by the performance of the hybrid loader in the simulation results.

Acknowledgments

The authors would like to acknowledge the support of the National Key R&D Program of China (Grant No. 2016YFC0802900) and the Fundamental Research Funds for the Central Universities.

References

[1]. Pattanayak P, Panigrahy N, and Mishra S 2015 *Int. Conf. on 2015 IEEE 9th International Conference on Intelligent Systems and Control (9-10 Jan 2015 Piscataway)* (IEEE).
[2] C Yu Huei, L Ching Ming and T Jiashen 2017 *Int. Conf. on 2017 IEEE 8th International Conference on Awareness Science and Technology (8-10 Nov 2017 Piscataway)* (IEEE).
[3] A Bertini, Ceraolo M and Lutzemberger G 2012 *Int. Conf. on 2012 IEEE International Energy Conference (9-12 Sept 2012 Piscataway)* (IEEE).
[4] Kumar S, Das S, Ghoshal S K and Das J 2017 *Int. Conf. on Mechanical, Material and Renewable Energy (8-10 Dec 2017 UK)* (IOP).
[5] Pan X, Yan W, Cheng Y H and Li G X 2016 Hybrid loader automatic shift strategy based on neural network *J. Donghua University (English Edition)* **33** 4.
[6] Li T Y, Zhao D X, Kang H L, Zhang Z W, Zhang Z F and Xu C B 2013 Parameter matching of parallel hybrid power loaders J. *of Jilin University (Engineering and Technology Edition)* **43** 4.
[7] Z Zhi Wen, Z Ding Xuan, Li Tian Yu and Z Min 2015 Control strategy based on automatictransmission for hybrid loader *J. Northeastern University. Natural Science* **36** 4.
[8] Zou N W, Dai Q L, Jia Y H, Zhang W and Ren Y C 2010 Modeling and simulation research of coaxial parallel hybrid loader *Advanced Materials Research* vol 29-32 pp 1634-9.
[9] Yan L J, Sun H, Liu W, Jiang J H and Han J W 2014 Hydraulic hybrid technology of moving construction machinery *J. Jilin University (Engineering and Technology Edition)* **44** 2.
[10] Lin M, Yu Z, Zhao L and Chen Y 2018 Working cycle identification-based braking control strategy and its application for hydraulic hybrid loader *Advances in Mechanical Engineering* **10** 5.
[11] Shi R, Zhao J and Sun H 2011 Energy-saving potential and influencing factors for hydraulic hybrid wheel loader *Transactions of the Chinese Society for Agricultural Machinery* **42** 3.
[12] Wang F, Mohd Zulkefli M A, Sun Z and Stelson K A 2016 Energy management strategy for a power-split hydraulic hybrid wheel loader. Proceedings of the Institution of Mechanical Engineers, Part D *J. Automobile Engineering* **230** 8.
[13] Duan B, Wang Q, Zeng X, Gong Y, Song D and Wang J 2017 Calibration methodology for energy management system of a plug-in hybrid electric vehicle *Energy Conversion and Management* **136**.
[14] Patel P S and Desai H P A 2015 *Int. Conf. on 2015 Electrical, Electronics, Signals, Communication and Optimization (24-25 Jan 2015 Piscataway)* (IEEE).

EEG Signal Radio Frequency Control Wheeled Robot Based on Bi-objective Chaotic Particle Swarm Optimization Algorithm

Shan Guan* **and Xingwang Liu**

School of Mechanical Engineering, Northeast Electric Power University 132012 Jilin

*865401811@qq.com

Abstract: Aiming at the EEG signal radio frequency control problem of EEG signal radio control wheeled robot traversing multiple objective points, a new method of EEG signal radio frequency control that integrates the bi-objective chaotic particle swarm optimization (BCPSO) is put forward. This method transforms the selection of objective points into the EEG signal radio frequency control problem and makes optimization by using the ant colony algorithm. The EEG signal control function between the two objective points is defined, and the particle swarm optimization algorithm is applied for optimization. Given the premature phenomenon of particle swarm optimization algorithm, the inverse learning strategy is introduced into the particle swarm optimization algorithm, and the inertia weight and learning factor of the particle swarm optimization algorithm are improved. The performance test results show that the bi-objective chaotic particle swarm optimization algorithm can effectively prevent the premature particle phenomenon and improve the optimization capacity and stability of the particle swarm optimization algorithm.

1. Introduction

Ant colony optimization (ACO) is a heuristic evolutionary algorithm evolved from ant foraging behavior. The ant colony algorithm has presented superiority in solving the TSP problem. The particle swarm optimization algorithm swarm optimization (PSO) is a swarm intelligence algorithm that is evolved from the bird's foraging behavior [1]. Compared with the general swarm intelligence algorithm, PSO has memory characteristics, which can obtain more information through self-learning and learning from others. Due to the advantages of fewer parameters and convenient calculation, PSO is widely applied in optimization problems and has achieved very excellent results. With the development of the EEG signal radio frequency control wheeled robot technology, domestic and foreign scholars have carried out a lot of studies on the EEG signal radio frequency control of the EEG signal radio frequency control wheeled robot, such as the multi-robot EEG signal radio frequency control [2-4], the free floating robot trajectory planning [5-7] and other application areas [8-9]. PSO is widely applied in EEG signal radio frequency control, and ACO is effective in solving the TSP problem. Hence, the combination of the two algorithms is very suitable for solving the multi-objective point EEG signal radio frequency control problem of the EEG signal radio frequency control wheeled robot [10-12].

In a space that contains obstacles, the PSO algorithm is used to control the EEG signal from the starting point to the objective point and between each objective point and the objective point. The starting point, the objective point and the obstacle position are known, but which route to be selected is unknown. In order to determine which path will be selected, ACO is applied to traverse the starting point and all objective points to obtain the shortest path through all points. In order to avoid premature PSO and improve the efficiency of the algorithm, a particle swarm optimization algorithm with fast convergence is put forward. The experimental results have verified the effectiveness of the hybrid algorithm and the superiority of the bi-objective chaotic particle swarm optimization algorithm.

2. Bi-objective chaotic particle swarm optimization algorithm

2.1. Iterative evolution process

The time when the particle identifies the optimal solution in the problem space is proportional to the distribution in the space when the particle is initialized. The particle near the optimal position identifies the optimal solution faster than the particle with the optimal position. However, particles are randomly distributed in the problem space, and the position of each particle relative to the optimal solution is unknown.

Combined with the above analysis, the introduction of the reverse learning strategy in the initialization of the bi-objective chaotic particle swarm optimization algorithm is conducive to the particle in identifying the optimal solution. Upon the initialization, firstly, the fitness value of the particle and its inverse particle fitness value are calculated, and their values are compared to select the particles with relatively good fitness value. Secondly, the particles with the optimal fitness value are selected from the population as the initial population.

2.2. Iterative evolution process

In the standard PSO algorithm,the inertia weight affects the global search and local search ability of the particle, and the learning factor affects the capacity of the particle to acquire information. In the process of evolution, the global search ability and local search ability should also change, that is, from global search to local search, to ensure that the particle can identify the optimal solution to the problem. Similarly, the particle should also be gradually enhanced social skills for the transition from "self" learning to learning from "others", so as to acquire more useful information.

According to the above analysis, the inertia weight should be kept dynamically changing during the optimization process. When the value taken for ω is relatively large, the global search ability of the particle is relatively strong; when the value taken for ω is relatively small, the local search ability of the particle is relatively strong. In the process, it will gradually approach the optimal point, and the inertia weight of the particle should change adaptively with the optimization process. Hence, the inertia weight update formula of the particle is ω_{max}

$$\omega = \omega_{max} - (\omega_{max} - \omega_{min})\left(1 - \frac{dist_i}{max\_dist}\right) \tag{1}$$

In the above equation: $\omega_{\max}$ and $\omega_{\min}$ stand for the maximum and minimum values of the particle inertia weight, respectively; $dist_i$ stands for the euclidean distance from the ith particle to the global optimal particle; $\max\_dist$ stands for the maximum distance from the particle to the global optimal particle. Hence, the expressions of $dist_i$ and $\max\_dist$ are as the following:

$$\begin{gathered} dist_i = \left(\sum_{d=1}^{D} g_{best_d} - x_{i,d} \right)^{\frac{1}{2}} \\ max\_dist = \underset{i}{arg\,max}(dist_i) \end{gathered} \tag{2}$$

The learning factorc $c_1$and c_2 control the relative influence between the particle's own memory and the companion memory: when the value of c_1 is relatively small, it is presented that the self-cognitive ability is insufficient; when the value of c_2 is small, it is presented that the social ability is insufficient.

In order to ensure that the self-cognition and social ability of the particle can vary from time to time in the iterative process, and the particle learning factor update formula is as the following:

$$c_1 = c_{1max} \times \left(\frac{c_{1min}}{c_{1max}} \right)^{\left(\frac{t}{T_{max}} \right)}$$
$$c_2 = c_{2min} \times \left(\frac{c_{1min}}{c_{1min}} \right)^{\left(\frac{t}{T_{max}} \right)} \quad (3)$$

In the above equation: c_{1max} and c_{2max}, as well as c_{1min} and c_{2min} stand for the maximum and minimum values of the learning factors c_1 and c_2, respectively; t stands for the current number of iterations; T_{max} stands for the maximum number of iterations.

2.3. Implementation process of the algorithm

Once the algorithm falls into premature, chaotic particle swarm optimization will search in the region around the optimal solution to replace some of the particles in the original population, leading the population to jump out of the local optimum. A chaotic variable has the following characteristics in a certain range: randomness, that is, its performance is as disorderly as random variables; ergodicity, that is, it can go through all States in space without repetition; regularity, the variable is derived from the determined iteration equation. Chaotic optimization is a novel optimization method, which uses the ergodicity of chaotic system to achieve global optimization, and it does not require the continuity and differentiability of the objective function.

In this paper, two different chaotic search spaces are used, one is to search in the neighborhood of space with global optimum gbest as the center and R as the radius, as shown in Formula (4); the other is to search in the spatial domain with origin as the center and R' as the radius, as shown in Formula (5).

$$R = \eta gbest \quad (4)$$

$$R' = \rho \left(x_{\max} - x_{\min} \right) \quad (5)$$

Among them: η is chaotic search coefficient, take 1.5; Xmax and Xmin represents the upper bound and lower bound of the position vector respectively, ρ is the shrinkage factor of chaotic search, which decreases linearly in the iteration process of the algorithm to reduce the scope of chaotic search and increase the search accuracy, taking 1.0-0.3 linear decline. The steps are as follows:

Step1: Initialize parameters, set the total number N, the maximum number of iterations, replacement probability, etc.

Step2: Update the velocity and position of the ith particle. Calculate the fitness value f_i of the ith particle.

Step3: Calculate the inverse solution of the position and velocity of the ith particle: $\tilde{x}_i = a_i + b_i - x_i$, $a_i = min\left(x_i\right)$, $b_i = max(\, x_i \,)$

Step4: Calculate the fitness value of the inverse solution

Step5: Select the particle with the best fitness to form the population

Step6: Update the velocity and position of the ith particle

Step7: Determine whether it falls into local optimal. If it gets stuck at the local optimum, it creates chaotic particles.

In Step7, D dimensional chaotic vector is generated according to formula (1), where P is the replacement probability, and the chaotic vector is linearly mapped to the set solution space to randomly replace the original population particles.

Step8: Determine whether the algorithm satisfies the termination condition, if so, terminate the algorithm and proceed to Step2 to continue execution.

3. Experimental results and analysis

3.1 BCPSO performance test

In order to verify whether the performance of the bi-objective chaotic particle swarm optimization algorithm is improved, the author compares it with other improved algorithms on four fitness functions. The four fitness functions are the Sphere function, Rastrigin function, Grewank function and Schaffer function, respectively, and the test function parameter settings are shown in Table 1 as the following.

Table 1 Test function parameter settings

Function	Value range	Optimal point	Optimal solution
Sphere (f_1)	[-100,100]	(0,0)	0
Bastrigin (f_2)	[-5.12,5.12]	(0,0)	0
Grewank (f_3)	[-600,600]	(0,0)	0
Schaffer (f_4)	[-100,100]	(0,0)	0

The BCPSO algorithm parameters are set as the following: the particle swarm size 40, the spatial dimension is 2, ω= [0.4, 0.9], c_1= [1.25, 2.75], c_2= [0.5, 2.25], and the maximum number of iterations is 100.

In Table 2, the mean value of the BCPSO algorithm on the four test functions is smaller than the values of the other three algorithms, especially in the Sphere function and the Rastrigin function. From the data in Table 2, it can be observed that the BCPSO algorithm has better convergence performance, and the solution thus obtained is more superior. The data in Table 3 reflects the stability of the improved algorithm. Comparison between the BCPSO algorithm and the IAPSO algorithm, between the IPSO algorithm and the WPSO algorithm shows that the improved algorithm has better performance in the stability.

Table 2 Comparison of the average optimal results

Function	BCPSO	IAPSO	IPSO	WPSO
f_1	0	0.132790	0.020476	0.034939
f_2	0	0.066379	0.066331	0.297488
f_3	0.002153	0.031692	0.017342	0.018080
f_4	0.003273	0.005328	0.006179	0.017561

Table 3 Comparison of the standard variance results

Function	BCPSO	IAPSO	IPSO	WPSO
f_1	0	0.714775	0.009423	0.006733
f_2	0	0.248174	0.248186	0.455948
f_3	0.004418	0.087505	0.021595	0.019151
f_4	0.0045600	0.004699	0.004650	0.032801

From the data in table 2 and table 3, it can be known that the reverse learning strategy has improved the diversity of the population, increased the probability of particle optimization success and saved the particle optimization time; the linearly varying inertia weight guarantees the particle from global search to local search. The linear transformation suppresses the premature phenomenon of the particles; the change of the learning factor ensures that the particles can fully complete the self-learning and social behavior and improve the convergence speed of the particles.

3.2 Performance comparison of EEG radio frequency contro

In order to verify the practicability and effectiveness of the new method, the author conducted the experiment in the simulation environment. The simulation experiment environment is set in an 18×16 two-dimensional rectangular space, which is further set into a simple environment and a complex environment. In the first experiment, the parameters of the BCPSO algorithm are set as the following:

ω_{max} =0.9, ω_{min} =0.4, S=10, c_{1max} =2.75, c_{2max} =2.25, c_{1min} =1.25, c_{2min} =0.5, v_{max} =1.9, v_{min} =0, the safety distance in the repulsive field $d_m = 2$, $k_r = 2.7$, λ =0.25, μ =0.25, and the maximum number of iterations is 100. The experiments for the above two tasks are repeated 10 times, and the mean value is obtained. The time consumption and moving distance performance of the multi-objective point EEG signal radio frequency control in the EEG signal radio frequency control wheeled robot are shown in table 4 as the following.

Table 4 Comparison of the path performance

Environment	Algorithm	Moving distance/m	Time consumed/s
Simple environment	BCPSO	40.943	8.866
	IAPSO	41.749	10.257
	IPSO	42.865	10.625
Complex environment	BCPSO	44.942	18.583
	IAPSO	43.127	23.265
	IPSO	45.336	25.183

From the data in Table 4, it can be concluded that with the increase of task complexity, BCPSO, IAPSO and IPSO are also increasing in time consumption. The more complex the task is, the more time it takes. However, when performing the same task, BCPSO takes time and IAPSO. In terms of moving distance, the moving distance of BCPSO is shorter than the other two comparison algorithms in simple tasks. In the complex task, although BCPSO has a longer moving distance than IAPSO, it is shorter than IPSO algorithm; in terms of the path security, the BCPSO algorithm is superior to the other two comparison algorithms. As the objective function of the EEG signal radio frequency control is designed, not only the path shortest problem is considered, but the safety problem is also taken into consideration. Hence, the experimental results can be obtained by combining the two environments. The comprehensive performance of the BCPSO algorithm is stronger than the IAPSO algorithm and the IPSO algorithm.

Figure 1 shows the convergence curves of the BCPSO algorithm, IAPSO algorithm and IPSO algorithm on the objective function in a simple environment. It can be observed that the BCPSO algorithm can be applied in the multi-objective point brain when iteration is performed for about 72 times. The optimal path is identified in the radio frequency control of the electrical signal, and the IAPSO algorithm and the IPSO algorithm still fail to obtain the convergence value of the BCPSO algorithm when iteration is performed for 100 times.

In summary, under the same task, the comprehensive performance of BCPSO algorithm is superior to that of the IAPSO algorithm and the IPSO algorithm. With the change of task complexity, BCPSO algorithm can effectively and properly complete the movement distance and time consumption and security performance, which indicates that the BCPSO algorithm can effectively and properly complete the multi-objective point EEG signal radio frequency control effect of the EEG signal radio frequency control wheeled robot.

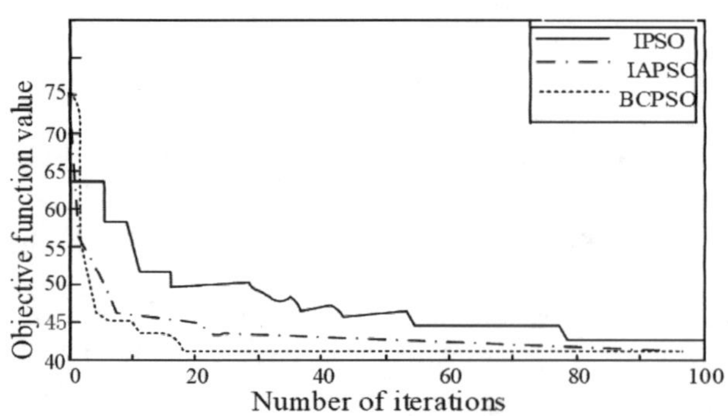

Figure 1. Comparison of BCPSO and IAPSO and IPSO objective function convergence curves

4. Conclusions

In this paper, a new method of multi-objective point EEG signal radio frequency control for the EEG signal-based wheeled robot based on the combination of ACO algorithm and BCPSO algorithm is put forward, and the feasibility of the new method is verified by experiments. The ACO algorithm presents relatively good superiority in the TSP problem. Hence, the ACO algorithm is applicable to the objective point selection problem of the EEG signal radio controlled wheeled robot. The BCPSO algorithm has not only inherited the simple calculation characteristics of the PSO algorithm but also introduced the reverse learning strategy during initialization to ensure the particle. In the iterative process, the BCPSO algorithm has suppressed the premature of the particles and improved the convergence speed. Four performance parameters are applied to evaluate the performance of the BCPSO algorithm, and excellent test results are obtained. The experimental results under the simulation experiment and real environment suggest that the new method proposed in this paper is a valid multi-objective point EEG signal radio frequency control method. The application of the new method proposed in this paper to the EEG signal multi-objective point EEG signal radio frequency control of the wheeled robot in the dynamic environment is the research content in the next step.

References

[1] Gao, H. , Song, X. , Ding, L. , Xia, K. , Li, N. , & Deng, Z. . (2014). Adaptive motion control of wheeled mobile robot with unknown slippage. International Journal of Control, 87(8), 1513-1522.

[2] Fateh, M. M. , & Arab, A. . (2015). Robust control of a wheeled mobile robot by voltage control strategy. Nonlinear Dynamics, 79(1), 335-348.

[3] Kralev, J. , Slavov, T. , & Petkov, P. . (2016). Design and experimental evaluation of robust controllers for a two-wheeled robot. International Journal of Control, 1-47.

[4] Alakshendra, V. , & Chiddarwar, S. S. . (2017). Adaptive robust control of mecanum-wheeled mobile robot with uncertainties. Nonlinear Dynamics, 87(4), 2147-2169.

[5] Héctor M. Becerra, Carlos Sagüés, Mezouar, Y. , & Hayet, J. B. . (2014). Visual navigation of wheeled mobile robots using direct feedback of a geometric constraint. Autonomous Robots, 37(2), 137-156.

[6] Ye, W. , Li, Z. , Yang, C. , Sun, J. , Su, C. Y. , & Lu, R. . (2015). Vision-based human tracking control of a wheeled inverted pendulum robot. IEEE Transactions on Cybernetics, 1-3.

[7] Chwa, & Dongkyoung. (2016). Robust distance-based tracking control of wheeled mobile robots using vision sensors in the presence of kinematic disturbances. IEEE Transactions on Industrial Electronics, 63(10), 6172-6183.

[8] Yang, H. , Fan, X. , Shi, P. , & Hua, C. . (2016). Nonlinear control for tracking and obstacle avoidance of a wheeled mobile robot with nonholonomic constraint. IEEE Transactions on Control Systems Technology, 24(2), 741-746.

[9] Rotondo, D. , Vicenç Puig, Nejjari, F. , & Romera, J. . (2014). A fault-hiding approach for the switching quasi-lpv fault tolerant control of a four-wheeled omnidirectional mobile robot. IEEE Transactions on Industrial Electronics, 62(6), 3932-3944.

[10] Fukushima, H. , Muro, K. , & Matsuno, F. . (2015). Sliding-mode control for transformation to an inverted pendulum mode of a mobile robot with wheel-arms. IEEE Transactions on Industrial Electronics, 62(7), 4257-4266.

[11] Juang, C. F. , Lai, M. G. , & Zeng, W. T. . (2017). Evolutionary fuzzy control and navigation for two wheeled robots cooperatively carrying an object in unknown environments. IEEE Transactions on Cybernetics, 45(9), 1731-1743.

[12] Korayem, M. H. , Yousefzadeh, M. , & Manteghi, S. . (2016). Dynamics and input–output feedback linearization control of a wheeled mobile cable-driven parallel robot. Multibody System Dynamics, 40(1), 1-19.

IOP Conf. Series: Materials Science and Engineering **686** (2019) 012009 doi:10.1088/1757-899X/686/1/012009

Design Method of Function-Structure Integrated Lattice Structure

G X Li, M Zhang, C Zhao*, X Jin and W P Luo

College of Intelligent Science, National University of Defense Technology, China

*Corresponding author: 1849559322@qq.com

Abstract. Structure lightweight has always been a hot topic in the field of engineering. Lattice structure has characteristics of light weight, high specific stiffness and high specific strength because of its high porosity and low relative density. It has been widely used in structural lightweight design. In addition, lattice structure has broad application prospects in advanced industrial equipment, due to the potential of anti-vibration, anti-impact, heat transfer and heat dissipation, zero/negative thermal expansion, electromagnetic wave absorption, sound absorption and noise reduction. In this paper, a design method of function-structure integrated lattice structure is proposed, topology optimization and variable density lattice filling technology are studied, the research methods of multi- functional characteristics of lattice structures are discussed, and the related simulation and experimental methods are introduced, which is of certain reference significance to the development of technology in this field.

Key words: lattice structure; topology optimization; lightweight structure; integration of function and structure

1. Introduction

The technology of structure lightweight aiming at reducing equipment self-weight, reducing driving energy and improving carrying capacity has always been a hot topic in the fields of engineering machinery, aerospace, rail transit and so on[1]. Lattice structure has good properties such as lightweight, high specific stiffness and strength because of its high porosity, low relative density and periodic distribution [2]. It has been well applied in structural lightweight. In recent years, researchers have gradually found that lattice structure has many functions such as anti-vibration and anti-impact, heat transfer and heat dissipation, electromagnetic wave absorption, sound absorption and noise reduction, zero/negative thermal expansion, etc. [3]. Lattice structure provides some new ideas on the development of multifunctional materials and integrated design technology. It is of great significance to study the basic theory, simulation and experiment of the multifunctional characteristics of lattice structures.

2. Design of function-structure integrated lattice structure

An design method of function-structure integrated lattice structure is proposed, which included three parts: topology optimization and variable density lattice filling, multi-functional characteristics of lattice structure, simulation and experimental verification. The flow chart of the function-structure integrated lattice structure is shown in Figure 1.

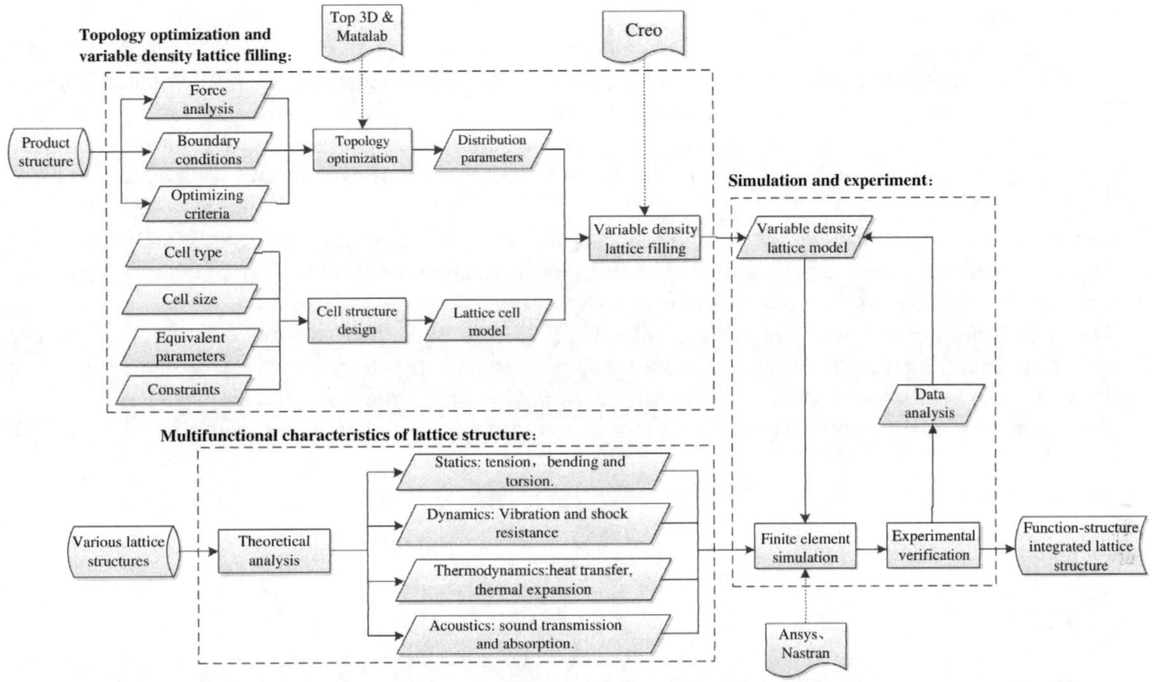

Figure 1. Flow chart of design for the function-structure integrated lattice structure

Firstly, the given product structure is optimized by topology and filled with variable density lattice. Top 3D & Matlab is used to optimize the topology and obtain the density distribution parameters of continuum materials. A single lattice cell model is obtained by the Creo, Solidworks and other three-dimensional modeling software. The secondary development application program for Creo is used to fill the continuum product structure with variable density to obtain the variable density lattice structure model.

Secondly, the multi-functional characteristics of lattice structure are studied. The multi-functional characteristic is the basis of integrated design of functional structure. It mainly includes the characteristics of statics, dynamics, thermodynamics and acoustics, involving basic theoretical research, mathematical and physical model building, calculation equation deduction and so on.

Then, the functional characteristics of lattice structure are simulated and verified by experiments. Finite element software such as ANSYS is used for simulation analysis, and the simulation results are obtained. The experimental platform is built to test the functional characteristics, and the experimental results are obtained. And the functional characteristics of lattice structure are improved continuously.

Finally, the function-structure integrated lattice structure with functional characteristics is obtained, that adapted to the specific working conditions.

3. Topology optimization and variable density lattice filling technology

3.1. Structural Topology Optimization

The topology optimization is an important method for lightweight structural design. A given product structure, with certain load conditions, constraints and performance indicators, is optimized and analyzed to remove redundant materials and retain key materials, so as to reduce the relative density of materials and achieve the goal of lightweight[4].

Common topology optimization software includes OptiStruct, Tosca, Top 3D, etc. The continuum structure is topologically analyzed to obtain the material density distribution parameters, which provide the basis for lightweight design and variable density lattice filling.

IOP Conf. Series: Materials Science and Engineering **686** (2019) 012009 doi:10.1088/1757-899X/686/1/012009

3.2. Lattice Structure and Cell Design

Lattice structure is also called multicellular structure or artificial periodic structure. According to cell structure and composition rules, it can be divided into: (1) two-dimensional lattice structure, such as grid structure, honeycomb structure (square, hexagon), etc.; (2) three-dimensional lattice structure, such as truss structure, sandwich structure (inclined column, pyramid), cubic lattice structure (face-centered FCC, body-centered BCC), etc.; (3) other complex lattice structure, such as wire-wound structure, etc. [5].

In practical application, the lattice type should be selected and lattice size should be designed according to different working conditions. The basic unit of lattice structure is lattice cell. The analysis and design of single cell is the basis of the follow-up research.

For two-dimensional lattice structures, Gibson method is often used to analyse lattice cells. The stress-strain equation is established according to the compatibility relation of deformation, and the equivalent elastic parameters of lattice structure, named Gibson formula[6], are calculated.

For example, Gibson's formula for equivalent elastic properties of square honeycombs:

$$\begin{cases} x = 2\dfrac{t}{a}\left(1-\dfrac{1}{2}\dfrac{t}{a}\right) \approx 2\dfrac{t}{a} \\ \dfrac{E_1^*}{E_s} = \dfrac{E_2^*}{E_s} = \dfrac{t}{a} = \dfrac{1}{2}x \\ \dfrac{G_{12}^*}{E_s} = \dfrac{1}{2}\left(\dfrac{t}{a}\right)^3 = \dfrac{1}{16}x^3 \\ \upsilon_{12}^* = \upsilon_{21}^* = \upsilon_s\dfrac{t}{a} = \dfrac{1}{2}\upsilon_s x \end{cases} \tag{1}$$

Among them, the elastic modulus, shear modulus and Poisson's ratio of cell are expressed as functions of relative density x, and the elastic modulus and Poisson's ratio of solid material of cell wall are expressed by E_s and υ_s, and the cell size and rod size are expressed by a and t.

For the three-dimensional lattice structure, the relationship between relative density x and cell size a and rod size t can be expressed by the ratio of lattice cell volume V_{entity} and cell outline volume V_{outline}.

Three-dimensional software such as Creo, UG and SolidWorks are often used to design and model lattice cells.

The manufacturing constraints of lattice structure are analyzed according to the characteristics of augmented material manufacturing process, which provides a basis for cell structure design, cell type and cell size optimization.

3.3. Variable Density Lattice Filling Technology

Lattice structure is a good material for lightweight design because of its porous and lightweight characteristics. Combining lattice structure with topology optimization technology leads to variable density lattice filling technology.

Because of the complex spatial structure of three-dimensional lattice cells and the variable density arrangement, it often takes a lot of time and is difficult to operate by using existing three-dimensional software to model. Therefore, based on Pro/Toolkit function library developed by REO or UG, a parameterized automatic assembly aided program for variable density lattice structure is developed. Each lattice cell is defined and assembled in the form of cell elements, and finally a variable density lattice structure model is formed[7].

The flow chart of variable density lattice structure modeling is shown in Figure 2.

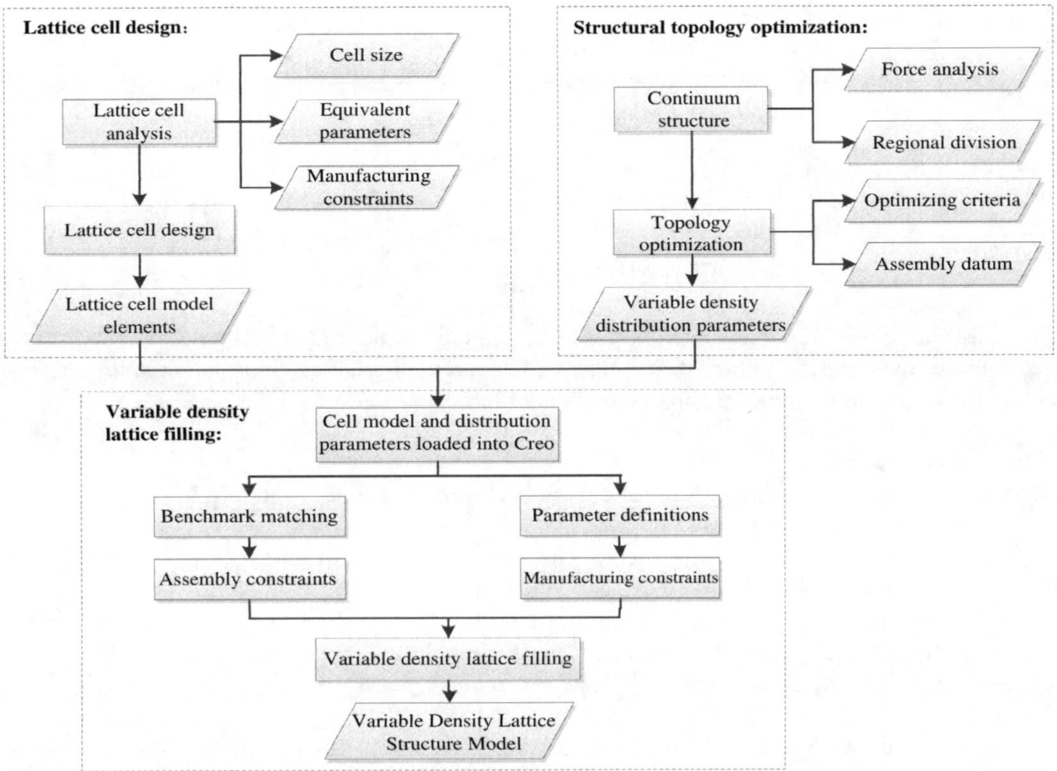

Figure 2. Flow chart of variable density lattice structure modeling

4. Research methods of multifunctional characteristics of lattice structures

The multifunctional characteristics of lattice structure mainly include static, dynamic, thermodynamic, acoustic and other characteristics.

4.1. Static characteristics

Static and mechanical properties of lattice structures include tensile, compressive, bending and torsional properties.

For the two-dimensional lattice structure, the simulation results are fitted by using MATLAB, and the fitting formula is obtained, and the Gibson formula is modified. Based on the modified Gibson formula, the equivalent elastic model, the equivalent Poisson's ratio and the equivalent shear modulus are calculated, and the two-dimensional elastic constitutive relationship is established.

For three-dimensional lattice structures, the equivalent elastic constitutive relation is established [8]. For the body-centered cubic lattice structure, based on the method of strain energy equivalence, the formula for calculating the equivalent parameters in the elastic constitutive relation of lattice structure is obtained as follows:

$$\begin{Bmatrix}\sigma_{11}\\ \sigma_{22}\\ \sigma_{33}\\ \sigma_{23}\\ \sigma_{31}\\ \sigma_{12}\end{Bmatrix}=\begin{bmatrix}\frac{E^H(1-\upsilon^H)}{(1+\upsilon^H)(1-2\upsilon^H)} & \frac{E\upsilon^H}{(1+\upsilon^H)(1-2\upsilon^H)} & \frac{E^H\upsilon^H}{(1+\upsilon^H)(1-2\upsilon^H)} & 0 & 0 & 0\\ \frac{E^H\upsilon^H}{(1+\upsilon^H)(1-2\upsilon^H)} & \frac{E^H(1-\upsilon^H)}{(1+\upsilon^H)(1-2\upsilon^H)} & \frac{E^H\upsilon^H}{(1+\upsilon^H)(1-2\upsilon^H)} & 0 & 0 & 0\\ \frac{E\upsilon^H}{(1+\upsilon^H)(1-2\upsilon^H)} & \frac{E\upsilon^H}{(1+\upsilon^H)(1-2\upsilon^H)} & \frac{E(1-\upsilon^H)}{(1+\upsilon^H)(1-2\upsilon^H)} & 0 & 0 & 0\\ 0 & 0 & 0 & G^H & 0 & 0\\ 0 & 0 & 0 & 0 & G^H & 0\\ 0 & 0 & 0 & 0 & 0 & G^H\end{bmatrix}\begin{Bmatrix}\varepsilon_{11}\\ \varepsilon_{22}\\ \varepsilon_{33}\\ \varepsilon_{23}\\ \varepsilon_{31}\\ \varepsilon_{12}\end{Bmatrix} \quad (2)$$

$$\begin{cases} E^H = \dfrac{E}{V_{bcc}} \left[\dfrac{\alpha_5 (Aa)^3 + \alpha_5 (A^2 Ia) + \alpha_5 \left(\dfrac{AI^2}{a} \right) + \alpha_5 \left(\dfrac{I}{a} \right)^3}{(Aa)^2 + \alpha_5 (Aa) + \alpha_5 \left(\dfrac{I}{a} \right)^2} \right] \\ \upsilon^H = \dfrac{\beta_1 (Aa)^2 + \beta_2 (Aa) + \beta_3 \left(\dfrac{I}{a} \right)^2}{(Aa)^2 + \beta_5 (Aa) + \beta_5 \left(\dfrac{I}{a} \right)^2}, \ G^H = \dfrac{E}{V_{bcc}} \left[\gamma_1 (Aa) + \gamma_2 \left(\dfrac{I}{a} \right) \right] \end{cases} \tag{3}$$

The constitutive relation, mechanical equilibrium equation and equivalent parameter expression of lattice structure material are obtained, by means of theoretical analysis and mathematical derivation, which are the key to study the static characteristics of lattice structure.

4.2. Kinetic Characteristics

The dynamic performance of lattice structure includes vibration, impact and fatigue.

(1) Vibration of lattice structures

Based on Reissner sandwich plate theory, the vibration equation and natural frequency solution formula of lattice truss sandwich plate are established[9].

The thin shell theory assumes that the strain and stress obey Hooke's law. Based on Love's thin shell theory, the natural frequency equation and the solution formula of natural frequency are established[10].

Based on the equilibrium differential equation of elastic body in space, the differential equation of elastic surface and the solution formula of natural frequency of thin plate lattice structure are established.

Vibration frequency response characteristics of several different lattice structures should be discussed, natural frequencies and modes should be calculated, and the effects of lattice type, cell size and relative density on anti-vibration performance should be studied.

(2) Impact of lattice structures

The three basic equations of lattice structure subjected to impact are established. The physical model of explosive shock is built, and the constitutive relation of structure and the propagation theory of shock wave under explosive shock are studied[11].

The dynamic response of several different lattice structures under a certain impact load should be studied. The deformation and energy absorption should be calculated. The influence of lattice type, cell size and relative density on impact resistance should be analyzed.

4.3. Thermodynamic properties

Thermodynamic properties of lattice structures include heat transfer, heat dissipation, zero/negative thermal expansion, etc.

(1) Heat transfer and heat dissipation of lattice structure.

The lattice structure has high porosity, high specific surface area and good permeability, which is convenient for convective heat transfer and can achieve good heat transfer and heat dissipation effect[12].

The physical model and thermodynamic equation of lattice heat transfer and heat dissipation should be established. The thermal stress distribution of different lattice structures under different thermal loads should be discussed. The effects of lattice type, cell size and density distribution on heat transfer and heat dissipation performance should be studied.

(2) zero/negative thermal expansion lattice structure.

Zero/negative thermal expansion structure can be formed by using different thermal expansion coefficients of different materials to design certain material ratio and micro-structure, which can realize the regulation of thermal expansion [13].The thermal expansion physical model of composite rod structure should be established, the thermal deformation mechanism under thermal load should be analyzed, the connection modes of rods with different materials and the filling process of zero/negative thermal expansion lattice cells should be studied. The thermal deformation laws of different zero/negative thermal expansion structures under different temperature loads should be discussed. The effects of lattice type, bar shape, connection mode and material properties on the zero/negative thermal expansion characteristics of structures should be studied.

According to the thermal expansion coefficients of different materials, with specific modulus and multiple symmetry as constraints, by topological optimization, the microstructures could be designed to achieve zero/negative thermal expansion[14], as shown in Figure 3.

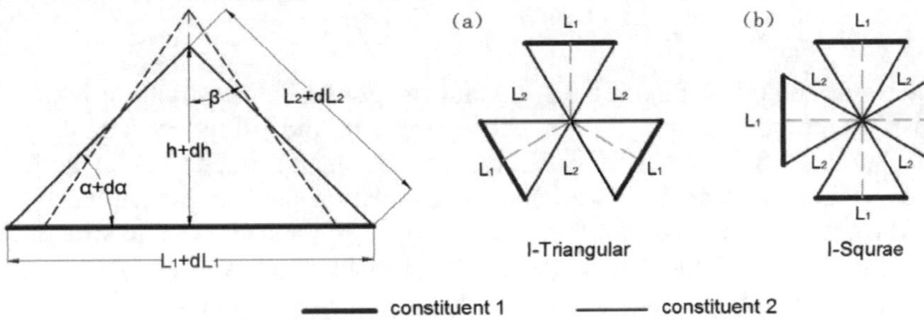

Figure 3. Zero/negative thermal expansion microstructure

4.4. *Sound transmission and noise reduction characteristics*

The sound absorption and noise reduction performance of lattice structure includes sound transmission/absorption, noise reduction and so on.

When noise propagates in rods, nodes and panels of lattice structures, sound energy dissipates in the form of heat energy, which can absorb sound waves and reduce noise [15].

The physical model of sound propagation in lattice structure should be established. Then, the phononic crystal of lattice structure should be designed. The sound propagation law in different lattice structures should be analysed, and the effects of lattice type, cell size and density distribution on sound transmission characteristics, phononic crystal band gap characteristics and noise reduction characteristics should be studied.

5. Simulation and experimental methods

According to the different functional characteristics of lattice structure, different modeling and simulation software are often used. Finite element software ANSYS is often used to analyze static and dynamic loads of lattice structures; Nastran is often used to analyze the frequency response characteristics of lattice structures; ANSYS and Thermo-Calc are often used to analyze the thermodynamic characteristics of lattice structures; MSC is often used to analyze the sound transmission and absorption characteristics of lattice structures.

For example, the deformation and displacement of two-dimensional honeycomb structure and three-dimensional cubic lattice structure under tension, compression and torsion loads were simulated and analyzed by finite element software ANSYS, and the compressive and bending properties of lattice structure are analysed [16], as shown in Figure 4.

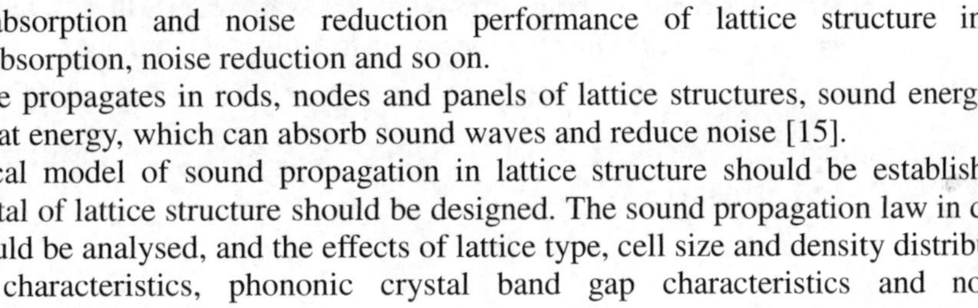

(a) hexagonal honeycomb lattice (b) body-centered cubic lattice

Figure 4. Three-point bending force simulation of lattice structure

On the basis of numerical simulation analysis of lattice structure and specific working conditions, the corresponding experimental platform is built to verify the functional characteristics of lattice structure and obtain experimental data. Finally, through data analysis and mining of theoretical, simulation and experimental results, the influence of lattice type, cell size and density distribution on functional characteristics will be summarized, and the coupling relationship between functional characteristics is analyzed. Then, the optimization design of lattice cell configuration, cell size, density distribution and other structural parameters is carried out continuously to improve the lattice structure performance and realize the lattice structure design of integration of work and structure.

6. Conclusion

In this study, a design method of functional-structural integrated lattice structure is proposed, which includes three parts: topology optimization of lattice structure and filling of variable density lattice, multi-functional characteristics of lattice structure, simulation and experimental verification. three research methods: theoretical analysis, numerical simulation and experimental verification, is combined, the influence rules of parameters is summarized by data analysis, and structural parameters is optimized. By filling with variable density lattice, the product structure is designed to be lightweight. At the same time, it also had multi-functional characteristics, such as anti-vibration, anti-impact, heat transfer and heat dissipation, zero/negative thermal expansion, etc. It realized the lattice structure design of power structure integration to meet the high performance requirements of industrial equipment. The research in this paper is meaningful for the development of the design of functional-structural integrated structure. However, The disadvantage is that some theories and techniques involved in this paper are still immature and need to be further studied.

Acknowledgments

The authors are grateful for the financial support of this work by the National Natural Science Foundation of China (No. 51505488).

References

[1] Q.X. Feng, Mechanical properties of metal porous structure based on selective laser melting and its variable density design method [D]. Chongqing: Chongqing University, 2017.
[2] Recep M. Gorguluarslan, Umesh N. Gandhi, et al. Design and fabrication of periodic lattice-based cellular structures [J]. Computer-Aided Design & Applications, 2016, 13: 50~62.
[3] B. Zhao, Z.Q. Li, et al. Advances in the fabrication of three-dimensional lattice structures of metals, [J], Rare Metals Materials and Engineering, 2016, 45 (8),
[4] Elnasri I, Pattofatto S, Zhao H, Tsitsiris H, Hild F, and Girard Y. Shock Enhancement of Cellular Structures Under Impact Loading: Part I Experiments [J]. J. Mech. Phys. Solids, 2007, 55(12): 2652~2671.
[5] C.Y. Yi, L. Bai, A review of the research and application of metal three-dimensional lattice structure topology, [J], functional materials, 2017, 48 (10):
[6] Gibson L J, Ashby M F. Cellular solids: structure and properties [M]. Cambridge: Cambridge University Press, 1999.
[7] X. Jin , G. X. Li , et al. Modeling and Analysis of Non-Uniform Honeycomb Structures Based on Topology Optimization [J]. Proc IMechE Part C:J Mechanical Engineering Science 2018, Vol. 232(7) 1280–1293
[8] X. Jin , Research on Optimum Design and Modeling of Variable Density Polycellular Structures for Laser Addition Manufacturing, [D], 2018
[9] S.J. Yan, Research on Vibration and Sound Propagation Characteristics in Light-weight Lattice Structures, [D], 2013
[10] Y. Wang, Z. Luo, et al. Study on high-order modal vibration characteristics of thin-walled cylindrical shells [J], Journal of Dynamics and Control, 2016, 14 (2): 131-137

[11] Sing Ying Choy, C.N. Sun, et al. Compressive properties of functionally graded lattice structures manu- factured by selective laser melting [J]. Materials & Design, 2017, 131: 112~120.

[12] Lehman J, Lakes R. Stiff lattices with zero thermal expansion and enhanced stiffness via rib cross section optimization. Int J Mech Mater Des, 2013, 9: 213–225

[13] Y.P. Chen, L. Ago, Topology Optimization Design of Heat Dissipation Structure Based on Variable Density Method, [J], Computer Integrated Manufacturing System, 2018, 24 (1)

[14] K. Wei, Y.M. Pei, Research Progress in Thermal Expansion Control Design of Light Composites and Structures, [J], Science Bulletin, 2017, 62 (1) 47-60

[15] Y. Xiao, Bandgap Regulation and Vibration and Noise Reduction Characteristics of Local Resonance Structures [D], 2016

[16] X. Jin , G. X. Li , et al. Optimal design of three-dimensional non-uniform nylon lattice structures for selective laser sintering manufacturing[J].Advances in Mechanical Engineering, 2018,10(7)1-19

A Study on the Characteristics of Road Landscape for 3D Design Developing Process

Chun Hyunjin[1,2,*] and Wang Yue[3]

[1] School of Arts, Nanjing University of Aeronautics and Astronautics, NJ 211106, China
[2] School of Architecture, Southeast University, NJ 210096, China
[3] School of Industrial Design, Nanjing University of the Arts, NJ 210013, China
No. 29 JIangjung Road, Jiangning District, Nanjing 211106, Jiangsu, China

[*]tough4324@naver.com

Abstract. Due to Asia's economic growth and industrialization, there has being a growing interest in road landscapes recently. In the meanwhile, aesthetic aspects in road design are becoming more and more important as 3D design technology advancing. With a view to the above, designers must consider both the structural engineering and the aspect of aesthetics when conceiving new urban roads blueprint. Much more in-depth studies are expected to focus on the aesthetics of roads by utilizing 3D design technology. In light of this, this study was centred on the landscape preferences of road in modern cities from the perspective of building height and planting type alongside. The main results of the study were summarized as follows: The images of H:W=1:3 and H:W=1:1 were more highly preferred than the rest seven images on the grounded that the lower buildings seemed to have better correspondence with the surroundings. And the image of planting with 7.5 meter-interval received the highest evaluation for visual preference on account that this interval provides road landscape with a visual image of stability. In conclusion, this study results illustrated that ratio of building height to road width and the planting interval on footpath were the crucial elements in landscape design. Furthermore, this study also make efforts to present a methodology for 3D design develoement.

1. Introduction

1.1. Research Background and Purpose

Due to economic development and recent rise of many nations in Asia, regional governments in Asia are increasingly interested in urban landscape. The cityscape consists of a variety of elements. Among these factors, roads play an important role in urban landscape. When designing roads, experienced designers would consider two aspects: structural engineering and aesthetics. Despite of the importance of road landscape, the road aesthetics in design hasn't been studied a lot up to now. More researches are called to analyze the aesthetics of roads so as to optimize road landscape during the design phase. Existing theses have formerly only suggested guidelines for design based on visual preference, but it has been as yet no research on the practical process of landscape design [1]. In response to this, this thesis analyzed the visual preference of urban roads to present theoretical foundations about road landscape using 3D design technology. Furthermore, this research hoped to present a methodology for

3D design developing process, to some extent.

1.2. Literature Review

The existing studies on road landscape were overwhelmingly conducted by the researchers from developed countries; prominent studies have not been found in developing countries. Additionally, most of the current studies on this topic were about structural stability and structural mechanics. Chen Yiling etc. (2015) analyzed the difference between the urban road landscape design and the highway landscape design [1]. Shin Jae Yun etc. (2014) analyzed the preference depending on street planting types and visual characteristics on Dongdaegu-ro located in Suseong-gu, Daegu Metropolitan City [2]. Jeong Dae young etc. (1996) figured out visual character and preference of roadscape for the main entrance road in Chongju [3]. Suh Joo Hwan etc. (1999) focused on exploring the relationship between the index of shape and the factor of perception about road landscape [4]. Almost all research about road landscape showed that the shape of the street trees and background were important factors in the image and visual preference of road landscapes. Background plays an especially significant role in visual preference, and more research is needed from the point of diversified backgrounds in the future. The background is comprised of a variety of factors, including buildings. So, this study surveyed the landscape preferences according to the height of the building and planting type.

2. Theoretical Background

2.1. Building Height

The height of the surrounding buildings along a road affects visual preferences. The ratio between the height of a building and the width of a road is particularly distinct within road landscape. If the height of a building is called H and the width of the road is W, then the ratio between the height of the building and the width of the road is H/W. If H/W is higher than 1, the closed image of the road increases and if H/W is lower than 1, the open image of the road increases [3]. So the ratio between the height of a building and the width of a road is an important factor in determining the image of the road (figure 1).

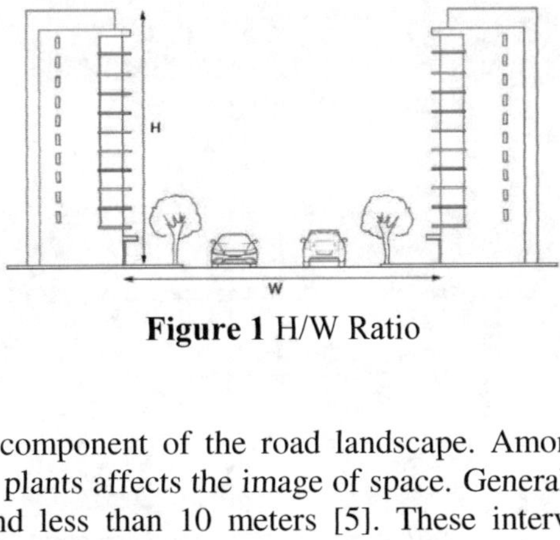

Figure 1 H/W Ratio

2.2. Planting Type

Planting is an indispensable component of the road landscape. Among such factors related to the planting, the interval between plants affects the image of space. Generally, the interval between plants is greater than 4.8 meters and less than 10 meters [5]. These intervals should be determined by considering both the engineering and aesthetic aspects (figure 2).

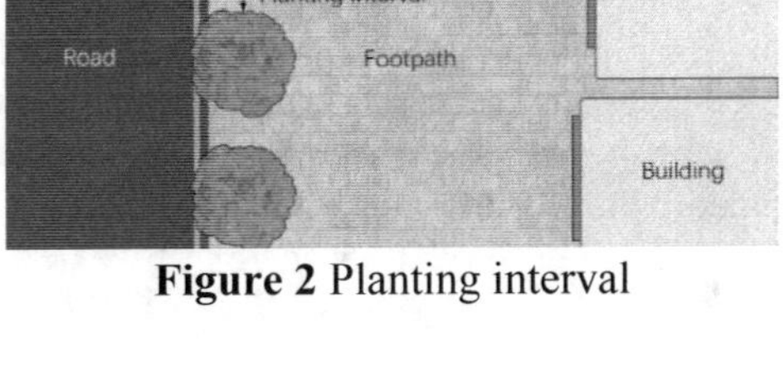

Figure 2 Planting interval

3. Research Method

3.1. Selection of Background

This view point should be the road's landscape surroundings along with the road itself, as observed from the centre of a road, where the road landscape can be seen clearly by drivers. The background was chosen as a typical road view for drivers when driving. In addition, this research chose view point on road, selected as a prospect point on the road landscape as seen from the center of the road.

3.2. Selection of Type and 3D Design Simulation

Table 1. 3D Design Simulation

Figure 3 Planting with 5 meter-interval (H:W=1:3)

Figure 4 Planting with 5 meter-interval (H:W=1:1)

Figure 5 Planting with 5 meter-interval (H:W=3:1)

Figure 6 Planting with 7.5 meter-interval (H:W=1:3)

Figure 7 Planting with 7.5 meter-interval (H:W=1:1)

Figure 8 Planting with 7.5 meter-interval (H:W=3:1)

Figure 9 Planting with 10 meter-interval (H:W=1:3)

Figure 10 Planting with 10 meter-interval (H:W=1:1)

Figure 11 Planting with 10 meter-interval (H:W=3:1)

This study completed a total of 9 computer-generated road landscape simulation images using 3D computer programs such as Auto-CAD (Autodesk, 2005), SketchUP (Google Inc., 2007), and Adobe Photoshop CS (Adobe Systems, 2003). Simulation images depicted three ratios between the height of a building and the width of a road, featuring H:W=1:3, H:W=1:1 and H:W=3:1 types. The plantings were respectively divided at intervals of 5, 7.5 and 10 meters (figure 3-11). A survey was then conducted by simulation produced from driver perspective [2]. These types were appropriate for road designs that can fit into this site, and thus were chosen for this experiment. Road shape was drawn according to the most common floor plan. Other elements of each simulated image, such as color and material, were kept constant (Table 1).

3.3. Selection of Survey Group and Method of Analysis

This research was conducted five times via survey from March 3rd, 2019 to May 10th, 2019. The size of all images was 210×297mm equally [6-7]. 60 students, who are the graduate students in Seoul national University and Konkuk University in South Korea, responded in the survey [6-7]. All students were landscape architecture majors. In addition, in order to analyze the differences in preference between the driving and non-driving groups, the subjects were sampled into 30 people who were driving and 30 people who were not driving. Before the survey began, each student were reminded of the evaluation method. The assessment time was 8 seconds. This survey asked an important question about visual preferences of the road landscape [6-7]. The evaluation of the visual preference used the 7-point likert scale. Statistical analysis, such as descriptive analysis, f-test, t-test [6-7], were conducted by employing such statistical programs like Excel and KESS (Korean Educational Statistics Software). This research also used two-way analysis of variance and Duncan's test (post hoc tests) to analyze preferential differences based on the variation of the building height and planting type [6-7].

4. Results and Discussion

4.1. Visual Preferences on a Road Landscape

Table 2. Visual Preferences on a road landscape

Landscape		Average	Standard deviation	F-Value	Significant probability	post hoc tests a=0.05
Planting with 5 meter-interval	H:W=1:3	3.82	1.32	8.322	0.001	b
	H:W=1:1	3.78	1.22			c
	H:W=3:1	3.46	1.12			d
Planting with 7.5 meter-interval	H:W=1:3	4.22	1.38	12.324	0.002	a
	H:W=1:1	4.16	1.34			a
	H:W=3:1	3.92	1.26			a
Planting with 10 meter-interval	H:W=1:3	3.94	1.32	9.148	0.003	b
	H:W=1:1	3.88	1.76			b
	H:W=3:1	3.72	1.46			c

The research also included conducting a study on visual preference on a road landscape. The results were summarized as follows: Significance level of results is less than 0.05. These significant differences of visual preferences on a road landscape can be measured to a given confidence level according to significant probability. As a result, in planting with 7.5 meter-interval, images of H:W=1:3 and H:W=1:1 were more highly preferred than other images. In planting with 5 meter-interval, images of H:W=1:3 and H:W=1:1 also received high evaluation for visual preferences. In planting with 10 meter-interval, image of H:W=1:3 received high evaluation for visual preferences. Because low buildings seem to have a good correspondence with the surroundings. And, the image of planting with 7.5 meter-interval received high evaluation for visual preferences. This was because this interval gave the road landscape a visual image of stability. The above results suggested that the variation in visual preference of road landscapes was profoundly linked to the ratio of building height to road width and planting interval on footpath. At the same time, this research conducted a Duncan test (post hoc tests) to analyze what the difference in landscape was according to the ratio of building height to road width and planting interval on footpath. As a result of the post-hoc tests, significance level of results was less than 0.05. The landscape preferences had significance according to the ratio of building height to road width and planting interval on footpath. The image of H:W=1:3 and the image of H:W=1:1 and the image of H:W=3:1 in planting with 7.5 meter-interval obtained similar evaluation performance in visual preferences. Furthermore, the H:W=1:3 image in planting with 5 meter-interval and the H:W=1:3 image, the H:W=1:1 image in planting with 10 meter-interval acquired similar evaluation in visual preferences. Afterward, the H:W=1:1 image in planting with 5 meter-interval and the H:W=3:1 image in planting with 10 meter-interval got paralleled performance in visual preferences. To sum up, visual preferences would change along with the building height and planting type (Table 2).

4.2. Visual Preference's Variation According to Driving

This research conducted a study on the difference of visual preference's variation according to driving through T-test. T-test was used to find the difference between the road landscapes of preference with the data of factor score [7-8]. The results were summarized as follows: Significance level of results was less than 0.05. T-value was a minus sign about images of H:W=3:1. T-value was a plus sign about images of H:W=1:3 and H:W=1:1.

Table 3. Visual Preferences According to Driving

Landscape		Subject	Average	Standard deviation	t-Value	Significant probability
Planting with 5 meter-interval	H:W=1:3	Driver	3.88	1.14	1.214	0.016
		Non-driver	3.76	1.26		
	H:W=1:1	Driver	3.80	1.36	0.416	0.012
		Non-driver	3.76	1.18		
	H:W=3:1	Driver	3.40	1.22	-1.214	0.022
		Non-driver	3.52	1.38		
Planting with 7.5 meter-interval	H:W=1:3	Driver	4.30	1.14	1.624	0.018
		Non-driver	4.14	1.18		
	H:W=1:1	Driver	4.18	1.26	0.412	0.014
		Non-driver	4.14	1.18		
	H:W=3:1	Driver	3.86	1.04	-1.216	0.018
		Non-driver	3.98	1.32		
Planting with 10 meter-interval	H:W=1:3	Driver	4.02	1.44	1.638	0.016
		Non-driver	3.86	1.12		
	H:W=1:1	Driver	3.91	1.36	0.614	0.018
		Non-driver	3.85	1.18		
	H:W=3:1	Driver	3.64	1.22	-1.626	0.022
		Non-driver	3.80	1.34		

In H:W=1:3, the preference of drivers was higher than that of non-drivers. In H:W=1:1, the preference of drivers was higher than the preference of non-drivers. But, in H:W=3:1, the preference of non-drivers was higher than that of drivers. The reason for these results was that the low building served to broaden the driver's horizons. Based on the analysis of visual preference's variation, there was a difference in landscape preference according to the ratio of building height to road width and planting interval on footpath. In conclusion, the results illustrated that the ratio of building height to road width and planting interval on footpath were also important elements in the landscape design. When designing the road, the ratio of building height to road width and planting interval on footpath should be regarded as important points and reflected as such in the design (Table 3).

5. Conclusion

Due to Asia's economic growth and industrialization, recently there has been a growing interest in road landscape. This study surveyed the landscape preferences of road landscapes, which was a special type from the angle of the ratio of building height to road width and planting interval on footpath. The main results were as follows: Images of H:W=1:3 and H:W=1:1 were more highly preferred than other images, because low buildings seemed to have better correspondence with the surroundings. And, the image of planting with 7.5 meter-interval received high evaluation for visual preferences, due to this interval brought the road landscape a visual image of stability. In H:W=1:3, the preference of the drivers was higher than the preference of non-drivers. In H:W=1:1, the preference of drivers was higher than the preference of non-drivers. In H:W=3:1, nevertheless, the preference of non-drivers was higher than that of drivers. The reason for these results was that the low building served to broaden the drivers' horizons. In conclusion, the results illustrated that ratio of building height to road width and planting interval on footpath were vital elements in landscape design. There are also other various elements on roads which show an obvious influence on the visual preference of the road landscape. But the results in this research only consider the three most essential factors, which were the ratio of building height to road width and planting interval on footpath, thus more research needs to be prepared and expected to get better and more accurate results in the field of road landscape in the future.

References

[1] Chen Yiling etc. 2015 *Difference Between the Urban Road Landscape Design and the Highway Landscape Design.* International journal of engineering research and applications. **7** Vol 5 pp 111-114

[2] Shin Jae Yun etc. 2014 *Evaluation of Streetscape by Street Planting Types using Dynamic Simulation - In the Case of Ginkgo Planted on the Dongdaegu-ro in Daegu -.* Journal of Korea Institute of landscape architecture. **42** Vol 1 pp 89-103

[3] Jeong Dae young etc. 1996 *A Study on the Visual Character and Preference of Roadscape -In Case of the Main Entrance Road in Chongju-.* Journal of Korea Institute of landscape architecture. **1** Vol 24 pp 15-31

[4] Suh Joo Hwan etc. 1999 *A Study on the Analysis of Landscape Preference in the Road landscape by Index of Shape - The case of Sorak National Park-.* Journal of Korea Institute of landscape architecture. **4** Vol 27 pp 87-93

[5] Park Tae Keun etc. 2000 *A study on the present status and improvement of planting method of street trees in Kimhae city.* Journal of Institute of Construction Technology, Dong-A University. Research Works. **24** Vol 1 pp 195-206

[6] Feimer N R 1979 *Personality and Environment Perception: Alternative Predictive System and Implications for Evaluative Judgement.* Ph.D. Dissertation. University of California. Berkely.

[7] Im Seung Bin 1991 *Landscape analytics.* Seoul: Seoul National University Press.

[8] Kevin Lynch 1960 *The image of the city.* Cambridge, MA: MIT press.

Research on Interactive System of Science Exhibition Based on Leap Motion

An Zhongliu[a] and Zou Xin[*,b]

School of Mechanical Engineering, Guizhou University, Huaxi District, Guiyang City, Guizhou Province

E-mail: [a]2747766929@qq.com; [*, b]2794976556@qq.com

Abstract. Objective In order to explore the application of non-contact interaction in Science exhibition, an interactive system of science exhibits based on leap motion technology is proposed. **Method** Based on the Leap motion-based Science exhibition interactive system, the design of a "planetary story" science exhibit with interactive experience is evaluated according to three interactive experience indicators: realism level, learning pressure level and pleasure level.In this way, the feasibility of Leap motion technology in the popular science interactive system is verified. **Result** The use of the Leap motion gesture interaction library, combined with Unity3D development tools, to design science exhibits. **Conclusion** By proposing an Science exhibition interactive system based on Leap motion, it provides a design theory for the development of popular science exhibits. At the same time, the combination of Leap motion and popular science exhibits improves the user's interactive experience and meets users' interest in popular science content.

Keywords: leap motion; interactive system; user experience; popular science exhibits

1. Introduction

Science exhibition as an important means of communication for understanding scientific knowledge, its interactive experience affects people's participation in popular science activities. With the rapid development of science and technology, the interactive technology of 3D gesture recognition has come one after another. Based on this, the interaction mode of Science exhibition should be considered from the perspective of user experience, and the gesture interaction technology is applied to the popular science exhibits [1]. The current Science exhibition is mainly presented to the audience through two-dimensional and three-dimensional methods. The audience interacts through direct contact in the process of participation. This form of interaction is relatively simple, and the user is difficult to distinguish in the participation process. Therefore, this study introduces the Leap motion somatosensory device by analyzing the Science exhibition interactive system based on Leap motion, which can enrich the interactive experience of the audience and solve the problem of poor user experience.

2. Related research on Leap motion technology

Leap motion sensors detect and track hands, fingers or rods, capture their position in real time with high precision, high tracking frame rate, and identify related gestures and actions. The modules of the Leap motion sensor include two high frame rate cameras, LED lights, infrared filters and a USB 3.0

chip. Through the tracking of the object, the instruction of the gesture to the PC is realized, and finally the interactive object is controlled, as shown in Figure 1.

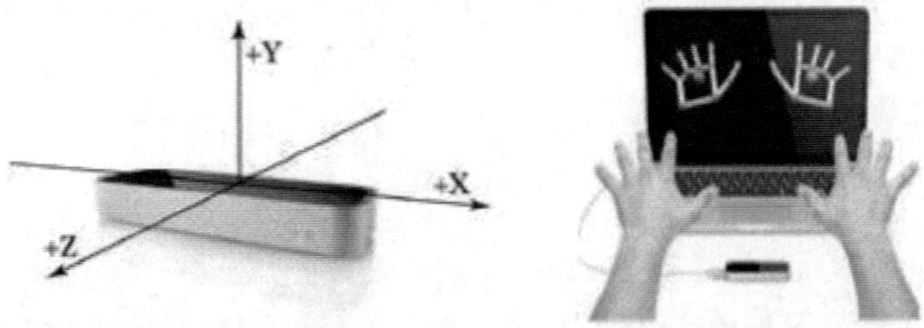

Figure 1. The Interactive demonstration of leap motion

In China, Li Yinghui et al [2] proposed a three-dimensional gesture recognition method based on the somatosensory device Leap Motion, and established a one-hand and two-hand static three-dimensional gesture library. In view of the traditional sand painting demonstration process, the performer needs to carry the sand table, fine sand and projection equipment, and has high requirements on the scene light and environment. Jin Yunhuan [3] designed a virtual sand painting based on Leap motion gesture interaction. Display system. In order to solve the problem of finger occlusion and improve the interactive experience, Sun Guodao et al. [4] proposed a method of setting dual Leap Motion at different angles, which has a significant effect on reducing finger aliasing and improving ease of use. In view of the shortcomings of automatic geometry-driven virtual cultural relics splicing, Lu Haoran et al. [5] proposed a method for implementing fragmentation and splicing using Leap motion. The traditional virtual plant human-computer interaction system generally runs on a specific operating system or mobile platform, and the interaction mode mostly involves human-computer interaction through the mouse and keyboard, which requires users to input more cumbersome parameters and commands, resulting in a lack of good interactive experience for system users. In view of the above situation, Wu Fuli et al [6] designed and developed a virtual crop somatosensory interaction system based on cloud computing and somatosensory interaction technology.

In foreign countries, Catalin et al [7] proposed a new five-finger humanoid gripper command and control scheme based on leap motion sensor, which is used in industrial machines to achieve low complexity and medium complexity. Manawadu et al. [8] developed an operator interface for a driver's vehicle based on Leap motion sensors, which sends commands to the car through gesture interaction to achieve automatic driving. Aashni Haria et al. [9] designed a robust marker-free gesture recognition system based on leap motion, which can effectively track static and dynamic gesture systems and improve the human-computer interaction experience. Haiyang et al. [10] used a new type of tracking sensor, Leap motion, to perform a non-contact demonstration of the robot.

In summary, Leap motion has a wide range of applications. In the process of human-computer interaction, the Leap motion sensor acts as a "bridge" between people and machines. By introducing Leap motion into Science exhibition, it helps to improve the user's interactive experience in Science exhibition. At present, the development and design of Leap motion is only to achieve the technical level of implementation, and does not establish a unified system design theory. By proposing a Science exhibition interactive system based on Leap motion, a user-centered system design theory is provided as a reference for the design and development of Leap motion.

3. Design of interactive exhibition system for science exhibits based on Leap motion

The Science exhibition interactive system based on Leap motion is mainly based on the user experience to improve the user's interactive experience. In the design of interactive systems, the scenes of science exhibition are mainly science and technology museums, museums and schools. Users usually include expert users and beginners. Science exhibits are the carriers of popular science content. The interactive method is the most important channel for users to obtain popular science content. The interactive system design is shown in Figure 2.

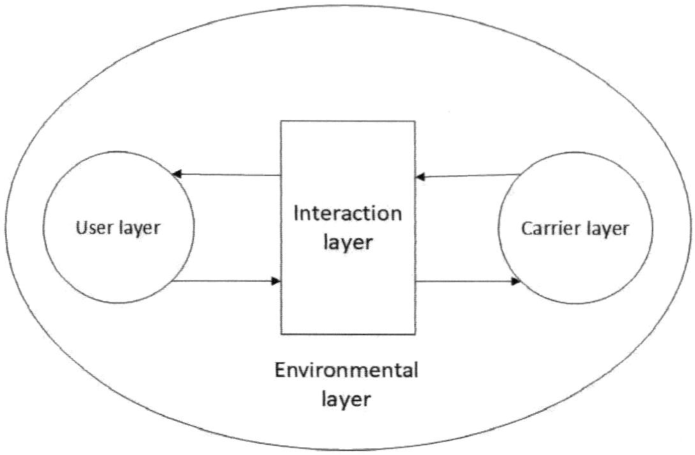

Figure 2. Science exhibition interactive system based on Leap motion

3.1. User layer
The user's proficiency in Leap motion can be divided into beginner users and expert users. The beginner user refers to the user who is used for the first time and is in the exploration stage, while the expert user refers to the cognitive and familiarity of the gesture interaction of Leap motion. Beginner users are not proficient in operation, resulting in unstable Leap motion recognition, which easily affects the user's interactive experience. Beginner users can continue to practice to reach the level of expert users.

3.2. Environmental layer
The Science and Technology Museum is a place for users to learn popular science knowledge. The use of Leap motion somatosensory equipment requires consideration of the environment. The science and technology museum has a large flow of people and the use of equipment is relatively frequent, so the stability of Leap motion sensing device identification is an important factor to enhance the user experience. Considering the special nature of the Science and Technology Museum, Leap motion needs to maintain high recognition performance.

3.3. Interaction layer
The gesture interaction of Leap motion is more and more complex, and the interactive experience is not obvious enough. Create a simple gesture interaction library and define related interaction events. Users can complete simple and correct interactions and improve the recognition performance of Leap motion, further enhancing the user experience. The design of the gesture interaction library is shown in Table 1.

Table 1. Gesture interaction library

Gesture interaction library	Definition of event
Gesture 1	Click
Gesture 2	Rotate
Gesture 3	Zoom
Gesture 4	Grab
Gesture 5	Mobile
....	....

3.4. Carrier layer

The main function of the Science exhibition is to show the science to the audience, popularize the scientific knowledge, and display the scientific knowledge vividly and visually [11]. Traditional science exhibits are books, exhibition boards, and models. They are usually only for users to learn and refer to. This interactive mode is that the user is in a passive receiving state, and the user can only passively accept the feedback of the popular science exhibits, so the user is interested in the traditional science exhibits. The cost of learning increases, as shown in Figure 3.

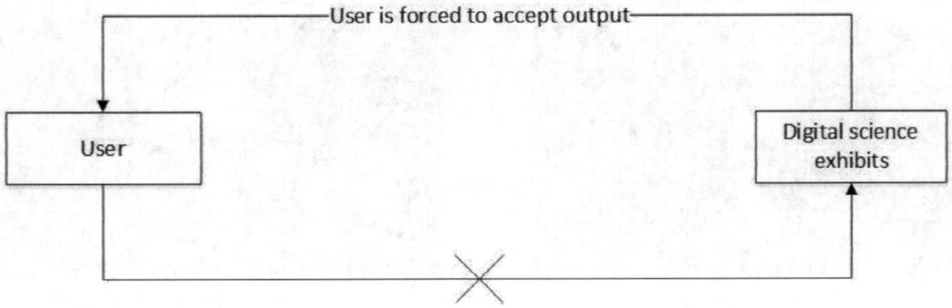

Figure 3. Traditional interactive mode of Digital science exhibits

With the continuous development of Internet technology, science exhibits gradually shift from traditional exhibits to digital science exhibits. The advancement and transformation of technology introduces various interaction methods, further enhancing the user's experience of popular science exhibits. Users can use their body language to interact with the exhibits in a spatial manner. This form of interaction allows the user's sense of participation and experience to be more satisfied [12]. The development of interactive design of popular science exhibits is closely related to interactive technology [13]. Somatosensory interaction technology is an important interactive technology to meet the design of interactive products centered on "people". For science exhibits, interaction is the most important part of the popular science experience [14]. Through the improvement of interactive technology, the user's interactive experience can be improved. Digital science exhibits mainly include mobile and PC. Users can interact in two-dimensional and three-dimensional ways. This interaction is two-way. Users input interactive behavior through Leap motion. Leap motion sends commands to PC, digital science exhibits. Feedback to the user in a timely manner, the user proceeds to the next operation based on feedback, as shown in Figure 4.

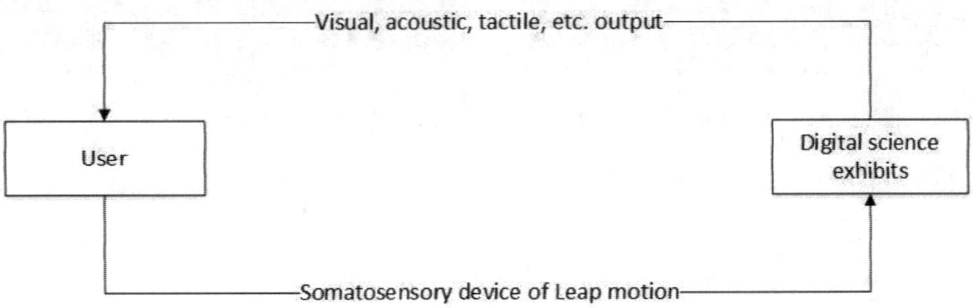

Figure 4. The interactive mode between leap motion and Digital science exhibits

4. "Planet Story" Science Exhibition Interactive System Design Case

The Unity3D platform was used to build the scene by combining the "Planet Story" project of the Science and Technology Museum of Guizhou Province with Leap motion.

Step 1, install the Leap motion SDK development environment and connect the Leap motion sensing device to the PC. If it is green, the connection is normal.

Step 2, the planet is created by adding a map to the sphere model, and the sun is created by means of textures and particle effects, forming a solar system.

IOP Conf. Series: Materials Science and Engineering **686** (2019) 012011 doi:10.1088/1757-899X/686/1/012011

Step 3, import the plugin developed by Leap motion in unity3D. Calls are made using the gesture control class of Leap motion.

4.1. Interaction design of the main scene

In the main scene, the principle of emitting rays is used to pick up the planets to solve the lack of interactive experience of Leap motion in the three-dimensional scene. Taking the Earth as an example, by clicking on the ray, you can rotate, zoom, and “long press” clicks on the earth. Enter the Earth's secondary scene interface by clicking on the Earth,as shown in Figure 5.

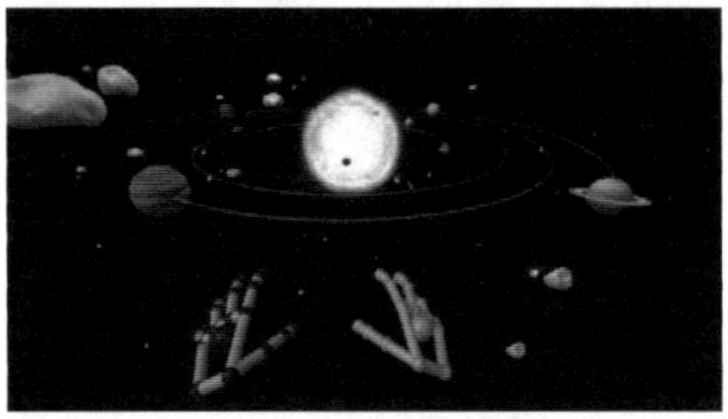

Initialized scene Launching a ray for grabbing

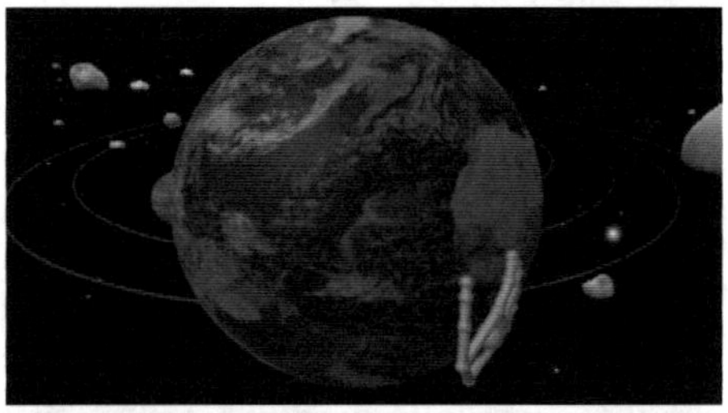

Grab the earth Rotating earth

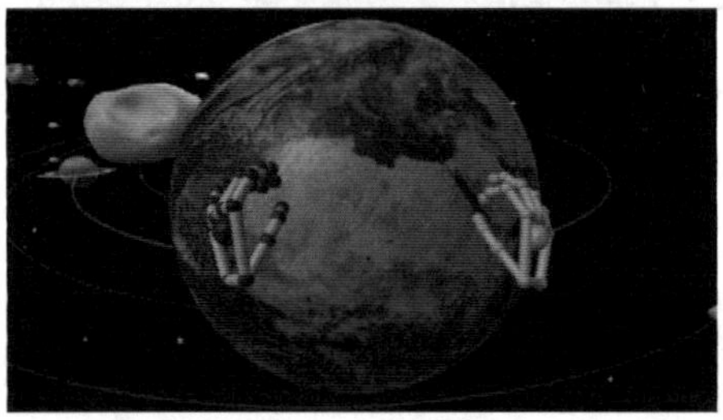
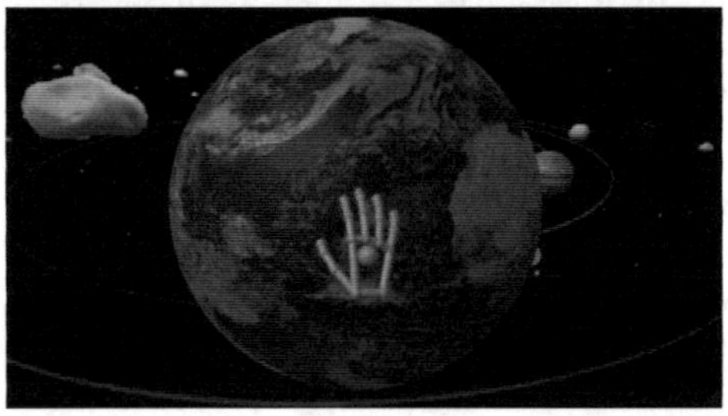

Zooming the earth “Long press” Click

Figure 5. Interaction case of main scene

4.2. Interaction design of secondary scenes

In the secondary interface of the Earth, a two-dimensional scene interface is built. Gestures can directly trigger the UI interface, generate UI interaction effects by clicking, and pop up the corresponding UI menu for introduction. You can rotate the earth by swiping left and right by gestures,as shown in Figure 6.

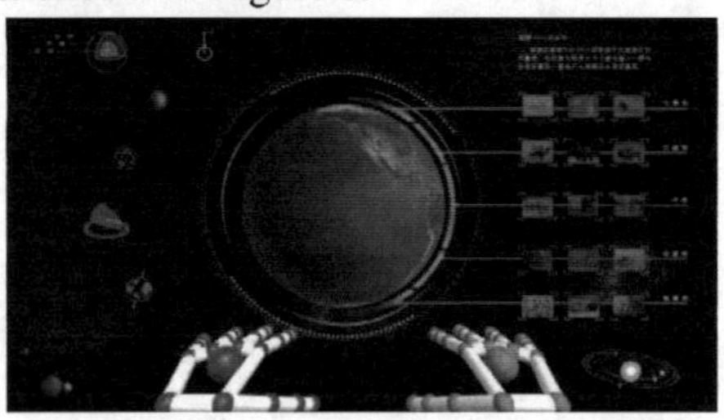
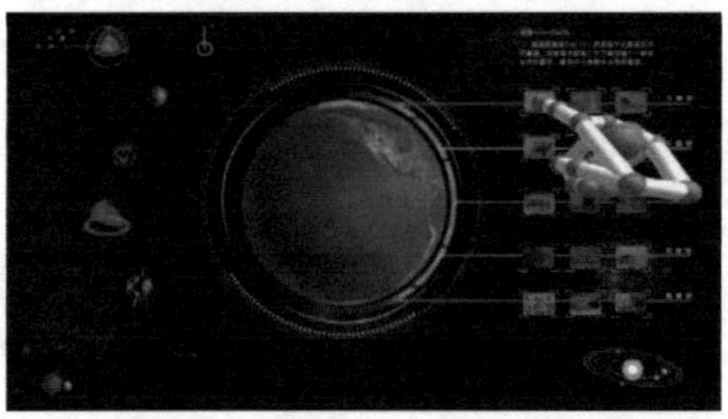

Entering the earth secondary interface Click on UI - Animal Introduction

IOP Conf. Series: Materials Science and Engineering **686** (2019) 012011 doi:10.1088/1757-899X/686/1/012011

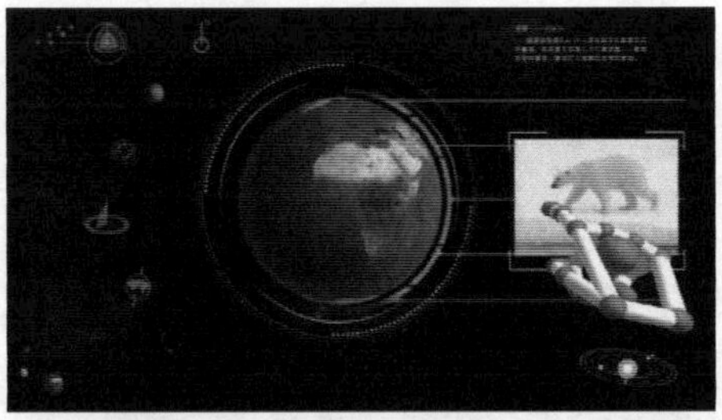

Pop up the introduction of the polar bear

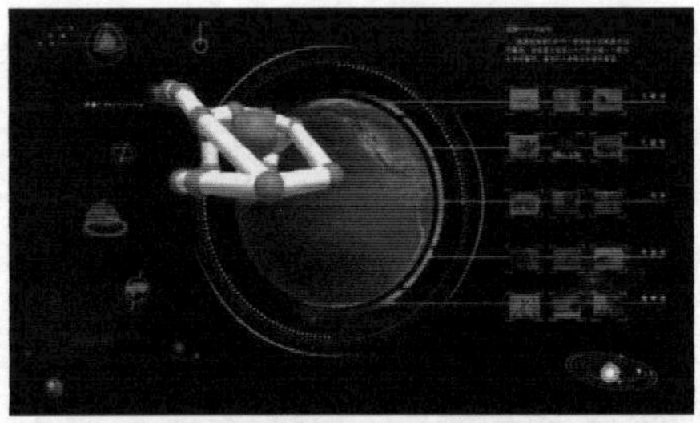

Click UI - Earth Quality

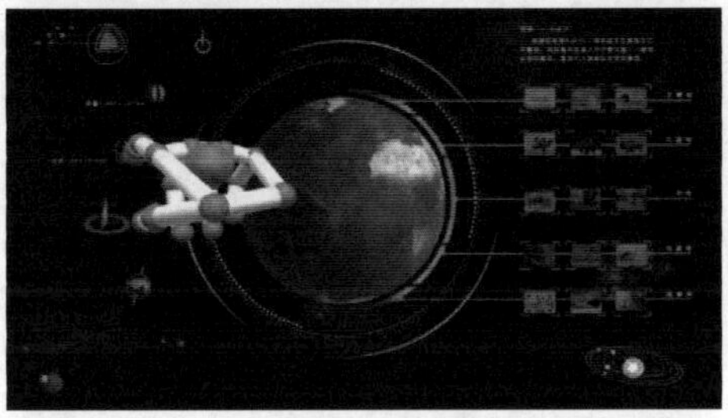

Click UI - Earth Volume

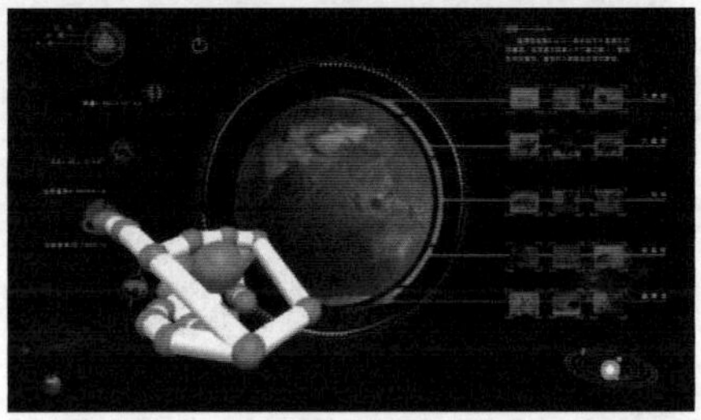

Click UI - Earth rotation, revolution rate

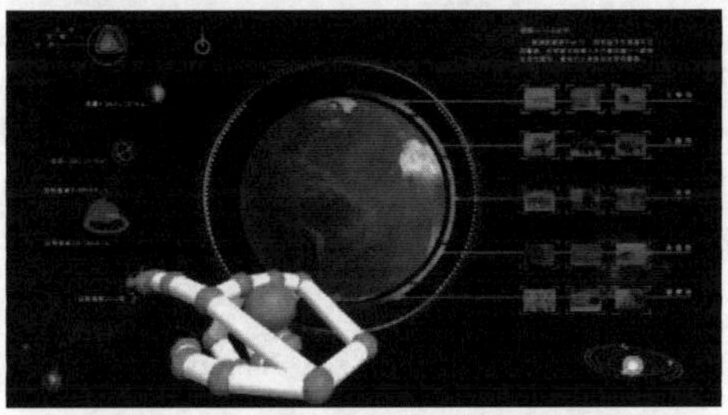

Click UI - Earth Rotation Cycle

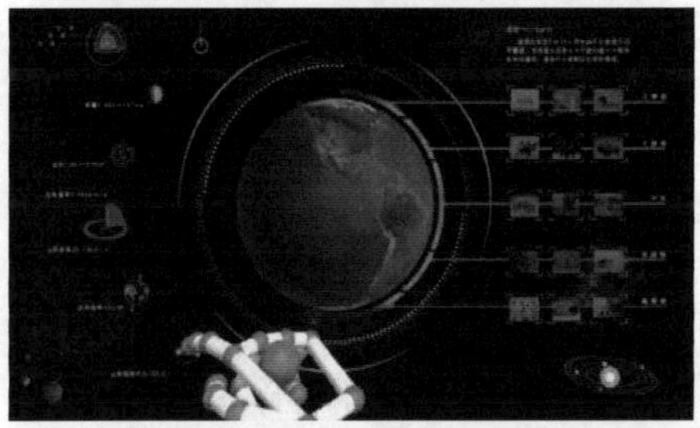

Click UI - Earth revolution cycle

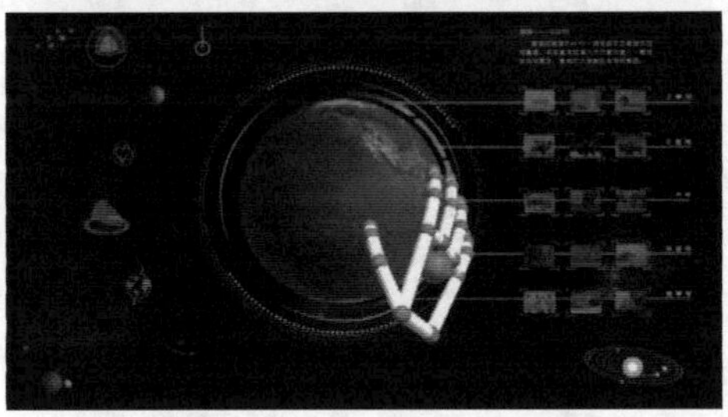

Rotating earth

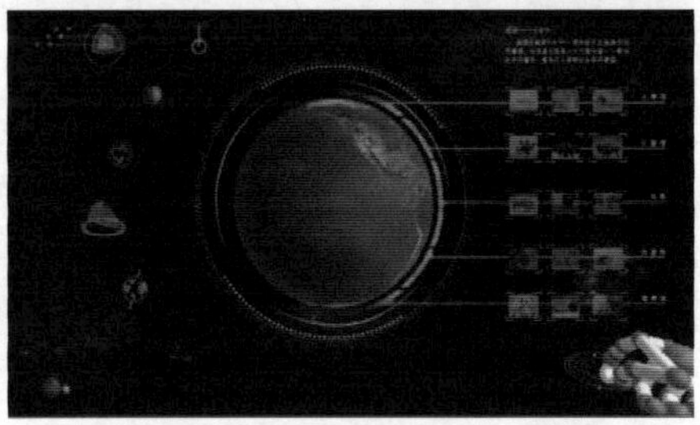

Click to exit (return to the main scene)

Figure 6: Interactive case of earth scene

5. User Evaluation of "Planet Story" Science Exhibition Interactive System

In this experiment, 50 users were selected for experiments, and tested according to three indicators of interactive experience: realism level, learning stress level and user pleasure level. The level is -3~3, the closer to -3, the smaller the level, the weaker the indicator. Conversely, the closer to 3, the greater the rating, the stronger the indicator of performance. The number of people at each level is displayed by selecting each level by 50 users. To verify the effectiveness of Leap motion in the Science exhibition interactive system by analyzing the two sets of data using Leap motion device and no interactive device, the formula of three indicators: Effect(Total)=index(positive)+index(negative) .

5.1. User interaction realism evaluation

IOP Conf. Series: Materials Science and Engineering **686** (2019) 012011 doi:10.1088/1757-899X/686/1/012011

Table 2. User interaction realism evaluation

realism level	Leap motion (number of people)				None (number of people)			
	Click	Rotate	Zoom	Grab	Click	Rotate	Zoom	Grab
-3	2	5	1	1	10	11	15	20
-2	6	3	2	3	8	9	7	12
-1	7	8	3	2	7	10	8	8
0	4	4	6	5	6	7	5	6
1	6	6	8	5	5	6	6	2
2	10	8	10	16	7	5	5	1
3	15	16	20	18	7	2	4	1
Trueness	16	14	32	33	-6	-17	-15	-36

According to the data in Table 2, Trueness (Leap motion)>Trueness(none), which is used in Science exhibition by Leap motion, can enhance the realism of user interaction.

5.2. User Learning Stress Assessment

Table 3. User Learning Stress Assessment

Learning stress level	Leap motion (number of people)				None (number of people)			
	Click	Rotate	Zoom	Grab	Click	Rotate	Zoom	Grab
-3	12	15	16	14	1	6	3	2
-2	10	12	10	11	5	8	4	3
-1	8	6	7	6	7	5	6	7
0	5	3	3	4	3	2	5	6
1	3	6	8	5	7	6	7	4
2	7	5	4	6	12	13	9	15
3	5	3	2	4	15	10	16	13
Pressure	-15	-19	-19	-16	11	10	19	20

According to the data in Table 3, Pressure(Leap motion) < Pressure(none) , which is used in Science exhibition by Leap motion, which can reduce the user's learning pressure.

5.3. User pleasure assessment

Table 4: User pleasure assessment

Pleasure level	Leap motion (number of people)				None (number of people)			
	Click	Rotate	Zoom	Grab	Click	Rotate	Zoom	Grab
-3	3	6	2	3	13	16	12	15
-2	5	9	3	2	11	13	14	12
-1	7	4	7	8	7	4	6	7
0	5	4	8	6	4	3	6	3
1	6	3	5	2	5	6	4	5
2	10	11	10	13	6	3	5	3
3	14	13	15	16	4	5	3	5
Pleasure	15	8	18	18	-16	-19	-20	-11

According to the data in Table 4, Pleasure (Leap motion)>Pleasure(none) , which is used in Science exhibition by Leap motion, can improve the user's pleasure.

6. Conclusion

Science exhibition is an important way for viewers to understand the contents of popular science. By introducing Leap motion gesture recognition technology to interact, the user's realism in the interactive system is enhanced, the user's learning cost is reduced, and the user's participation in the process is improved. The interactive development of Science exhibition injects new forms of interaction. While the Leap motion device improves the single, boring interactive experience of the past, the Leap motion sensing device limits the number of participants, and is used by only one user at a time. In the future development, increase the number of people involved in Leap motion, so that more users can participate at the same time.

Acknowledgments

This work is supported by Guizhou Science and Technology Fund Project “Science and Technology Museum Exhibition Engineering and Science Exhibition Teaching Innovation Design Theory and Methodology” (Qiankehe J word [2014] No. 2057)

References

[1] YANG Jian, CHEN Yang, WANG Dan-dan, ZHONG Fang-xu. Research on interactive interface design of popular science exhibits based on emotional concept[J]. Packaging Engineering, 2016, 37(06): 109-113.

[2] Li Yinghui, Shi Zhuo, An Yalei. Three-dimensional gesture recognition method based on Leap Motion[J].Modern Computer (Professional Edition),2016(14):74-76+80.

[3] Mou Yunhuan.Design and implementation of virtual sand painting display system based on gesture recognition[J].Electronic Design Engineering,2018,26(20):169-173.

[4] Sun Guodao, Huang Puyong, Liu Yipeng, Liang Ronghua. Research on 3D Visualization Interaction Method Based on Double Leap Motion[J]. Journal of Computer-Aided Design & Computer Graphics,2018,30(07):1268-1275

[5] LU Haoran, ZHANG Yu, NIU Miaomiao, TAI Wuyang, ZHOU Mingquan. A virtual mosaic method for 3D artifacts based on Leap motion[J]. Journal of System Simulation, 2015, 27(12): 3006-3011+3017.

[6] Wu Fuli, Ding Wei, Ding Weilong, Xie Tao. Design of virtual crop human-computer interaction system based on Leap Motion[J]. Transactions of the Chinese Society of Agricultural Engineering, 2016, 32(23): 144-151.

[7] Catalin Constantin Moldovan, Ionel Staretu. An Anthropomorphic Hand with Five Fingers Controlled by a Motion Leap Device [J]. Procedia Engineering, 2017, 181.

[8] Manawadu U E, Kamezaki M, Ishikawa M, et al. A hand gesture based driver-vehicle interface to control lateral and longitudinal motions of an autonomous vehicle [C]/ /IEEE International Conference on Systems, Man, and Cybernetics. IEEE, 2016: 001785-001790.

[9] Aashni Haria, Archanasri Subramanian, Nivedhitha Asokkumar, Shristi Poddar, Jyothi S Nayak. Hand Gesture Recognition for Human Computer Interaction [J]. Procedia Computer Science, 2017, 115.

[10] Haiyang Jin, Qing Chen, Zhixian Chen, Ying Hu, Jianwei Zhang. Multi-LeapMotion sensor based demonstration for robotic refine tabletop object manipulation task[J]. CAAI Transactions on Intelligence Technology,2016,1(1).

[11] Liang Henglong. Analysis of the function and form of science exhibits[J].Science and Technology,2015(04):41.

[12] Peng Mengjun, Peng Li. Research on the design of popular science exhibits integrating augmented reality and user experience. Design. 2018

[13] Yan Jingyan.Research on the Influence of Artificial Intelligence on Interaction Design[J].Packaging Engineering,2017,38(20):27-31.

[14] TAN Jie,HAN Jinghong,YANG Tao,WANG Jingchun.Design and Implementation of Internet-based Remote Operation and Experience System for Science Exhibitions[J].Experimental Technology and Management,2016,33(09):134-136.

Creative Design of Transmission Device for Workpiece Transfer in Production Line

Rongyu Ge[1,*], Miaomiao Xin[2], Jialiang Wu[2] and Xiaomei Ni[2]

[1]School of Mechanical Engineering, University of Jinan, Jinan, 250022, China
[2]School of Mechatronics Engineering, Qilu Institute of Technology, Jinan, 250200, China

[*]me_gery@ujn.edu.cn

Abstract. It is necessary for the workpiece transfer system in the production line to realize periodic indexing and reciprocating oscillation motion. In view of the complicated structure of the traditional transmission device and the need of multiple driving motors, a creative design integrating the indexing transmission and the oscillating transmission is made based on the globoidal cam mechanism. Taking the material transfer system in the LED beam splitter as a design example, structural design and 3D modeling were carried out, and the 2D drawings were obtained. The virtual prototype of the overall transmission is established by the dynamic simulation software ADAMS and the angular displacement of the three output shafts is simulated. The simulation results verify that the scheme can meet the requirements of the workpiece transfer system.

1. Introduction

In the automatic production line, the workpiece transfer system must realize the coordination between various technological actions, mainly including periodic indexing motion and reciprocating oscillation [1]. The former realizes intermittent rotation of multi-station rotary table, while the latter realizes grasping and handling of workpieces between different worktables. For the existing workpiece transfer system, indexing and oscillating actions are performed by different intermittent mechanisms, and driven by different motors, as shown in figure 1, which has disadvantages such as complex structure, multiple sensors feedback required, difficulty to control the coordination from multiple power sources. An innovative transmission device using single power source to meet the multiple technological actions is designed in this paper, which is much simpler and more compact.

2. Functional principle of implementation scheme

According to the process requirements of the workpiece transfer system, the transmission device needs to convert the rotating motion from the motor not only into intermittent indexing motion of the work table but also intermittent reciprocating oscillation for the feeding manipulator. There are many mechanisms to fulfil indexing movement like cam indexing mechanism, ratchet mechanism, groove wheel mechanism and incomplete gear mechanism, and also many mechanisms to complete oscillating movement like crank-rocker mechanism, crank-slider mechanism, cam mechanism with oscillating followers and so on. Therefore, the cam mechanism can meet two requirements simultaneously. The globoidal cam mechanism, because of its compact structure, high load capacity, low noise, high transmission accuracy and high reliability, is widely used in various production lines [2]. According to the motion form of follower, it is divided into oscillating globoidal cam and indexing globoidal cam mechanism. Based on the globoidal cam mechanism, a combined transmission device integrating the indexing drive device and oscillating drive device is accomplished to transform the single input motion into multiple output motion. In other words, multiple globoidal cams are installed on the same input shaft and different followers on different output shaft mesh with the corresponding cams respectively, so diverse indexing or intermittent reciprocating output motion can be achieved through the design of cam profile, as is shown in figure 2. This innovative design requires only one motor to directly transfer torque to multiple groups of cam mechanisms, so indexing motion and oscillating motion are controlled by a power input, and the coordination of output action is just controlled by cam profile. This control method is very simple and reliable.

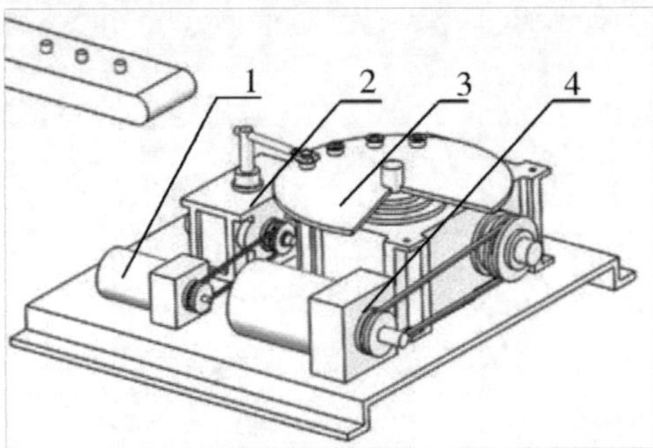

1-motor I; 2-oscillating transmission device; 3-indexing transmission device; 4-motor II

Figure 1. Exiting transmission scheme

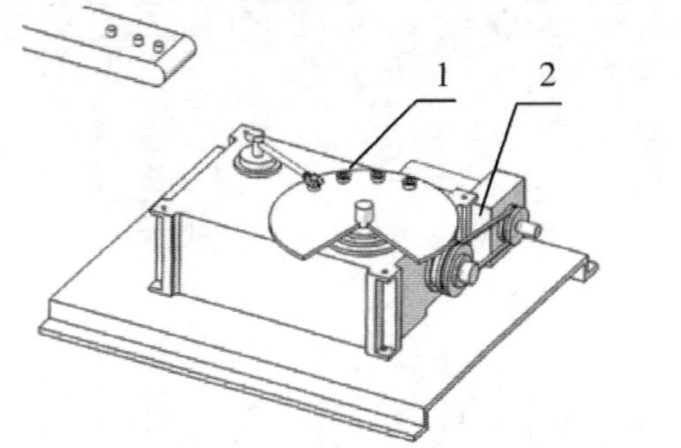

1-combined transmission device; 2-motor

Figure 2. Innovative design scheme.

3. Design example

3.1. Structure design

LED light splitter is a kind of equipment to classify LED according to the wavelength, light intensity and voltage of the diode. At present, the combination of compound cam device and indexing cam device is generally adopted in the mechanical system of LED light splitter [3]. The disk cam in the compound cam is used to control the reciprocating motion of the mechanical gripper in the X direction, and cylindrical cam in the compound cam is applied to control the reciprocating motion in the Y direction. The indexing cam device intermittently transfers the LED to the designated position. The disadvantages of this system are that multiple motors are required to control each intermittent mechanism separately, there are many maintenance failure points and complex structure, and it is difficult to be assembled. When the LED market becomes increasingly competitive, higher and higher requirements are put forward for the sorting speed of splitter. It is problematical for the existing mechanical structure to meet the high-speed condition. The motion cycle diagram of the mechanical drive system of a LED light splitter on the market is shown in figure 3.

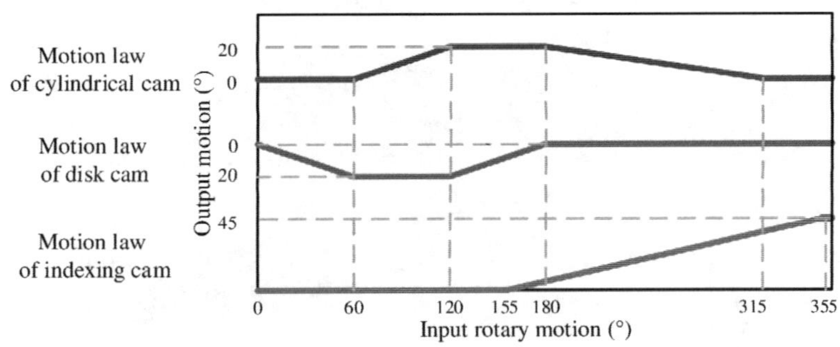

Figure 3. Motion cycles of three cam mechanisms.

A driving device fulfilling the indexing and oscillating motion based on the globoidal cam mechanism is designed, which replaces the original material transfer system of LED splitter. As shown in figure 4, the device consists of a power input shaft and three output shafts. The oscillating globoidal cam Ⅰ, oscillating globoidal cam Ⅱ and the indexing globoidal cam are installed on the input shaft in turn. As a result, the power input can actuate the three output shafts to achieve the specified movement law shown in figure 3. Both the oscillating output shaft I and output shaft Ⅱ with the swing angle of 20°, connected with the material receiving plate through the curved arm, will control the movement of the gripper in the X direction and Y direction. In this way, the material transmission of LED is executed replacing the original cylindrical cam and disk cam mechanism. The indexing output shaft with the indexing angle of 45°, connected with the material feeding plate through the synchronous belt transmission, will accomplish the material feeding movement replacing the original indexing cam mechanism.

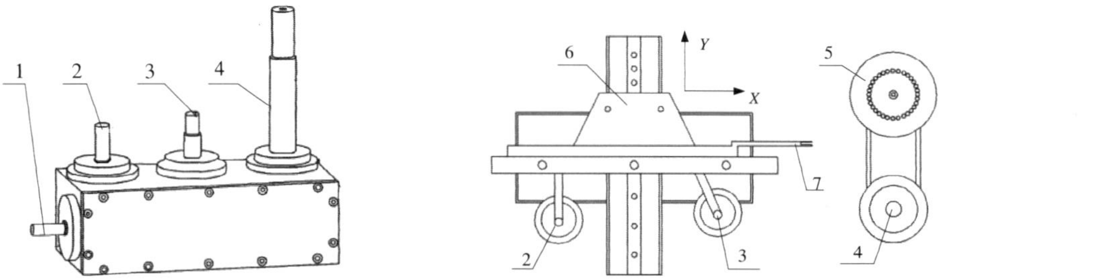

1-input shaft; 2-swing shaft I; 3-swing shaft II; 4-indexing shaft;
5- material feeding plate; 6-material receiving plate; 7-gripper

Figure 4. Schematic diagram of innovative material transmission system in LED light splitter.

3.2. Three-dimensional modeling

In order to kinematics simulation, part and assembly modelling for the transmission device containing indexing and oscillating motions should be made. The working surface equation between the indexing cam and oscillating cam is similar but the motion law function of follower [4]. Working surface equation of the globoidal cam is especially complex [5], so that it is difficult to directly establish its mathematical model in 3D software. Some software are integrated with the modeling method of the offset surface, and it can indirectly realize the precise modeling of complex regular surface. Here, Pro/E software is used as a tool for the model of the globoidal cam mechanism.

According to the theoretical surface equation, two benchmark curves are firstly created by software command "by equation". Then, "variable cross section scanning" is used to generate the theoretical surface and then "offset" to generate two working surfaces of globoidal cam with offset distance R, the radius of rollers, as shown in figure 5. At last the overall 3D entity of cam body is finished through commands such as "boundary blending", "surface merging", "materialization", etc.

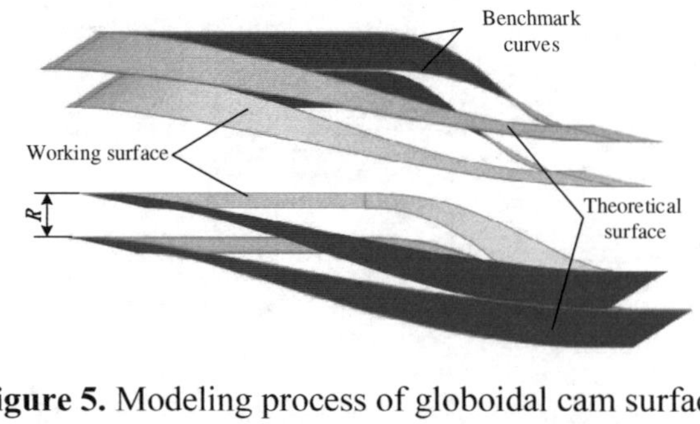

Figure 5. Modeling process of globoidal cam surface.

In the assembly module of Pro/E, two vertical input axes and output axes are established according to the center distance between cam and following plate, as shown in figure 6. Taking two axes as assembly datum, the globoid cam and the follower are connected by a pin for assembly. The "pin connection" and "groove connection" are used to fully define the driven plate connection for limiting the roller of the following disk. For the definition of groove connection, take a certain point P of roller axis as "point on the line", and select the theoretical line corresponding to point P as "groove axis". Further using the mechanism module of Pro/E, the servo motor is defined with the globoidal cam as the active part, so as to realize the motion simulation and collision interference inspection of the globoidal cam mechanism.

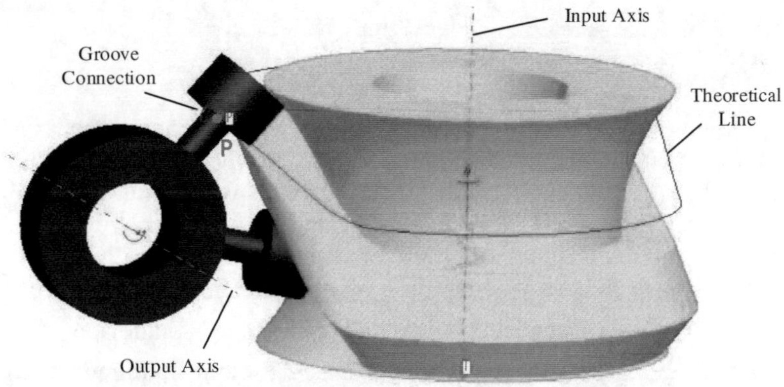

Figure 6. Assembly process of arc cam mechanism

After the assembly of the main transmission system is completed, the other parts of the device are assembled, including the box body, the fastener on the shaft and so on. Figure 7 shows the overall 3D model of the combined transmission device.

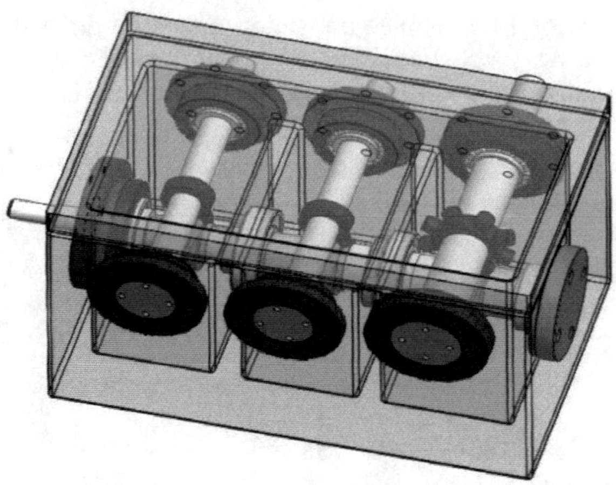

Figure 7. 3D model of combined drive.

3.3. Creation of engineering drawings

After 3D model creation is completed, engineering drawings need to be generated to meet production requirements, as shown in figure 8.

As can be seen from figure 8(a), three sets of globoidal cams are mounted on the input shaft, which are two oscillating cams and one indexing cam. The driven turntable meshes with the cam to drive the output shaft to rotate according to the defined motion. Two sections of the box wall with holes are added inside the box body for mounting deep groove ball bearings, which ensure the stability of the input shaft with the tapered roller bearings at both ends of the shaft together.

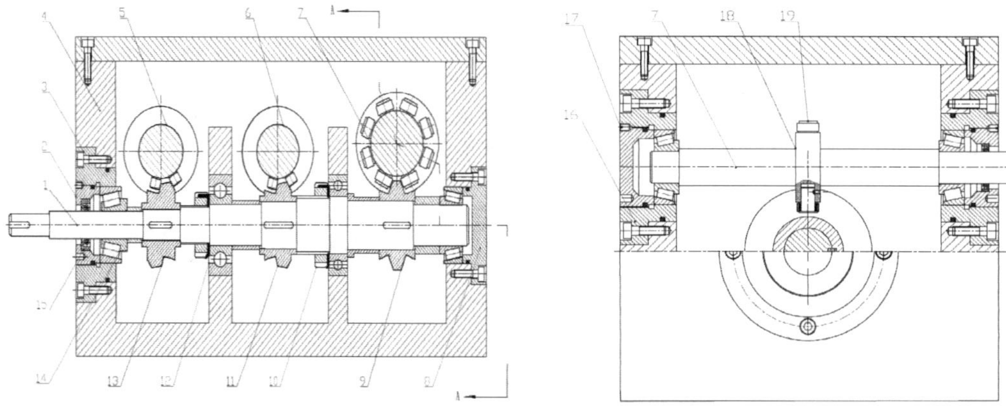

(a) main visual profile (b)A-A cutaway view
1-input shaft; 2-adjustment cover; 3-input support set; 4-box body; 5-oscillating shaft I; 6- oscillating shaft II; 7-indexing shaft; 8-bearing end cover; 9- indexing globoidal cam; 10-circular nut; 11- oscillating globoidal cam I; 12-deep groove ball bearing; 13-oscillating globoidal cam II; 14-tapered roller bearing; 15-lip sealing ring; 16-eccentric end cap; 17-clamping bolt; 18-driven follower; 19-roller

Figure 8. Engineering drawings of transmission device.

As can be seen from figure 8(b), an eccentric end cap is used at both ends of the output shaft to adjust the center distance between the input shaft and the output, that is, the gap between the cam ridge and the driven roller. After the eccentric end cap is rotated, its relative position is fixed by a set screw. At one end of the input shaft and at both ends of the output shaft, a screw-on adjustment cover is applied to adjust the clearance of the tapered roller bearing and achieve axial positioning.

4. Dynamic simulation

A partial 3D assembly model of the combined transmission was introduced into the mechanical system dynamics analysis software ADAMS. Each part is applied with a fixed pair or a rotating pair. The rotary drive of the input shaft is set to 300r/min, and a contact force is added between each roller and the cam. Stiffness coefficient is $K = 7.5\times10^5 N / mm^2$, with force linear index e=1.5 and damping $C = 750\ N\cdot S / mm$. As shown in figure 9, a virtual prototype of the device is built.

Figure 9. Virtual prototype model of transmission device with ADAMS software.

Dynamic simulation of the model in the ADAMS/view module should be operated to verify the correctness of the transmission design. The actual angular displacement of the three output shafts is illustrated in figure 10. The simulated angular displacement curve is consistent with the given theoretical angular displacement curve, shown in figure 3. The two oscillating plates have a swing angle of 20°, the indexing turntable has an indexing angle of 45°, and the motion coordination relationship between the three output shafts is also consistent with the design requirements.

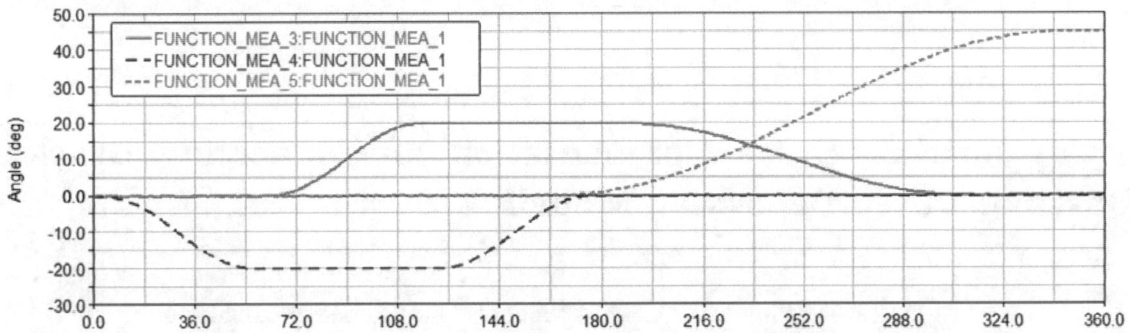

Figure 10. Simulation of angular displacement of three output motions.

5. Conclusions

In the automatic production line, it is often necessary to achieve the indexing or oscillating movement due to the requirements of the production process. However, most existing devices or institutions can only achieve a single movement. At the same time, the intermittent globoidal cam mechanism on the automatic production line gradually replaces the traditional intermittent mechanism because of its high precision and good dynamic performance. Therefore, a combination device of indexing and oscillating transmission based on globoidal cam is designed and its 3D model is constructed to realize the motion transmission from single input to three outputs. The virtual prototype model of the device is established by ADAMS software, and the simulation analysis is carried out, which provides the reference for the design of this type of device in the future.

Acknowledgement

The authors gratefully acknowledge that the present research is supported by the Key Research Project of Shandong Province in China (No. 2019GNC106123). In addition, valuable suggestions from the reviewers in improving the quality of the paper are highly appreciated.

References

[1] Liang Y F and Tan W M 2008 *Automatic Machinery Automatic* production line (Beijing: High Education Press) p52
[2] Robert L Norton 2009 *Cam design and manufacturing handbook* (Newyork: Industrial Press) p9
[3] Zhou J and Zhao Y 2010 Kinematics design on positioning and testing instrument in LED sorting machine *Mech. Trans.* **34** 48-51
[4] Ge R, Mi Z and Gao C 2018 Model and analysis of globoidal cam mechanisms with oscillating and dual stopping followers *J. Mach. Des.* **35** 62-66
[5] Zhang Y, Ji S and Zhao J 2016 Study on the geometric characteristics of mating surfaces of globoidal cam mechanisms *Mech. Mach. Theory* **100** 44-62

IOP Conf. Series: Materials Science and Engineering **686** (2019) 012013 doi:10.1088/1757-899X/686/1/012013

Study on Size Error Compensation of Connecting Bracket Based on Fused Deposition Modeling

Shibo Li, Tingting Liu*, Xingzhi Xiao, Weiyue Hu and Wenhe Liao

School of Mechanical Engineering, Nanjing University of Science and Technology, Digital Forming Technology and Equipment National Local United Engineering Laboratory, Nanjing, Jiangsu 210094

* E-mail: liutingting@mail.njust.edu.cn

Abstract: In the integrated additive manufacturing of electronic products, the connecting bracket is a typical fixed structure for embedding electronic components such as PCBs, sensors, batteries, etc. The dimensional accuracy of column spacing affects the stability of embedded components directly. In this paper, aiming at the printing error of connecting bracket column spacing during the Fused Deposition Modeling (FDM), the printing error compensation is studied via the single factor experiments. The impact of design size on printing error is analyzed, and the printing error compensation model is established. Based on the FDM printing process of the connecting bracket, the main influencing factors of the printing error are analyzed. The results show that the column spacing printing error is the result of the combination of the dimension error of column printing, the dimension error of base printing and the motion error of the machine. Finally, based on the error compensation model, the compensated connection bracket is printed and the PCBs embedding verification is performed. The PCB can be embedded steadily, which indicates that the established model has higher precision.

1. Introduction

With the development of 3D printing, the additive manufacturing of structure parts have gradually evolved into integrated additive manufacturing of electronic products. During the printing, the structures are manufactured by FDM, the Printed Circuit Boards (PCBs), sensors, and other electronic components are embedded in the connecting brackets fabricated by FDM. The connection electric circuits are printed subsequently, which ultimately facilitates the integrated manufacturing of "structure-circuit-electronic component" for electronic products [1]. The process is shown in figure 1.This technology subverts the traditional manufacturing method of electronic products, and it also provides more space for designing and manufacturing, as well as simplifying the manufacturing process. It is expected to provide a brand new idea and solution for the development of innovative products, and to become one of the research hotspots in 3D printing [2-4].

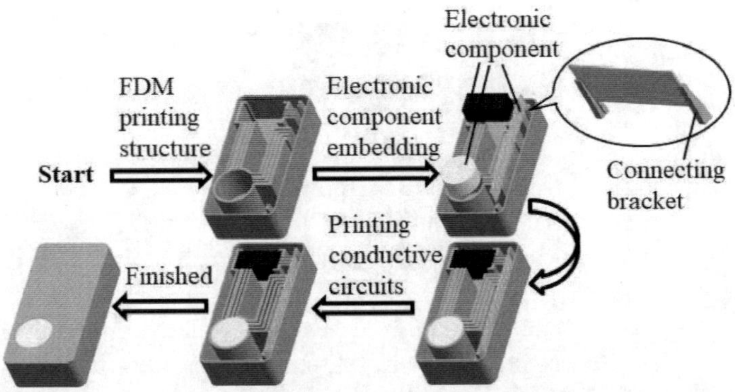

Figure 1. Flow chart of embedded electronic product manufacturing.

The effective and stable embedding of PCBs, sensors and other electronic components is the key to realize the integrated additive manufacturing of electronic products. Currently, the electronic components are embedded into the reserved cavities of FDM printing connecting brackets by setting the pause program [5-7].However, in the embedding process, due to the printing error of the dimension between two columns of the connecting brackets, it is easy to cause the unstable embedding of components and other problems. Therefore, it is necessary to carry out an error compensation study on the size of FDM printed connecting brackets.

Numerous studies have been conducted to solve the size error compensation for FDM parts. Tong et al put forward an error compensation method for slicing software, which can reduced the average volume error of printed parts by about 30% [8]. Yaman et al proposed a method of the internal structure of parts, which can control the shrinkage of circular and square holes and make error compensation for the size of characteristic holes [9]. Zou et al set up a compensation model of ideal contour line from the FDM's microscopic size, and verified that this method could compensate the outer contour size of the model correctly [10]. Dong et al analyzed the printing error caused by heating and contract on cooling of materials with the example of 3D printed threaded connector, and proposed that the printing error can be reduced by reducing the material shrinkage rate and establishing a compensation model [11]. However, the existing studies are mostly based on the analysis of size error caused by a single factor. On the one hand, the comprehensive influence of multiple factors such as model size interaction and nozzle movement form are not taken into account, and the dimensional accuracy after compensation still needs to be improved. On the other hand, there is still less research on the printing error compensation of the connecting brackets for components embedding. Therefore, for the FDM printing of connecting brackets, it is still necessary to study dimensional error compensation in-depth, and to reveal the law that how multiple factors affect the dimensional accuracy, which will provide technical support for the effective and stable embedding of electronic components.

In this paper, the single-factor experimental design is utilized to carry out the FDM printing connecting brackets experiment. The correlation between the design size and the printing size error are analyzed and to build an error compensation model. Then, the causes of size error are deeply studied to reveal the influence mechanism of multiple factors on size error. Finally, a verification experiment is carried out to verify the validity of the error compensation model.

2. Experimental design

The FDM printer used in this experiment is Mass Portal Pharaoh ED 20, as shown in figure 2. The forming space size of device is ϕ200mm×200mm (bottom diameter ×height). This printer using Delta architecture technology, can achieve fast printing with less vibration and noise. The printing material is PolyPlus PLA supplied by Polymaker. In order to ensure the consistency of different experimental process, printing parameters are set to machine default parameters. The parameters as follows: Nozzle diameter is 0.4 mm, Contour width is 0.4 mm, Scanning times is 1, Layer thickness is 0.2 mm, Printing speed is 2400 mm/min, Filling rate is 5%, Nozzle temperature is 210°C.

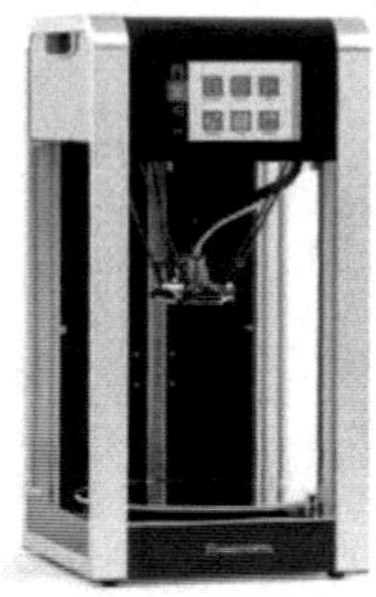

Figure 2. Mass Portal desktop 3D printer.

According to the basic structure of the connecting brackets in figure 1, the experimental parts are designed, shown as figure 3. The height of the base *h* is 2mm, and the spacing of the columns in Y direction *m* is 5mm. The dimension *L* is the column spacing, which is the measured size of this experiment. L_2 is the base length, L_1 is the column length, and the column design size is 5mm×5mm×10mm (length × width × height). The spacing size *L* of the experimental variables is shown in table 1, a total of 26 levels are set. Each group printed four samples, that is, print size data of 8 column spacing are obtained for each groups.

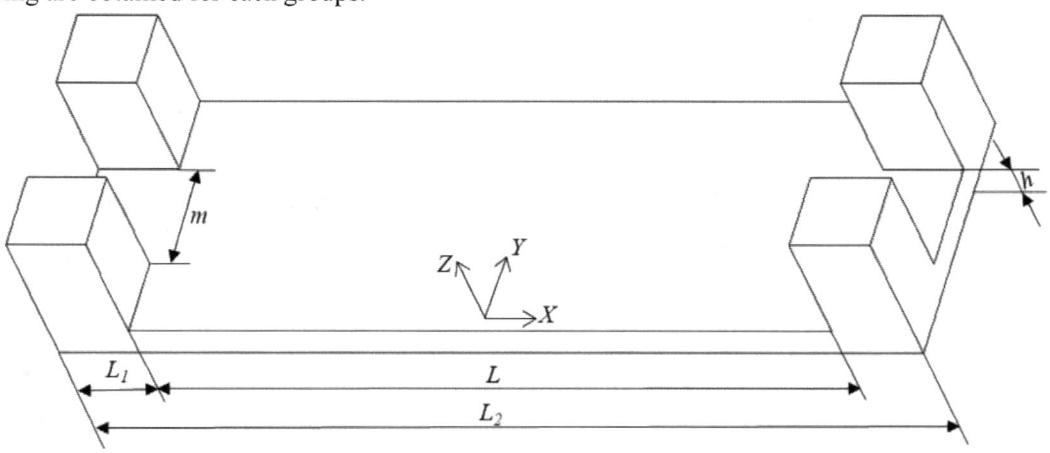

Figure 3. 3D model of the experiment piece.

3. The experimental results

The FDM printed connecting brackets experiment is carried out. By using the vernier caliper with a precision of 0.02 mm to measure the column spacing dimension *L*, and to record the data and calculate the average print size. The printing average size subtracted from the column spacing design size is the printing size error *Δ*. The specific data is shown in table 1.

Table 1. Column spacing design size, printing size and printing error.

Design size (mm)	Printing size (mm)								Printing average size (mm)	Printing size error (mm)
3	2.80	2.80	2.86	2.82	2.86	2.78	2.84	2.82	2.82	0.18
5	4.84	4.90	4.84	4.68	4.88	4.80	4.80	4.86	4.83	0.18
9	8.80	8.80	8.82	8.86	8.78	8.82	8.86	8.84	8.82	0.18
14	13.84	13.80	13.82	13.86	13.84	13.80	13.82	13.80	13.82	0.18
18	17.78	17.82	17.80	17.80	17.82	17.88	17.84	17.80	17.82	0.18
22	21.86	21.82	21.94	21.80	21.82	21.82	21.86	21.82	21.84	0.16
24	23.92	23.86	23.86	23.86	23.94	23.86	23.88	23.88	23.88	0.12
28	27.88	27.88	27.86	27.86	27.92	27.84	27.86	27.88	27.87	0.13

32	31.86	31.90	31.96	31.86	31.94	31.86	31.88	31.86	31.89	0.11
36	35.86	35.80	35.88	35.78	35.86	35.86	35.90	35.88	35.85	0.15
40	39.90	39.84	39.90	39.92	39.90	39.86	39.88	39.86	39.88	0.12
44	43.94	43.90	43.82	43.86	43.94	43.90	43.88	43.88	43.89	0.11
48	47.94	47.88	47.98	47.98	47.94	47.94	47.94	47.86	47.93	0.07
52	51.92	51.92	51.92	51.88	51.86	51.92	51.90	51.88	51.90	0.10
56	55.94	55.86	55.88	55.84	55.92	55.90	55.96	55.94	55.91	0.10
60	59.92	59.86	59.89	59.88	59.94	59.90	59.96	59.94	59.91	0.09
65	64.98	64.94	64.98	64.94	64.98	64.96	64.98	64.88	64.96	0.05
69	68.90	68.88	68.88	68.86	68.90	68.86	68.96	68.94	68.90	0.10
73	72.94	72.92	72.96	72.92	72.96	72.94	72.98	72.94	72.95	0.06
77	76.92	76.96	77.00	76.94	76.96	76.88	76.98	76.94	76.95	0.05
81	80.96	80.94	80.96	80.92	80.96	80.96	80.96	80.92	80.95	0.05
85	85.02	84.98	84.94	84.96	84.88	84.88	84.94	84.94	84.94	0.06
89	88.94	89.04	88.98	88.98	88.94	88.90	89.00	89.00	88.97	0.03
93	92.92	92.92	92.98	92.88	93.00	92.98	92.92	92.90	92.94	0.06
97	96.90	96.98	97.06	97.00	96.96	97.04	96.96	96.86	96.97	0.03
100	99.94	99.90	99.94	99.94	100.08	100.08	100.08	99.96	99.99	0.01

The scatter diagram of printing size error for column spacing size as shown in figure 4, where the abscissa is the design size of column spacing, and the ordinate is the printing size error. Linear regression analysis was carried out on the printing size error data, which got design error compensation model of printing size error Δ and size L:

$$\Delta = -0.0016L + 0.1867,\ R^2 = 0.894 \quad (1)$$

Where, R^2 is the coefficient of determination, which is used to measure the goodness of fit of the regression equation. The closer the value is to 1, the higher the degree of interpretation of the dependent variable by the independent variable in the regression equation, and here is the degree of fitness is good.

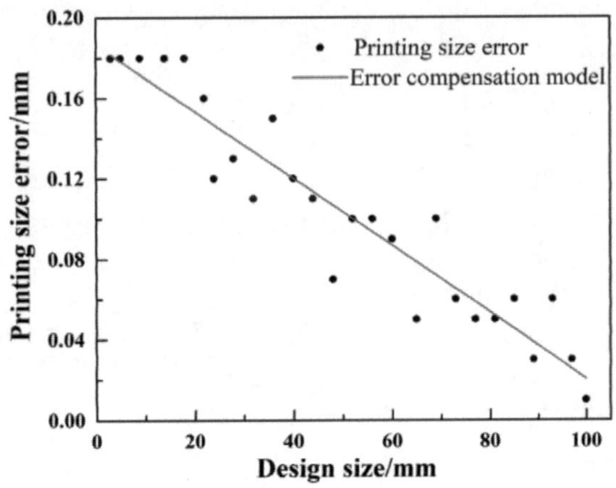

Figure 4. The relation between printing size error and design size.

4. Error analysis

As can be seen from figure 4, when the design size is within 100mm, the printing average size of the column spacing formed by FDM is all smaller than the design size, and the printing size error gradually decreases with the increase of the design size. According to the structure of the experimental parts as shown in figure 2, it can be found that the spacing L of the column is actually indirectly formed by printing the base L_2 and two columns L_1, and its measured size $L= L_2-2L_1$.It can be seen that the printing size error of column spacing L is affected by the printing size error of base size L_2 and column itself L_1. In this section, the size error of the column itself, the size error of the base and other causes of the error will be studied further, and the comprehensive influence of each factor in the final printing size error of the column spacing will be clarified.

4.1 Error analysis of column size L_1

Select one of the experimental parts of each size, and measure the printing size of the four columns by the vernier caliper. The average value was calculated. The specific data is shown in table 2.

Table 2. Parts design size, column printing size and printing size error.

Design size (mm)	Printing size (mm)				Printing average size (mm)	Printing size error (mm)
3	5.10	5.08	5.14	5.10	5.10	-0.10
5	5.06	5.12	5.12	5.06	5.09	-0.09
9	5.10	5.12	5.10	5.14	5.12	-0.12
14	5.16	5.14	5.12	5.2	5.15	-0.15
18	5.16	5.12	5.12	5.14	5.13	-0.13
22	5.18	5.14	5.12	5.16	5.15	-0.15
24	5.08	5.14	5.14	5.18	5.14	-0.14
28	5.22	5.18	5.16	5.16	5.18	-0.18
32	5.16	5.20	5.14	5.22	5.18	-0.18
36	5.08	5.16	5.18	5.12	5.14	-0.14
40	5.12	5.16	5.20	5.20	5.17	-0.17
44	5.16	5.22	5.16	5.20	5.18	-0.18
48	5.16	5.16	5.12	5.18	5.15	-0.15
52	5.16	5.22	5.20	5.26	5.21	-0.21
56	5.16	5.18	5.16	5.20	5.17	-0.17
60	5.16	5.20	5.18	5.16	5.17	-0.17
65	5.24	5.22	5.20	5.18	5.21	-0.21
69	5.22	5.20	5.20	5.26	5.22	-0.22
73	5.20	5.18	5.22	5.2	5.20	-0.20
77	5.18	5.24	5.22	5.14	5.20	-0.20
81	5.26	5.18	5.24	5.20	5.22	-0.22
85	5.16	5.16	5.18	5.14	5.16	-0.16
89	5.16	5.18	5.22	5.14	5.18	-0.18
93	5.20	5.18	5.28	5.24	5.22	-0.22
97	5.20	5.22	5.24	5.18	5.21	-0.21
100	5.18	5.20	5.20	5.24	5.21	-0.21

The column design size minus the printing average size gets the printing error ΔL_1, the column size print error scatter diagram as shown in figure 5, where the abscissa is design dimension of the parts spacing, and the ordinate is the printing error of L_1.

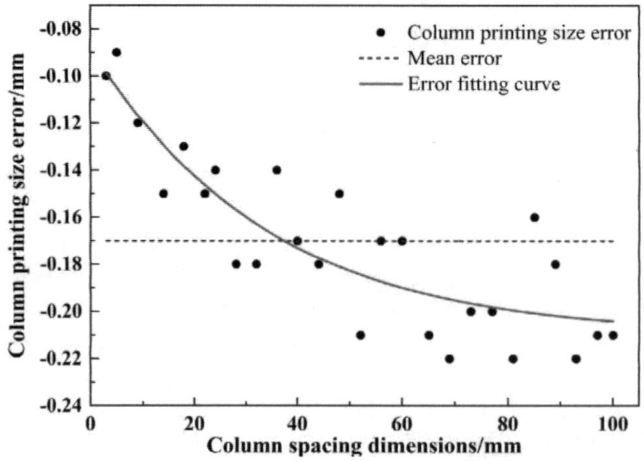

Figure 5. Printing size error of 5mm×5mm (length × width) column.

The data is fitted to the printing error data to obtain an error compensation model of the printing error ΔL_1 and the design size L:

$$\Delta L_1 = -9e^{-10}L^4 + 3e^{-7}L^3 + 5e^{-5}L^2 - 0.0036L - 0.0089,\ R^2 = 0.733 \quad (2)$$

As shown in figure 5, the print size of the column is larger than the design size, and the deviation is between +0.09mm~+0.23mm. As the column spacing increasing, the printing error of L_1 increases gradually. The mean error is 0.17mm. This data shows that the printing size error of column size L_1 will lead to an average decrease of 0.17mm in the column spacing size L of the experimental parts compared with the designed size. Comparing the error compensation model shown in (1), when L=0, Δ=0.1867, which is the print size error when the column spacing L is designed to be 0. It can be seen that when L=0, the error of the column size L_1 accounts for 91.05% of the error of the column spacing dimension L, which is the main reason that the design spacing of the column is larger than the printing size. But at the same time, it can be seen that due to the size of the column tends to be gentle as the column spacing increases, and the trend of printing error do not match the linear decrease of the design size. Therefore, the base size L_2 needs to be analyzed.

4.2 Error analysis of base size L_2

According to the formula $L= L_2-2L_1$, the size of the base L_2 is another factor affecting the dimensional accuracy of column spacing. It can be seen from figure 3 that theoretical base size $L_2=L+2L_1$. According to the analysis in section 4.1, there is a printing error in column size L_1. In order to avoid the influence of L_1 when measuring the base size L_2, according to the structure of the base, the design of the printing test part is shown in figure 6. The size of experimental parts in the Y and Z directions are fixed, which are 5 mm and 10 mm respectively. The dimensions in X direction are set to 5 groups of different levels, which are 5mm, 20mm, 40mm, 60mm and 80mm. After printing, the vernier caliper was used to measure the printing size of the experimental parts in the X direction, and the results are shown in table 3.

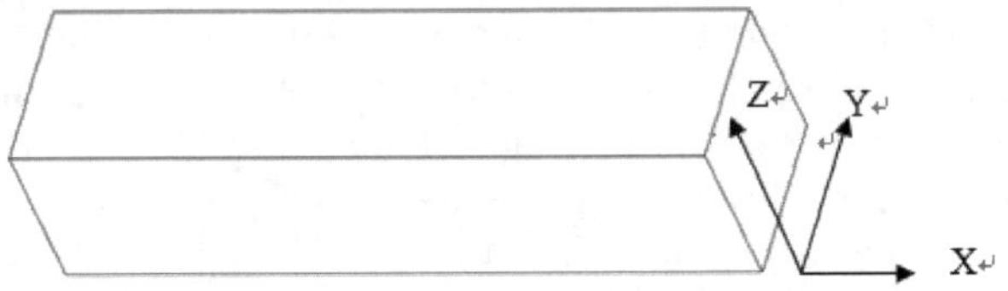

Figure 6. Printing test part of base size.

Table 3. Printing size, printing average size and printing size error of the test parts.

Design size (mm)	Printing size (mm)					Printing average size (mm)	Printing size error (mm)
5.00	5.00	5.02	5.02	5.04	5.00	5.02	-0.02
20.00	20.16	20.08	20.12	20.10	20.12	20.12	-0.12
40.00	40.18	40.18	40.20	40.20	40.16	40.18	-0.18
60.00	60.20	60.26	60.16	60.24	60.22	60.22	-0.22
80.00	80.32	80.34	80.30	80.26	80.36	80.32	-0.32

Plot the scatter diagram of printing average error changing with design size L_2, as shown in figure 7.

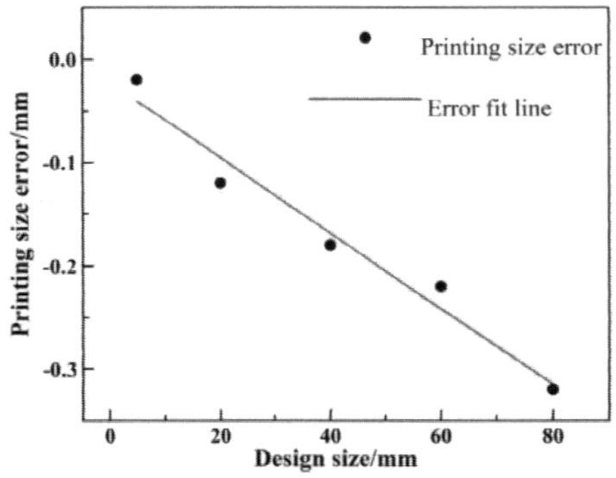

Figure 7. Relationship between printing size error and design size.

Linear regression analysis is performed on the printing size error Δ_2 to obtain the relationship between the design size L_2 and the printing size error Δ_2:

$$\Delta_2 = -0.0037L_1 + 0.02, \ R^2=0.95 \quad (3)$$

As shown in figure 7, with the increasing of base design size L_2, the print size error Δ_2 increases gradually. According to formula $L= L_2-2L_1$, base printing size error will cause the column spacing print size error to decrease with the design size increases, which is an important reason for the variation trend of the printing size error of the column spacing as shown in figure 4. But at the same time, it can be seen that the reduction degree of printing error in figure 4 and figure 7 is not completely consistent, indicating that there are other influencing factors.

4.3 Analysis of other error factors

The structure of FDM printer used in the experiment is Delta parallel mechanism. Some studies have pointed out that the operating accuracy of parallel mechanism is not evenly distributed in its workspace. The farther the position points are from the central axis, the larger the motion error will be [12]. Comparing the printing size which design size is 5mm in table 2 and table.3. In table 2, it can be find that the distance between center position of the columns and the center of the printing platform is $(L+L_1)/2$, the actual print size is 0.09 mm larger on average. Differently, the experimental parts size which in table 3 are placed at the center of the printing platform, and the actual print size is 0.02 mm larger on average. Therefore, it is necessary to consider the impact of the machine's motion error on the printing size of the connection brackets.

There are five experimental parts are designed, the size of parts are 5mm×5mm×5mm(length×width×height), as shown in figure 8. The experimental parts No. 3 is placed in the center of the printing platform, and other experimental parts are placed symmetrically along the Y direction of the platform. After printing is completed, the printing size in the X direction is shown in table 4.

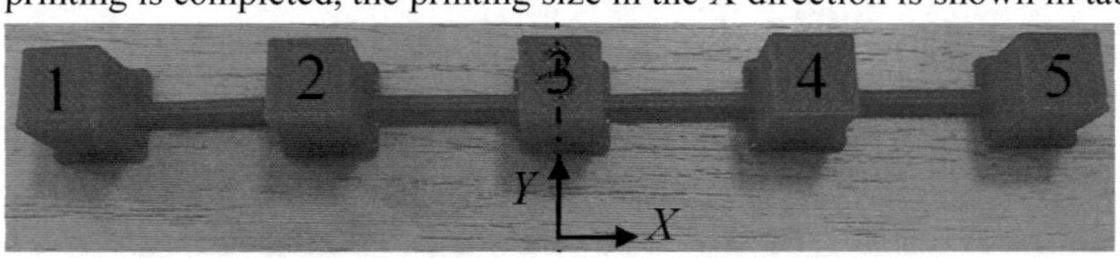

Figure 8. Part printing experiments at different positions.

Table 4. Experimental part size at different positions.

Part numbers	Printing size (mm)					Printing average size (mm)
1	10.10	10.12	10.10	10.10	10.12	10.11
2	10.18	10.20	10.18	10.20	10.20	10.19
3	10.04	10.04	10.06	10.08	10.06	10.06
4	10.22	10.24	10.20	10.24	10.24	10.23
5	10.22	10.24	10.22	10.20	10.22	10.22

As can be seen from table 4, the printing error of column size at the center of the printing platform is the smallest, and the size error of both sides is larger. This result also further explained the experimental results in Section 3.1, and why the printing size of the column increases with column spacing increasing. It is verified that the motion accuracy of the Delta mechanism is not evenly distributed in its workspace, and the closer the position point is to the central point, the higher the motion accuracy is, that indicating the printing size error is related to the printing position. This factor combined with the size error of the base causes the size error of the column spacing to decrease as the design size increasing.

5. Experimental verification

In this paper, the size error compensation model of desktop FDM printer is obtained through experimental research, as shown in formula (1). In order to verify the validity of the model, 5 PCBs of different sizes were selected for verification experiments. Firstly, the design size was compensated according to the compensation model, the original PCBs size were brought into the formula (1) to obtain the compensation value, and plus the original size value as the final design dimension of the column spacing. The original size of PCBs and the size after compensation are shown in table 5.

Table 5. PCBs original size and compensation size.

Number	1		2		3		4		5	
	O[a] size (mm)	C[b] size (mm)	O size (mm)	C size (mm)	O size (mm)	C size (mm)	O size (mm)	C size (mm)	O size (mm)	C size (mm)
Length	28.08	28.22	90.24	90.27	45.14	45.25	40.14	40.26	20.12	20.27
Height	33.12	33.25	95.36	95.38	30.06	30.19	40.20	40.32	24.16	24.30
Width	1.64	1.82	1.64	1.82	1.64	1.82	1.64	1.82	1.64	1.82

a The letter O represents the original size.

b The letter C represents the compensation size

The design validation model is shown in figure 9.After the printing is completed, embedding PCBs into the connection brackets to test the embedding situation, the result as shown in figure 10.

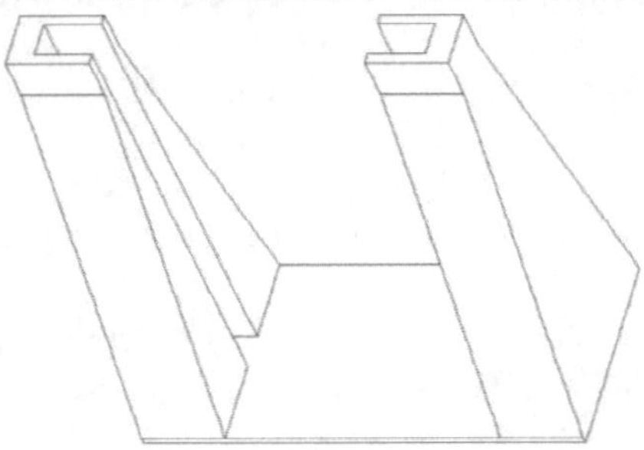

Figure 9. Varification model.

Figure 10. PCBs embedding verification.

Experimental verification shows that based on the error compensation model, designing of dimensional error compensation for connection brackets can ensure PCBs embedded smoothly, and there is no shaking phenomenon after embedded or unable embedding phenomenon.

6. Conclusion

In this paper, aiming at the problem of unstable component embedding due to the printing error of the FDM printed connecting brackets, based on the FDM research on error compensation of column spacing is carried out, and the error compensation model of the column spacing size was established. After in-depth analysis of the cause of the error, the main conclusions are summarized as follows:

(1) The average print size of the column spacing of the FDM forming connection brackets are smaller than the design size, and the print size error decreases linearly with the increase of the design size.

(2) Column spacing size error is the result of the combination of column size error, base size error and machine motion error. Among them, the column size error is the main reason that why the column

spacing design size is larger than the print size. The combination of the base size error and the machine motion error causes the column spacing size printing error to decrease with the increasing of the design size.

In the following work, the error compensation model of FDM printing column spacing size will be further studied. It is expected to establish a theoretical error compensation model which consider the forming direction, layer thickness and other process parameters, and provide theoretical guidance for the integrated additive manufacturing of electronic products.

Acknowledgement
This work was supported by The National Key Research and Development Program of China (No.2017YFB1103000), JiangSu Key Research and Development Program (No. BE2016182) and The Key Technology Breakthrough program in Advanced Industry of Nanjing (2018003)

References

[1] Kim R, Bae M, Kim D G, Cheng H, Kim B H, Kim D, Li M, Wu J, Du F, Kim H, Kim S, Estrada D, Hong S W, Huang Y, Pop E and Rogers J A 2011 Stretchable, Transparent Graphene Interconnects for Arrays of Microscale Inorganic Light Emitting Diodes on Rubber Substrates *NANO LETT* **11** 3881-86

[2] Lewis J A and Ahn B Y 2015 Device fabrication: Three-dimensional printed electronics. *NATURE* **518** 42

[3] Wicker R and Macdonald E 2012 Multi-material, multi-technology stereolithography *Virtual & Physical Prototyping* **7** 14

[4] Muth J T, Vogt D M, Truby R L, Mengüç Y, Kolesky D B, Wood R J and Lewis J A 2014 Embedded 3D Printing of Strain Sensors within Highly Stretchable Elastomers *ADV MATER* **26** 6307-12

[5] Espalin D, Muse D W, MacDonald E and Wicker R B 2014 3D Printing multifunctionality: structures with electronics *The International Journal of Advanced Manufacturing Technology* **72** 963-78

[6] Lu Y, Vatani M and Choi J 2013 Direct-write/cure conductive polymer nanocomposites for 3D structural electronics *J MECH SCI TECHNOL* **27** 2929-34

[7] Shemelya C, Banuelos-Chacon L, Melendez A, Kief C, Espalin D, Wicker R, Krijnen G and MacDonald E 2015 Multi-functional 3D printed and embedded sensors for satellite qualification structures.: IEEE) pp 1-04

[8] Tong K, Joshi S and Lehtihet E A 2008 Error compensation for fused deposition modeling (FDM) machine by correcting slice files *RAPID PROTOTYPING J* **14** 4-14

[9] Yaman U 2018 Shrinkage compensation of holes via shrinkage of interior structure in FDM process *The International Journal of Advanced Manufacturing Technology* **94** 2187-97

[10] G Zou 2002 Research on manucfacturing precision of fuesd deposition modelling and rapid mold manufacturing technology : Dalian University of Technology p 132

[11] X Dong,D Yin, H Fang and W Zhang 2016 FDM 3D Printing Dimension Error and Its Compensation Method Research *Modular Machin Tool & Automatic Manufacturing Technique,* 39-41

[12] Y Fang,H Fang and J Mou 2000 Error sensitive index of paraller Robots and its Distribution in Workspace *Journal of northern jiaotong university* 38-42

Thermodynamic Prediction Model of Single-stage Gradational Lead Screw Vacuum Pump

Fan Zhao*, Feng Han, Kun Sun, Shi-wei Zhang, Tai Yue and Jin-ming Li

School of Mechanical Engineering & Automation, Northeastern University, Shenyang 110004, China

*Corresponding author's E-mail: 1834228767@qq.com

Abstract. As a kind of fluid machinery, screw vacuum pumps are widely used in material engineering and semiconductor industry. Nowadays, new requirements for screw vacuum pumps are put forward on low energy consumption and low manufacture cost. Compared with the constant lead screw (CLS) vacuum pump, the gradational lead screw (GLS) vacuum pump has many advantages. Taking the SGLS vacuum pump as the research object, this paper divides the pumping process into four small stages and proposes a thermodynamic model to analyse the performance. The characteristic curves under diverse operating conditions are drawn and the energy consumptions are illustrated at the end of this paper. The results show that a thermodynamic model can provide a well explanation on the phenomena of pressure pulsation and power consumption, which will help us to understand the outstanding performance of SGLS vacuum pump better.

1. Introduction

The oil-free screw vacuum pump is a kind of positive displacement screw machines to obtain a clean vacuum environment, widely used in the material engineering and semiconductor industry. As a rough vacuum pump, the constant lead screw (CLS) vacuum pump was invented earlier than the gradational lead screw (GLS) vacuum pump (figure 1). The problems of CLS vacuum pump, such as high energy consumption, are also excessive and apparent when the pump system is running. In response to solve these issues, the GLS vacuum pump emerged at the right time [1].

The pitch of GLS rotor is variable along the axis direction. If the pitch of GLS rotor continuously decreases in the axial direction, this kind of rotor is known as a SGLS rotor (figure 2), and the pump comprising the SGLS rotors is referred to as a SGLS vacuum pump. A cylindrical coordinate system can be established to describe the helix of SGLS rotor. In this paper, if we assume that the pitch at outlet is λ and the varying pitch coefficient is α, the expression of helix and its pitch can be given orderly:

$$z(\theta) = \frac{\lambda}{2\pi} \cdot (\theta + \alpha\theta^2) \quad (1)$$

$$\lambda(\theta) = \frac{\lambda}{2\pi} \cdot (1 + 2\alpha\theta) \quad (2)$$

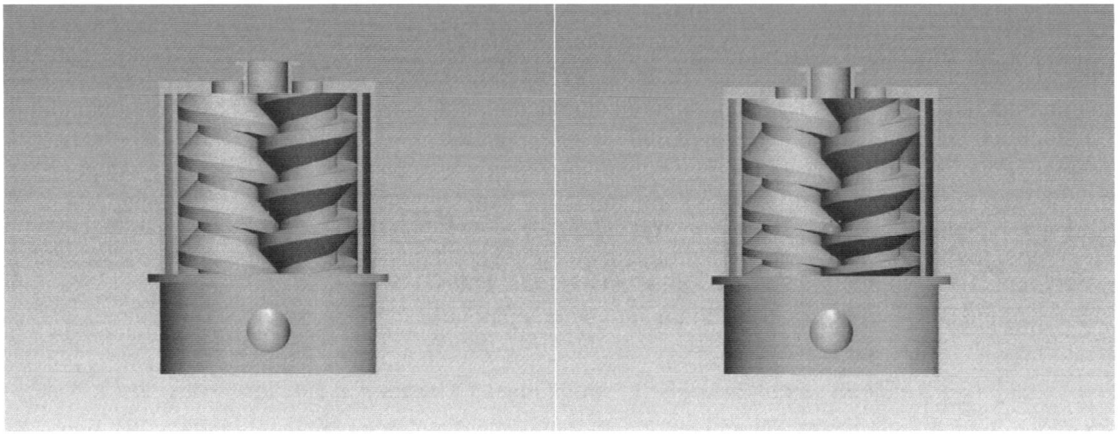

(a) (b)

Figure 1. The structure diagram of (a) CLS and (b) GLS vacuum pump

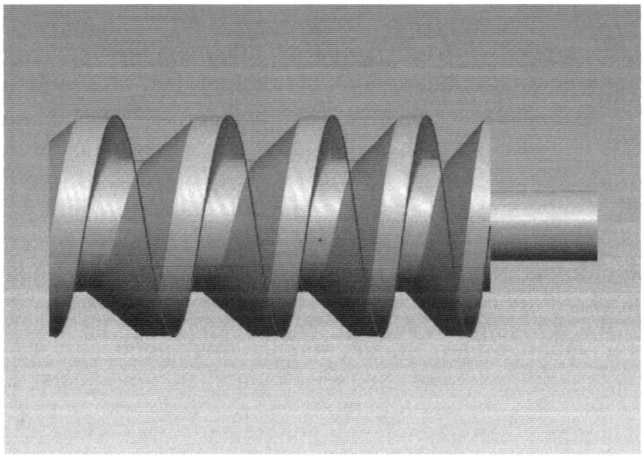

Figure 2. The outline of SGLS rotor and its helix

Considerable efforts have been carried out on the gradational screw machines, concerning novel product design [2], structure optimization [3], CFD simulation [4], etc. Although a number of efforts have been concentrated on the area of screw vacuum pump, few studies have been put into the thermodynamic field of SGLS vacuum pump [5]. In this paper, we divide the pumping process into four small stages according to the structure feature of screw rotor, and we establish a thermodynamic model based on the thermodynamics principle to analyze the performance quantitatively. The characteristic curves under diverse operating conditions are drawn and the energy consumptions are illustrated at the end of this paper. The results will help us to understand the outstanding performance of SGLS vacuum pump.

2. Pumping process

As the gas moves along the travel path from inlet side to the outlet side, the pumping process (figure 3(a) to 3(d)) can be split up into four successive stages (sucking, compressing, backlashing and exhausting stage) according to their respective feature. The gas in the said space is marked with the black dot in the drawing.

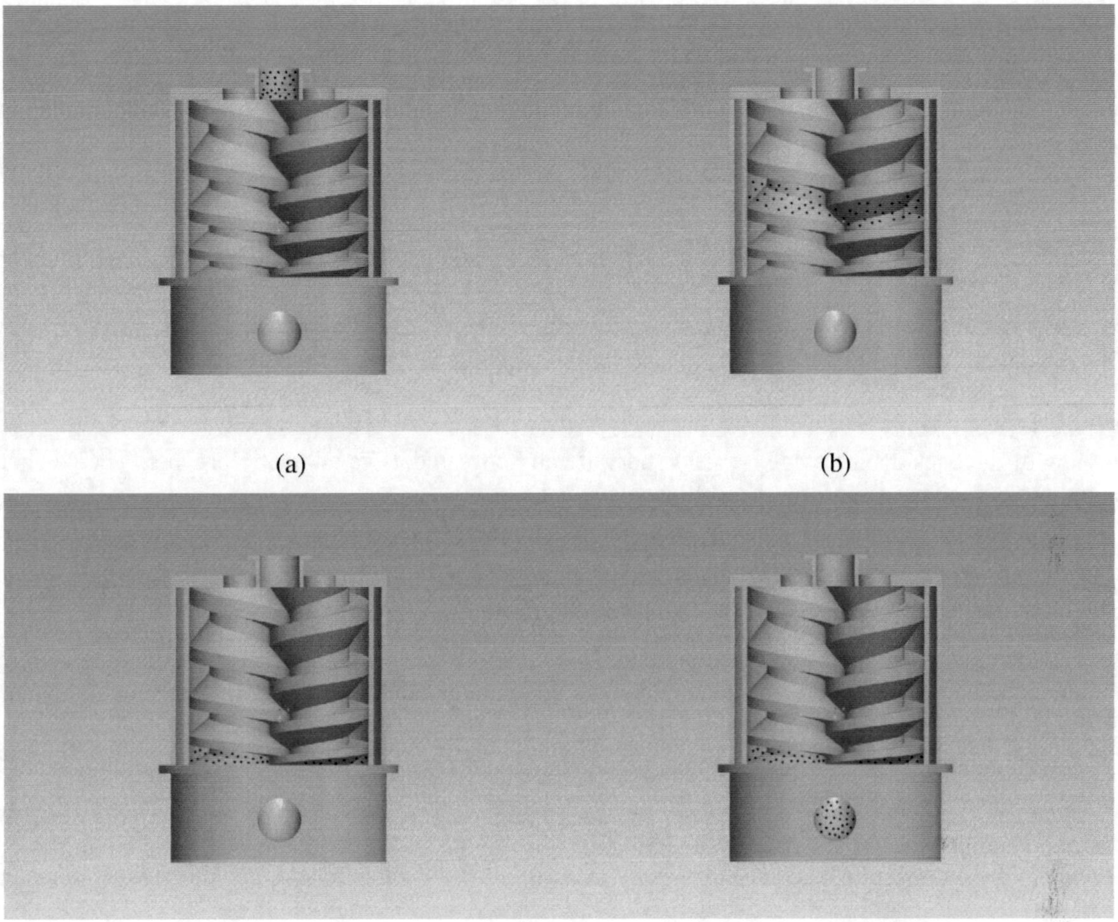

(a) (b) (c) (d)

Figure 3. Four successive stages occurred during the gas pass through the pump

3. Mathematical model

The thermodynamic model can be established according to the under-mentioned assumption:

(1) The gas is deemed as ideal gas and it is equally distributed in the space at any time
(2) The leak and back-diffusion through the gap are neglected
(3) The outlet pressure is assumed to be constant

3.1 Sucking stage

Owing to the connection between the sucking space and the inlet during this stage, the temperature and pressure of the gas can be seen as isothermal-isobaric process in the interval, thus,

$$T(t_1) = T_i \tag{3}$$

$$P(t_1) = P_i \tag{4}$$

The volume of sucking space is the product of the length and the cross section area, namely,

$$V(t_1) = A \cdot l(t_1) = \frac{A\omega\lambda}{2\pi} \cdot [(1 + 2\alpha n\pi)t_1 - \alpha\omega t_1^2] \tag{5}$$

The mass of gas are deduced by the following formula:

$$M(t_1) = \frac{P_i V(t_1)}{R_g T_i} = \frac{P_i A\omega\lambda}{2\pi R_g T_0}[(1 + 2\alpha n\pi)t_1 - \alpha\omega t_1^2] \tag{6}$$

3.2 Compressing stage

During compressing stage, volume of gas is lessened and the gas is compressed gradually:

$$V(t_2) = A \cdot l(t_2) = A\lambda\{1 + 2\alpha[(n+1)\pi - \omega t_2]\} \tag{7}$$

$$M(t_2) = M_1 = \frac{A\lambda P_i}{R_g T_i}(1 + 2\alpha n\pi - 2\alpha\pi) \tag{8}$$

$$P(t_2) = P_i\left[\frac{V_0}{V(t_2)}\right]^{\kappa} = P_i\left\{\frac{1+2\alpha n\pi-2\alpha\pi}{1+2\alpha[(n+1)\pi-\omega t_2]}\right\}^{n} \tag{9}$$

$$T(t_2) = T_i\left[\frac{V_0}{V(t_2)}\right]^{\kappa-1} = T_i\left\{\frac{1+2\alpha n\pi-2\alpha\pi}{1+2\alpha[(n+1)\pi-\omega t_2]}\right\}^{n-1} \tag{10}$$

3.3 Backlashing stage

In view of the backlashing stage is fairly short, it can be treated as an adiabatic process. The gas both in and out of the space conform the ideal gas law, namely,

$$PV = MR_gT \tag{11}$$

If the gas flowing back into the space is written as M_o, and the mixture gas in the space is written as M_3, the mixture mass of gas can be expressed as follow.

$$M_3 = M_1 + M_o = \frac{V_1P_i}{R_gT_i} + \frac{C_VV_2(P_o-P_2)}{C_PR_gT_o} = \frac{C_PT_oV_1P_i+C_VV_2T_i(P_o-P_2)}{T_iC_PR_gT_o} \tag{12}$$

$$T_3 = \frac{P_oV_2}{M_3R_g} = \frac{C_PT_oT_iP_oV_2}{C_PT_oV_1P_i+C_VT_iV_2(P_o-P_2)} \tag{13}$$

3.4 Exhausting stage

Owing to the connection between the space and the outlet during this stage, the temperature and pressure of the gas can be seen as isothermal-isobaric process in the interval, thus,

$$P(t_4) = P_o \tag{14}$$

$$T(t_4) = T_3 \tag{15}$$

$$V(t_4) = A \cdot l(t_4) = \frac{A\lambda}{2\pi}\{(n+2)\pi - \omega t_4 + \alpha[(n+2)\pi - \omega t_4]^2\} \tag{16}$$

$$M(t_4) = \frac{P_oV(t_4)}{R_gT_3} = \frac{P_oA\lambda}{2\pi R_gT_3}\{(n+2)\pi - \omega t_4 + \alpha[(n+2)\pi - \omega t_4]^2\} \tag{17}$$

4. Result and discussion

The above formulas can be used to predict the thermodynamic performance of SGLS vacuum pump. The structure parameters of SGLS vacuum pump are listed in Table 1. Since the inlet pressure and volume ratio are the main factors affecting pumping process, the impacts of these factors will be illustrated in this part.

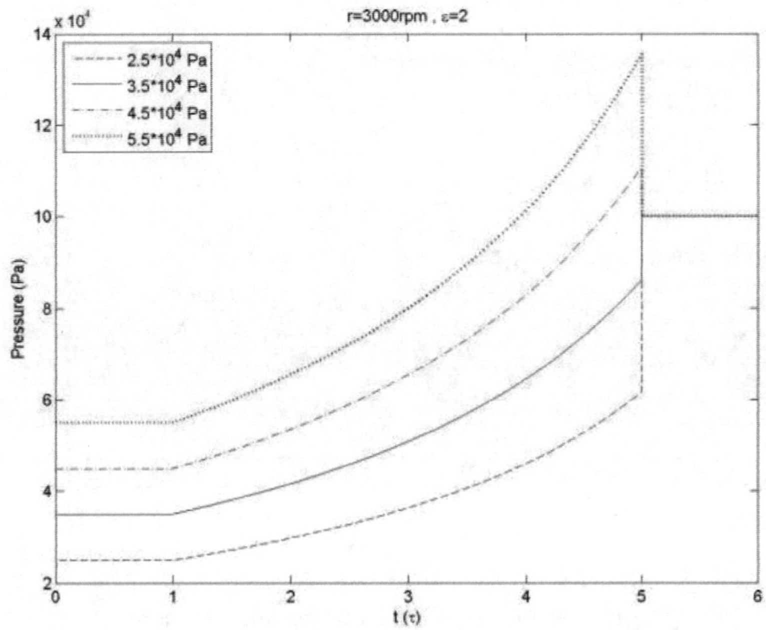

Figure 4. P-t diagram for various inlet pressures

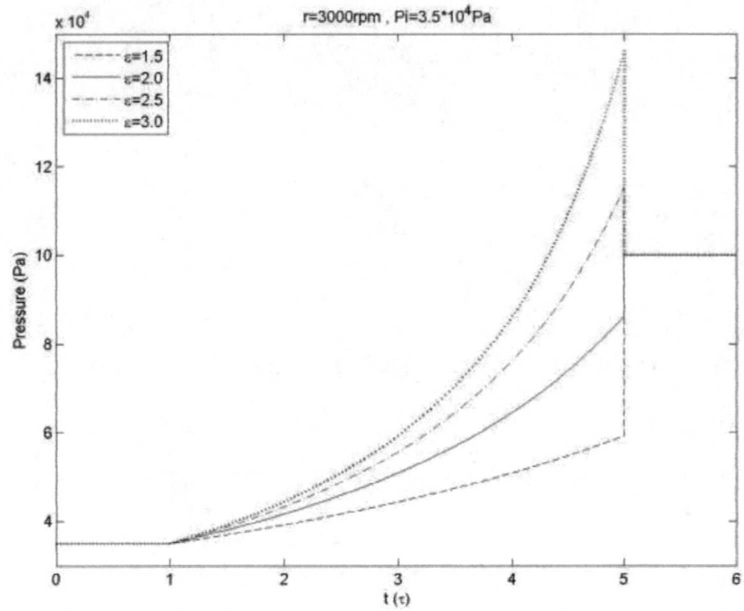

Figure 5. P-t diagram for various volume ratios

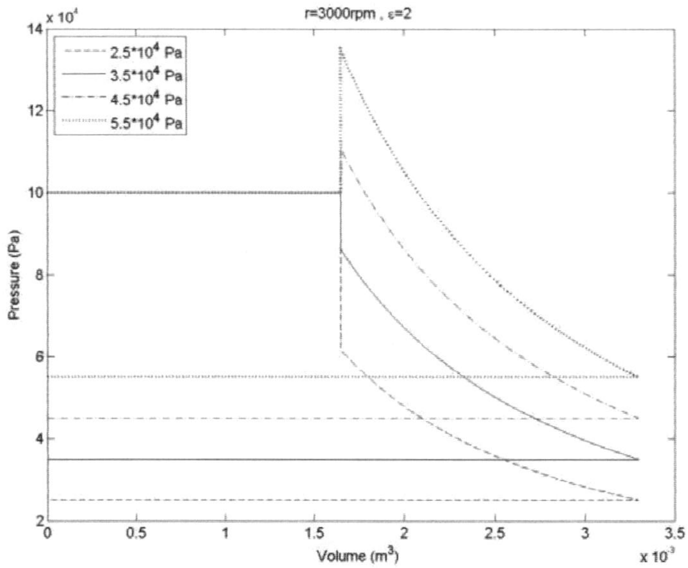

Figure 6. P-V diagram for various inlet pressures

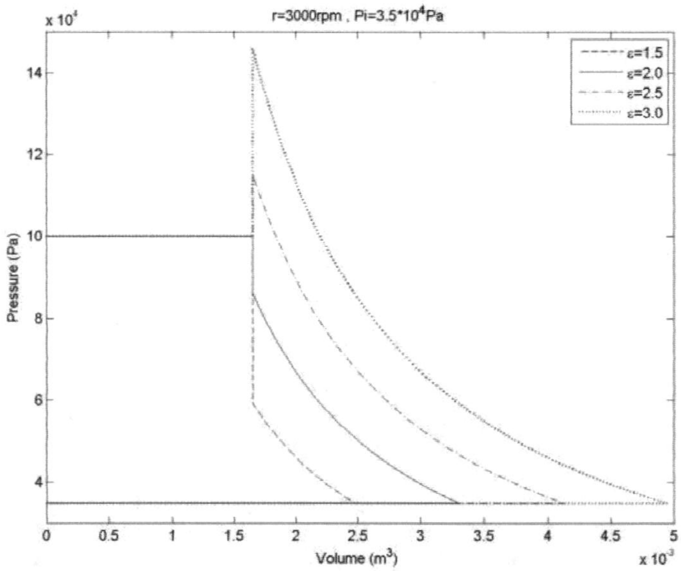

Figure 7. P-V diagram for various volume ratios

Figure 4 reveals the impact of inlet pressure on the pressure changing. The pressure curves in figure 4 can be roughly divided into the under-compression stage and the over-compression stage. The over-compression stage occurs at the initial operation of the pumping system, and later it will transit into the under-compression stage due to the decrease of inlet pressure. Meanwhile, the gas pressure in the space will decline with the decrease of the inlet pressure except the exhausting stage. Figure 5 reveals the effect of volume ratio, which has a visible impact on the pressure especially at the compressing stage. The compressing effect grows with the increase of volume ratio and the pressure can be raised up to a high level, even resulting in the over-compression. Therefore, both the inlet pressure and volume ratio have influences on pressure pulsation.

Table 1. Structure parameters of SGLS vacuum pump with the same discharge volume.

Structure parameter	Dimension
The total wrap angle	10π
The varying pitch coefficient	0.011,0.023,0.037,0.053
The initial pitch	56.25mm,52.50mm,48.75mm,45.00mm
The volume ratio of SGLS vacuum pump	1.5,2.0,2.5,3.0

The amount of work can be indicated by the area under the curve in the indicator diagram. Figure 6 reveals the impact of inlet pressure on the work. Since the discharge pressure, the intake volume and discharge volume stay constant, it is obvious that the thermodynamic work descends with the decrease of inlet pressure with the volume ratio of 2. Figure 7 reveals the impact of volume ratio on the work. Owing to the discharge pressure and discharge volume fixed, the compression governs the amount of work. That is to say, the higher volume ratio is, the more compressing work does. Thus, both the inlet pressure and volume ratio have influences on power consumption.

5. Conclusion

In our study, we introduced the structure and four thermodynamic stages of SGLS vacuum pump at first. Then the pumping process was divided into four successive stages a thermodynamic model was established according to the basic physics laws. The characteristic curves of SGLS pump were plotted. The results show that the thermodynamic model can provide a well explanation on the phenomena of pressure pulsation and power consumption, which will help us to understand the outstanding performance of SGLS vacuum pump.

Reference

[1] Lessard P A. Dry vacuum pumps for semiconductor processes: Guidelines for primary pump selection. *J Vac Sci Technol A*, 2000, **18(4)**:1777-1781.

[2] Akutsu I, Matsuoka T, Ozaki M, et al. A gradational lead screw dry vacuum pump. *J Vac Sci Technol A*, 2000, **18(3)**:1045-1047.

[3] Pfaller D, Brümmer A, Kauder K. Optimized rotor pitch distributions for screw spindle vacuum pumps. *Vacuum*, 2011, **85(12)**:1152-1155.

[4] Rane S, Kovacevic A, Stosic N, et al. Deforming grid generation and CFD analysis of variable geometry screw compressors. *Comput Fluids*, 2014, **99(5)**:124-141.

[5] Zhao F, Zhang S, Sun K, et al. Thermodynamic Performance of Multi-stage Gradational Lead Screw Vacuum Pump. *Appl Surf Sci*, 2018, **432**:97-109.

Preparation and Evaluation of Mefenamic Acid Transdermal Patches Prepared from Pressure Sensitive Adhesive

Jirapornchai Suksaeree[1,*], Yupaporn Sampaopan[1,2], Wiwat Pichayakorn[3] and Ekwipoo Kalkornsurapranee[4]

[1] Department of Pharmaceutical Chemistry, College of Pharmacy, Rangsit University, Muang, Pathum Thani 12000, Thailand
[2] College of Oriental Medicine, Rangsit University, Muang, Pathum Thani 12000, Thailand
[3] Department of Pharmaceutical Technology, Faculty of Pharmaceutical Sciences, Prince of Songkla University, Songkhla 90110, Thailand
[4] Department of Materials Science and Technology, Faculty of Science, Prince of Songkla University, Songkhla 90110, Thailand

[*] E-mail: jirapornchai.s@rsu.ac.th;

Abstract. The pressure sensitive adhesive is interesting to apply in pharmaceutical and medical products. It composed of STR 5L block rubber which has high uniformity and very low impurity. Thus, it was used to prepare the transdermal patches for mefenamic acid delivery. The mefenamic acid powder was loaded in pure STR 5L block rubber and pressure sensitive adhesive. The pressure sensitive adhesive was the STR 5L block rubber mixed with hydroxyethyl cellulose and white oil which were used as tackifier and softener, respectively. The preparation of mefenamic acid loaded in transdermal patches were crushed by melt blending technique with two-roll mill machine. The mefenamic acid could be mixed into pure STR 5L block rubber and pressure sensitive adhesive to be the homogeneous transdermal patches. The percentages of moisture uptake and swelling ratio were less than 5% and 15%, respectively. The mefenamic acid was entrapped in the range of 75-90%. However, the content of mefenamic acid from transdermal patches was a low cumulative release (less than 60% of cumulative drug release). Therefore, this preparation of mefenamic acid loaded in transdermal patches might be developed to increase the release of mefenamic acid from transdermal patches in further study.

1. Introduction

The natural rubber exhibits various outstanding properties such as excellent resilience, high elongation at break and tensile strength because it presents the high stretching and ease of patch formation but it has a low adhesion and tack properties. The main component of natural rubber latex is rubber particle in the concentration of 30–40% in aqueous serum that is difficult to storage because it has the bulk volume or microbial contamination in their serum [1, 2]. The various publication reported the usage natural rubber latex as a controlled-release membrane in pharmaceutical and medical products for controlled drug release of nicotine [3-8], ketoprofen [9, 10], naproxen [11], sulindac [12], etc.

The block rubber is one choice as interesting material for the controlled-release membrane. Because it is formed as a solid stage and high stability when compared to natural rubber latex. The

Standard Thai Rubber (STR) of block rubber in Thailand are graded into 8 types that are STR 5, STR 5L, STR 5CV, STR 10, STR 10CV, STR 20, STR 20CV, and STR XL [13]. However, the STR 5L is the best material with the great potential to be used as a controlled-release membrane in pharmaceutical and medical products due to its quality latex and low impurity. Recently, the STR 5L block rubber is reported that can be prepared for transdermal patches preparation [14-16]. Thus, this work interested to use the STR 5L block rubber as a controlled-release membrane for transdermal patches preparation of mefenamic acid delivery.

This aim of the study was to formulate mefenamic acid loaded in transdermal patches using pure STR 5L block rubber and pressure sensitive adhesive. The pressure sensitive adhesive was the STR 5L block rubber mixed with hydroxyethyl cellulose and white oil which were used as tackifier and softener, respectively and to evaluate the moisture uptake, swelling ratio, and *in vitro* drug release from the transdermal patches.

2. Experimental Works

2.1 Preparation of mefenamic acid loaded in transdermal patches
For mefenamic acid loaded in a transdermal patch made from STR 5L, 100 phr of STR 5L block rubber was crushed with a two-roll mill machine (191-TM, Yasuda Seiki Seisakusho, Japan). The gap of the machine was set at 0.5 mm and the mixing time was 15 min. The mefenamic acid powder was added at 5% w/w and crushed to combine with STR-5L block rubber for 15 min.

For mefenamic acid loaded in a transdermal patch made from pressure sensitive adhesive, 100 phr of STR 5L block rubber was crushed with a two-roll mill machine (191-TM, Yasuda Seiki Seisakusho, Japan). The gap of the machine was set at 0.5 mm and the mixing time was 15 min. The white oil and hydroxyethyl cellulose were used as a softener and tackifier, respectively. The mefenamic acid powder was added at 5% w/w and crushed to combine with pressure sensitive adhesive for 15 min.

2.2 Percentage of moisture uptake
The study of moisture uptake of mefenamic acid loaded in transdermal patches stored the samples in a desiccator which was equilibrated with saturated NaCl solution at room temperature under the condition of 75% relative humidity. Every week, the samples of each mefenamic acid loaded in transdermal patches were weighed until constant. The percentage of moisture uptake was calculated as the difference of the percentage between the increased weight of the sample and the initial weight of the sample. The percentage of moisture uptake was tested in triplicate.

2.3 Percentage of swelling ratio
The study of the swelling ratio of mefenamic acid loaded in transdermal patches used the distilled water as a solvent. The samples were immersed in 10 mL of distilled water and kept at room temperature. The samples of each mefenamic acid loaded in transdermal patches were weighed every hour until constant. The percentage of swelling ratio was calculated as the difference of the percentage between the increased weight of the sample and the initial weight of the sample. The percentage of swelling ratio was tested in triplicate.

2.4 Determination of drug entrapment efficacy in transdermal patches
The transdermal patches were cut in small pieces and extracted in ethyl acetate under sonication for 3 hrs. The extracted solution was diluted and adjusted to the determined volume. The content of mefenamic acid that entrapped in transdermal patches was assayed by a high performance liquid chromatography (HPLC) instrument. The analyzed mefenamic acid content was compared with the peak area of the linearity of the standard curve. The determination of drug content in transdermal patches was assayed in triplicate. The percentage of drug entrapment efficacy of mefenamic acid in transdermal patches was calculated as the percentage of the analyzed mefenamic acid content in the

patch compared to the theoretical mefenamic acid content. The percentage of drug entrapment efficacy was carried out in triplicate.

2.5 In vitro release evaluation of mefenamic acid from transdermal patches

The cumulative release of the drug from the transdermal patches was studied under USP dissolution apparatus V. 900 mL of 2% w/v sodium dodecyl sulfate in distilled water was selected as a receptor medium. Because the previous study reported the 2% w/v of sodium dodecyl sulfate could improve the solubility of mefenamic acid in water at 37°C [17]. The temperature of the water bath was set at 37±0.5°C with stirring constantly at 100 rpm. At sampling time 0, 0.5, 1, 2, 3, 4, 5, 6, 8,10, 12, and 24 hrs, the receptor medium was withdrawn and then replaced with the equal volume of fresh receptor medium. The content of mefenamic acid in each sampling time was analyzed by HPLC assay and compared to the calibration curve. The samples for each sample were evaluated in triplicate.

2.6 Analytical Method

The HPLC instrument from Agilent Technologies, USA, model Agilent 1260 Infinity system was used for drug analysis. A reverse-phase ACE Generix5 C18 with 4.6 mm × 150 mm, 5 μm particle size from DV12-7219, USA was used to separate the mefenamic acid. The ratio between methanol and acetate buffer pH 4.1 at 95:5 was used as a mobile phase with 5 minutes run-time. The flow rate was 1 mL/min. The solution of the sample was injected at 10 μL. The UV detection was set at 285 nm. The required concentration of pure mefenamic acid for validation was in the range of 5-50 μg/mL. The limits of detection and quantification, linearity, accuracy, and precision were observed. The limit of detection and limit of quantification were 1.10 μg/mL and 3.33 μg/mL, respectively. The calibration curve was good linearity with R^2 more than 0.9997. The accuracy was found in a range of 103-107% and had a good precision (less than 2% RSD for both intra- and inter-day analysis).

3. Results and discussion

Figure 1 and 2 showed the controlled-release membrane for mefenamic acid loaded in transdermal patches that made from STR 5L block rubber and pressure sensitive adhesive, respectively. Figure 1 showed the unevenness of the patch while Figure 2 showed the smoothness of the patch. The mefenamic acid powder could homogeneously disperse in a controlled-release membrane made from pressure sensitive adhesive more than controlled-release membrane made from STR 5L block rubber. This was due to the transdermal patch in Figure 1 composed only STR 5L block rubber while the transdermal patch in Figure 2 had various fillers such as white oil and hydroxyethyl cellulose as a softener and tackifier, respectively.

Figure 1. Mefenamic acid loaded in a transdermal patch made from STR 5L block rubber.

Figure 2. Mefenamic acid loaded in a transdermal patch made from pressure sensitive adhesive.

The moisture uptake and swelling ratio of mefenamic acid-loaded STR 5L block rubber and -loaded pressure sensitive adhesive are shown in Figure 3 and 4, respectively. The mefenamic acid-loaded pressure sensitive adhesive patch showed the high moisture uptake value than mefenamic acid-loaded STR 5L block rubber patch. This was due to the mefenamic acid-loaded pressure sensitive adhesive patch composed the ingredient in formulation more than mefenamic acid-loaded STR 5L block rubber patch. However, the swelling ratio values of both mefenamic acid-loaded STR 5L block rubber and -loaded pressure sensitive adhesive were not significant.

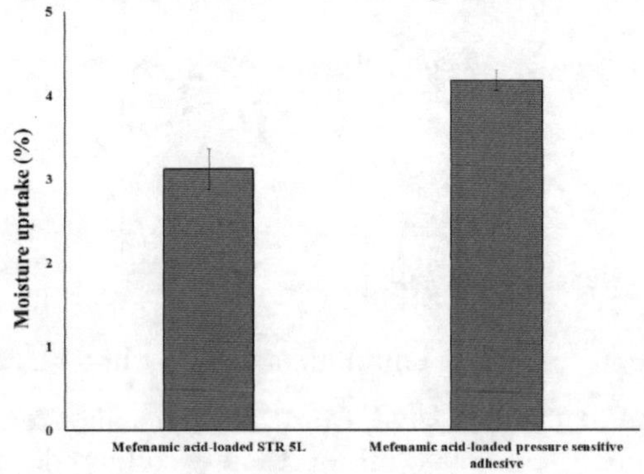

Figure 3. Percentage of the moisture uptake of mefenamic acid loaded in transdermal patches.

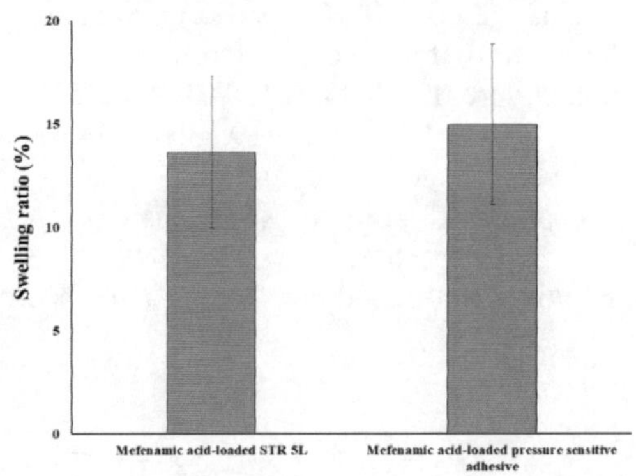

Figure 4. Percentage of the swelling ratio of mefenamic acid loaded in transdermal patches.

The mefenamic acid could be loaded in transdermal patches at 1.90±0.13 and 2.27±0.13 mg/cm^2 for mefenamic acid-loaded STR 5L block rubber and -loaded pressure sensitive adhesive, respectively (Figure 5). The percentages of drug entrapment efficacy in STR 5L block rubber and pressure sensitive adhesive were 76.16±5.11% and 90.67±5.03%, respectively (Figure 6). Thus, the pressure sensitive adhesive could entrap the mefenamic acid in their patch more than STR 5L block rubber.

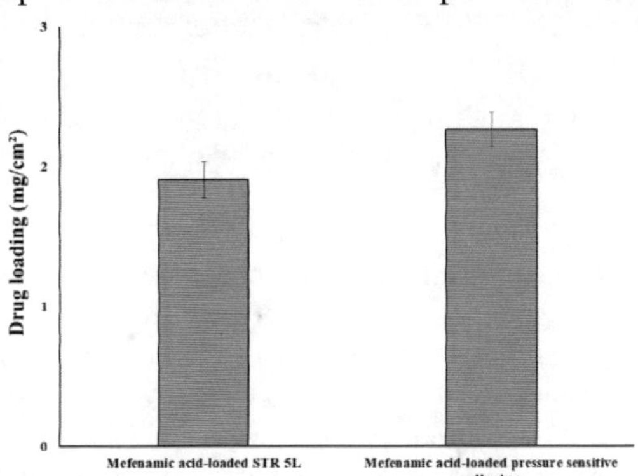

Figure 5. Drug loading in transdermal patches.

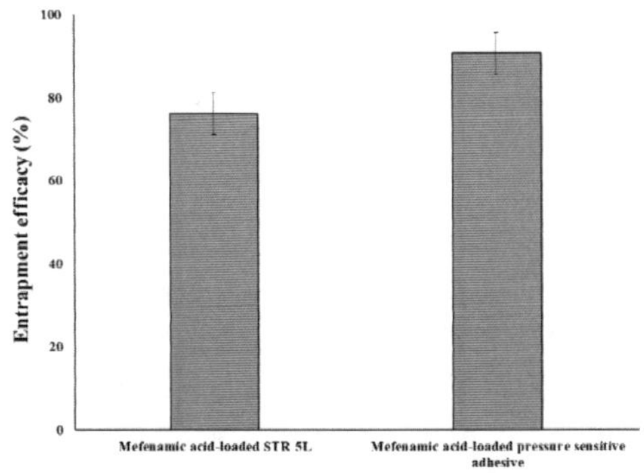

Figure 6. Percentage of drug entrapment efficacy in transdermal patches.

The cumulative release of mefenamic acid from the controlled-release membrane was studied under USP dissolution apparatus V using 900 mL of 2% w/v sodium dodecyl sulfate in distilled water as a receptor medium [17]. The times for sample collection was 0, 0.5, 1, 2, 3, 4, 5, 6, 8,10, 12, and 24 hrs. Each sample collection was analyzed by HPLC assay and compared to the calibration curve. The percentages of cumulative release of mefenamic acid from the controlled-release membrane that prepared from STR 5L block rubber were 0%, 5.33±0.12%, 9.96±2.63%, 15.68±3.68%, 19.89±2.65%, 25.63±5.53%, 29.32±9.64%, 34.36±8.94%, 42.59±9.33%, 44.95±8.30%, 48.63±9.78%, and 52.36±8.32%, respectively. The percentages of cumulative release of mefenamic acid from the controlled-release membrane that prepared from pressure sensitive adhesive were 0%, 6.98±0.22%, 11.69±2.70%, 19.62±3.85%, 22.65±3.36%, 28.31±7.89%, 33.67±6.88%, 39.87±5.97%, 44.29±11.64%, 49.65±7.64%, 53.96±12.63%, and 60.39±8.63%, respectively (Figure 7).

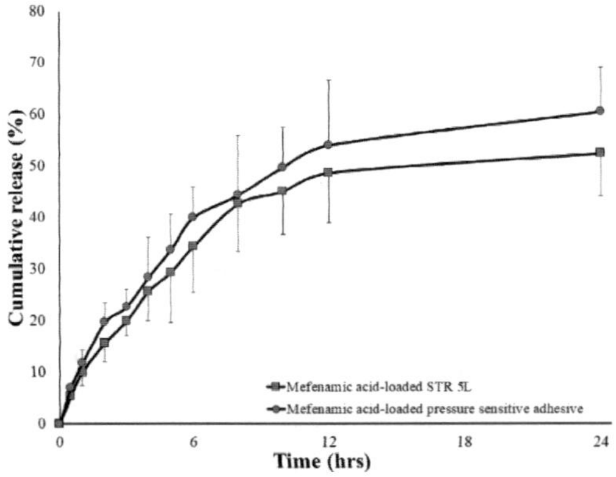

Figure 7. Percentages of the cumulative release of mefenamic acid from the controlled-release membrane that prepared from STR 5L block rubber and pressure sensitive adhesive.

4. Conclusions

The mefenamic acid could be successfully loaded into pure STR 5L block rubber and pressure sensitive adhesive to be the homogeneous transdermal patches. The percentages of moisture uptake and swelling ratio for both preparations were less than 5% and 15%, respectively. The percentage of mefenamic acid entrapment efficacy was in the range of 75-90%. However, the cumulative drug release of mefenamic acid from transdermal patches was a low cumulative release (less than 60%). Therefore, this preparation of mefenamic acid loaded in transdermal patches might be developed to increase the release of mefenamic acid from transdermal patches in further study.

Acknowledgments

The financial supporting of this work was the College of Pharmacy, Rangsit University.

References

[1] Suksaeree J, Pichayakorn W, Monton C, Sakunpak A, Chusut T and Saingam W 2014 *Ind. Eng. Chem. Res.* **53** 507-513.
[2] Pichayakorn W, Suksaeree Jand Boonme P. Mechanical and Viscoelastic Properties of Natural Rubber Based Blends and IPNs. Natural Rubber Materials: Volume 1: Blends and IPNs. vols. 1: The Royal Society of Chemistry; 2014, p. 481-500.
[3] Pichayakorn W, Suksaeree J, Boonme P, Amnuaikit T, Taweepreda W and Ritthidej CG 2013 *Pharmaceut. Dev. Tech.* **18** 1111-1121.
[4] Pichayakorn W, Suksaeree J, Boonme P, Amnuaikit T, Taweepreda W and Ritthidej GC 2012 *Ind. Eng. Chem. Res.***51** 8442-8452.
[5] Pichayakorn W, Suksaeree J, Boonme P, Amnuaikit T, Taweepreda W and Ritthidej GC 2012 *J. Membr. Sci.* **411-412** 81-90.
[6] Pichayakorn W, Suksaeree J, Boonme P, Amnuaikit T, Taweepreda W and Ritthidej GC 2012 *Chem. Eng. Res. Des.* **91** 520-529.
[7] Suksaeree J, Karnsopa P, Wannaphruek N, Prasomkij J, Panrat K, Monton C and Pichayakorn W 2018 *Journal of Polymers and the Environment* **26** 3531-3539.
[8] Suksaeree J, Karnsopa P, Wannaphruek N, Prasomkij J, Panrat K and Pichayakorn W 2018 *Journal of Polymers and the Environment* **26** 3216-3225.
[9] Suksaeree J, Charoenchai L, Monton C, Chusut T, Sakunpak A, Pichayakorn W and Boonme P 2013 *Ind. Eng. Chem. Res.* **52** 15847-15854.
[10] Floriano JF, de Barros NR, Cinman JLF, da Silva RG, Loffredo AV, Borges FA, Norberto AMQ, Chagas ALD, Garms BC, de Oliveira Graeff CF and Herculano RD 2018 *Journal of Polymers and the Environment* **26** 2281-2289.
[11] Kaewchingduang R, Paradee N, Sirivat A and Niamlang S 2019 *International Journal of Pharmaceutics* **561** 296-304.
[12] Banpean A, Paradee N, Sirivat A and Niamlang S 2018 *Journal of Polymers and the Environment* **26** 3745 - 3753.
[13] Astlett Rubber. Standard Thai Rubber (STR). Ontario, Canada. Available at: https:// www.astlettrubber.com/pdf/nr/str.pdf. Accessed 25 March 2019.
[14] Pichayakorn W, Boontawee H, Taweepreda W, Suksaeree J and Boonme P2014 *Industrial & Engineering Chemistry Research* **53** 1672-1677.
[15] Kalkornsurapranee E, Waiprib R and Pichayakorn W 2017 *Key Engineering Materials* **751** 236-241.
[16] Boonme P, Boontawee H, Taweepreda W and Pichayakorn W 2013 *Advanced Materials Researc* **747** 103-106.
[17] Patil PB, Gupta VRM, Udupi RH, Srikanth K and Prasad BSG 2010 *Res. J. Pharm. Biol. Chem. Sci.* **1** 544–549.

IOP Conf. Series: Materials Science and Engineering **686** (2019) 012016 doi:10.1088/1757-899X/686/1/012016

Analyzing Resistance of Al-10Si-5Mg Alloy from Stress Corrosion Cracking for Ocean Current Turbine Applications

Andre Amba Matarru [1, a], **Muhammad Syahid** [2, b], **Donny Yoesgiantoro** [1, 3, c] **and Syahril Hadi** [1, 4, d]

[1] Energy Security Graduate Program, Indonesia Defense University (UNHAN), Indonesia Peace and Security Centre (IPSC), Bogor 16810, Indonesia
[2] Departement of Mechanical Engineering, Hasanuddin University, Malino Street, Bontomarannu, Gowa, South Sulawesi 92119, Indonesia
[3] The Kirana Tembok (Pioneer of Eco Edu Tourism), Tejakula, Buleleng Regency, Bali 81173, Indonesia
[4] Directorate General of New Renewable Energy and Energy Conservation, Ministry of Energy and Mineral Resources of Energy, Pegangsaan Timur Street No.1, Menteng, Jakarta Pusat 10320, Indonesia

E-mail: [a]andreambamatarru@gmail.com ,[b] syahid.arsjad@gmail.com , [c]energyprogram@gmail.com , [d] hadi.syahriel21@gmail.com

Abstract. Stress Corrosion Cracking in Marine Applications, especially Ocean Current Turbines must be overcome. Al-10Si-5Mg Alloy which is based on the nature of flexibility and strength, is expected to survive from the SCC. The purpose of this paper is to examine the indications of SCC from immersed specimens in the 3.5% NaCl Solution + Aquades solution, then give varying deflections on the 3 Point Bending Testing. The other tests are harness testing, tensile testing, and macro visualization. The method is experimental for Al-10Si material in addition to a 5% Mg at casting. There are 9 specimens, namely 3 specimens from each of the 3 load values, namely 252 MPa (deflection (y)=2mm), 378 MPa (y=3mm), and 504 MPa (y=4mm). The results provide evidence that when the specimens were submerged with varying stresses, the absence of broken material was obtained even though there were indications of cracking and corrosion. Then the specimen that gets a high stress, becomes more susceptible to fracture due to the predominant cracking effect when compared to low standing specimens ($\sigma_{y=4\ mm} < \sigma_{y=3\ mm} < \sigma_{y=2\ mm}$). So by the resistance to SCC, the Al-10Si-5Mg alloy can be recommended as an Ocean Current Turbine Blade material.

1. Introduction

The steps taken by the government to anticipate the energy crisis in Indonesia include National Energy Policy, National Energy Management Blueprint 2005 - 2025, National Strategic Policy for Science and Technology Development, and the National Ocean Exploitation Policy which emphasizes energy sustainability through the creation and utilization of renewable energy sources. In Blue Print Energy Management, energy management will be optimized, so that by 2025 the composition of energy is expected to be 33% coal, 30% gas, 20% petroleum and 17% new renewable energy. Indonesia with a total ocean area of almost 8 million km^2 seeks to increase inventory of non-living resources, one of which is the energy potential of ocean currents [1]. Components that are of concern in the development of Marine Flow Turbines in the future are made from Aluminium, where the first generation turbines whose blades are made of Polymer (steel-frame fiberglass) are not able to withstand the instability of unstable underwater currents [2].

However, it is undeniable that aluminium alloys are also susceptible to fragmentation by hydrogen, although at the time of reaching the FCC microstructure they explain that the transfer of hydrogen is lower than in High Strength Steel (HSS) and hence, the average growth of cracks is slow. Cracking is usually intergranular. Just like steel, susceptibility increases with increasing alloy strength [3]. So that it is obtained that the indication of the emergence of SCC is something that harms certain alloys.

The design criteria needed include strength, fatigue resistance, corrosion resistance, potential long-term and short-term fabrication costs. Consideration includes the availability of several blade structures as shown in Figure 1 below.

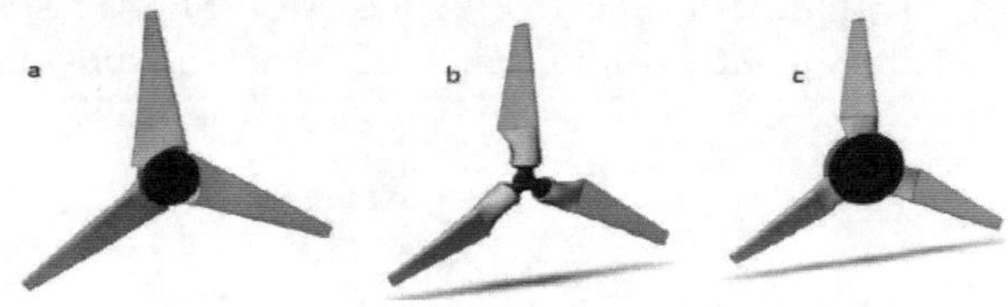

Figure 1. AlMg blade with bolt-pierced nap, same as Gen4c Verdant blade (a) Fiberglass Composite Blade with axial blade blade (b) Fiberglass Blade Flanked with Flat Side Towering or Broken Bolt (c)

The concept has been evaluated on the basis of price and structural performance for both class 5 and 10 m standards as shown in Table 1 below.

Table 1. Characteristics of Composite Blade Design

Rotor Diameter - Resource	5m - Tidal	7m - River	10m - River	10m - Tidal
Generator Class	56 kW	110 kW	110 kW	500 kW
Peak Water Velocity	2.5 m/s	2.5 m/s	2.0 m/s	3.3 m/s
RPM	40 rpm	31 rpm	22 rpm	31 rpm
Root - Thickness/Twist	Proprietary Verdant Power Design Values			
Tip - Thickness/Twist	Proprietary Verdant Power Design Values			

Fiberglass, resin and synthetic foam finishes are selected and the fabrication, bonding and filling fabrication stages have been terminated for resin filling of high and low pressure layers and shear layers. A variety of materials for bonded frame fillings are necessary including Nickel Bronze Aluminium considering balance of strength, durability, and cost [4]. However, it is identified that filling is an important factor in the development of blade manufacturing especially Aluminium that has a density of only 2.7 g/cm^3 indicates that only about one third of the steel density is 7.83 g/cm^3. In addition, as the Al-Mg alloys, character of Magnesium can be preferentially hardened on grain boundaries as a high anode phase (Mg_5Al_3 or Mg_5Al_8) which results in susceptibility to intergranular cracks and to Stress Corrosion [5].

Stress Corrosion Cracking (SCC) is the combination between Corrosion and Stress which could appear on aggressive maritime environment and generally on the structure that made from steel and aluminium alloy which has enormous structure and mostly applied as the other external components of the power transmission and control systems. During the second exploitation, the metal alloy is known to be intended for an aggressive maritime environment that directs the process of corrosion (ordinary corrosion, pitting, and SCC). On the other hand, rusted construction is the reason for component changes that are faster power transmission components and unexpected impacts cause the potential for dangerous accidents. In maritime applications, SCC is also the most dangerous type of corrosion caused by stress and the electrolyte environment arises simultaneously [6].

Clear rifts arise as a result of hydrogen embrittlement that supports a number of mechanisms and becomes a potential link to the SCC environment. Some environments that support hydrogen give rise to SCC for the susceptibility of alloys, usually in moist air to saline liquids [3]. This is the basic idea of the need for early learning about the treatment of SCC vulnerability arising from 3 determinants,

IOP Conf. Series: Materials Science and Engineering **686** (2019) 012016 doi:10.1088/1757-899X/686/1/012016

namely a material component that is vulnerable, an electrolytic environment, and sufficient stress on renewable energy technologies. In this study, the author intends to examine how Analysis of Stress Corrosion Cracking Al-10Si-5Mg Alloys for Maritime Applications especially preventing at Ocean Current Turbine further.

2. Research Methods

2.1. Test material

In this study the material used was Al-10Si in the form of bars with the addition of 5% wt Mg, which was cast with a melting temperature of 700° C which was then poured into metal molds and formed based on the size of ASTM G 39 buckling test specimens with long dimensions 120 mm and a width of 25 mm with a height of 3 mm. And the size of the Al-10Si-5Mg specimen Hardness test with dimensions of 20 mm long, 20 mm wide, with a height of 10 mm as shown in Figure 2 below.

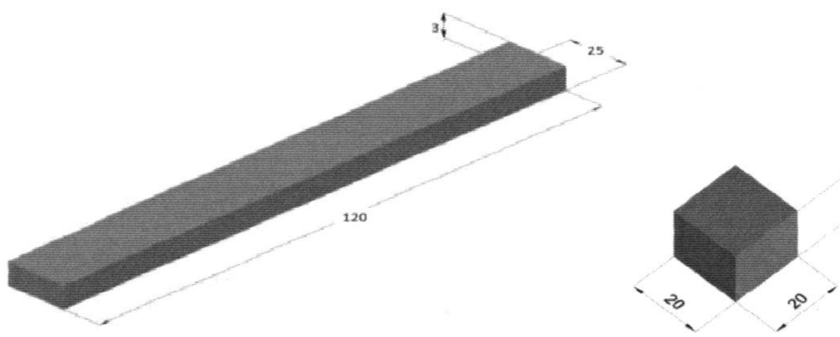

(a) (b)

Figure 2. Bending Test Specimens (a) Hardness Test Specimens (b)

2.2. Corrosion Media

Corrosive Environment in the form of Solution 3.5% NaCl + Aquades (distilled water). The description needed according to ASTM G39 is a description of the temperature and pressure during testing.

2.3. Bending Testing

- Specimens must be made of material that is not affected by the shape (for example plastic) of the corrosive environment or coating the support with a coating material which is usually used in electricity that shown in the Figure 3 below. If necessary, Hydrophobic Fillers can be added to protect against corrosion of cracks at the touch point between support and specimens (Grease or Wax) [7].

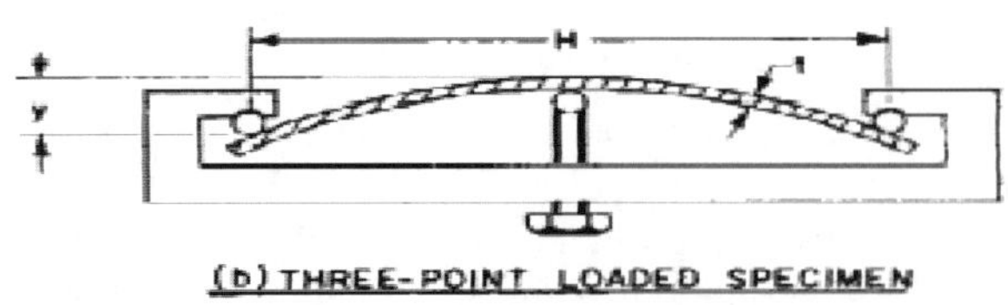

Figure 3. Scheme of Specimen Supporting Configuration

- Deflection Gauges, the presence of deflection in the specimen is determined by a measuring device which is separate or together on the completeness that shown in the Figure 4 below. In the design of the deflection gauge for adjustment of the appropriate individual conditions must weigh the provisions of the 3 Point Bending specimen.

IOP Conf. Series: Materials Science and Engineering **686** (2019) 012016 doi:10.1088/1757-899X/686/1/012016

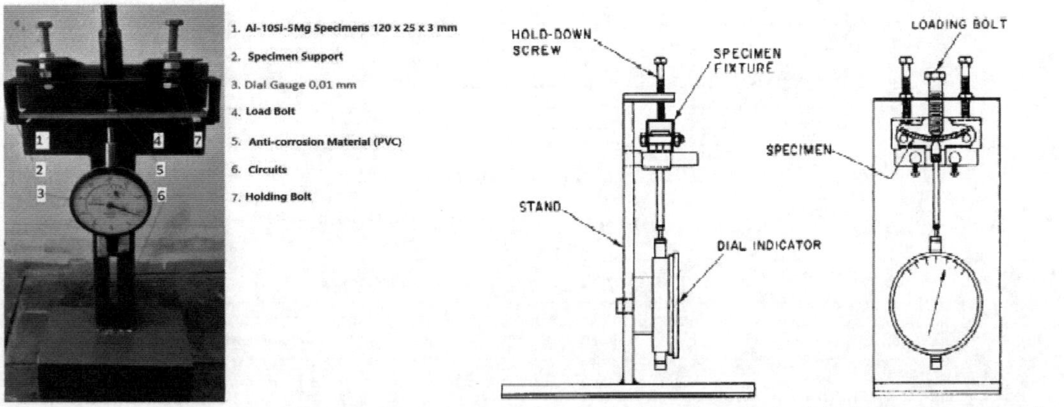

Figure 4. Specimen loading equipment for beam specimens 3 point bending with derivative deflection gauge

Testing is done by giving the specimen loading to deflection. The tools used are Support 3 Point Bending. The maximum deflection tested on specimens until broken without immersion is 4 mm, 5 mm and 7 mm. So that taken as a parameter is 4 mm as the maximum deflection. The maximum load is based on the maximum deflection. The maximum load equation with the formula based on ASTM G 39:

$$\sigma = 6Ety/H^2 \tag{1}$$

σ = *Maximum Aluminum Tension Voltage (Elastic Limit)*
E = *Aluminum Elastic Modulus*
t = *specimen thickness*
y = *Maximum deflection*
H = *Distance between 2 edge supports*

Then if the value of deflection is entered

σ = 6 (70,000,000 kPa) (3mm)(4mm)/(100mm)^2
= 252.000 kPa = 504 MPa

So that the variation of loading given is based on multiples of deflection integers, defined:

y_1 = 2 mm σ_1 = 252 MPa
y_2 = 3 mm σ_2 = 378 MPa
y_3 = 4 mm σ_3 = 504 MPa

2.4. Hardness Testing

Hardness testing is done by the Brinell method. Hardness can be defined as the ability of a material to resist plastic deformation. Hardness also has a correlation with strength. Hardness testing is one of the many tests used because it can be carried out on small test objects without difficulty regarding specifications. The most widely used test is to emphasize a certain emphasis and by measuring the size of the stresses formed above it, this method is called the method of suppression hardness.

2.5. Tensile Testing

Tensile Testing was carried out at the Physical Metallurgy Laboratory of the Universitas Kristen Indonesia Paulus. The procedure performed is:

- Removing the test specimens that were previously treated with bending and soaking tests in a mixture of aquades and NaCl of 3.5%.
- Cut specimens similar to the standards in accordance with ASTM B557M as shown in Figure 5 and following the requirements which shown in Table 2.

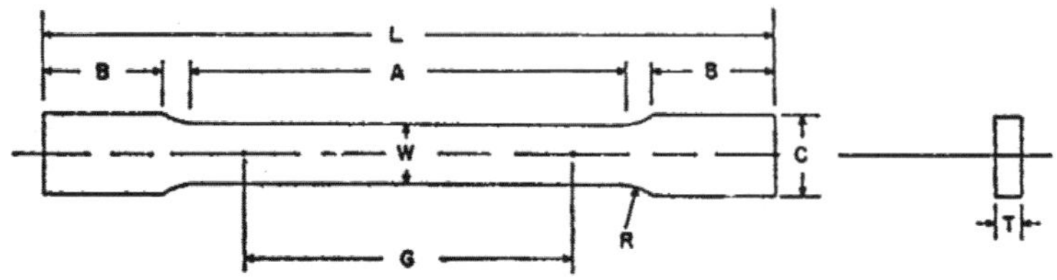

Figure 5. Image of Tensile Test Specimen Size

Table 2. Tensile Test Size Tables

	Dimensions, mm
	Subsize Specimen 6 mm Wide
G—gage length	25.0 ± 0.1
W—Width (Note 1 and Note 2)	6.0 ± 0.1
T—Thickness (Note 3)	3
R—Radius of fillet, min	6
L—Overall length, min (Note 4)	120
A—Length of reduced section, min	32
B—Length of grip section, min (Note 5)	40
C—Width of grip section, approximate (Note 2 and Note 6)	10

3. Testing Result

3.1. Results of Bending Testing

Bending Testing after immersion in a corrosive environment. The load given to the test specimens is:

3.1.1 Deflection 4 mm with σ = 504 MPa

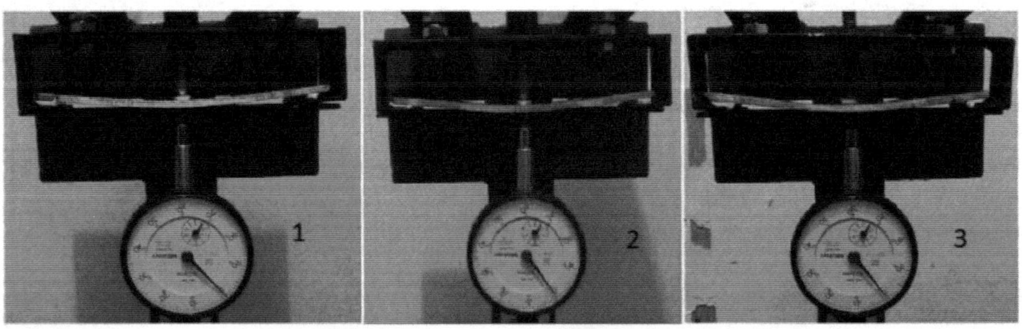

Figure 6. Dial gauge needle for 4mm deflection

Giving numbers to specimens applies to all three specimens with a thickness of 4 mm, Specimen 1, Specimen 2, and Specimen 3 as shown in the Figure 6 above. The appearance of the Dial Gauge small needle shows the number 4 which means 4 mm, where the long needle in the three specimens has reached number 0 which further clarifies the absolute magnitude to be achieved.

3.1.2 Deflection 3 mm with σ = 378 MPa

Figure 7. Dial gauge needle for 3mm deflection

IOP Conf. Series: Materials Science and Engineering **686** (2019) 012016 doi:10.1088/1757-899X/686/1/012016

Giving numbers to specimens applies to the three specimens with a defect of 3 mm namely Specimen 4, Specimen 5, and Specimen 6 as shown in the Figure 7 above. The appearance of the Dial Gauge small needle shows the number 3 which means 3 mm, where the long needle on the three specimens has reached the number 0 which further clarifies the absolute magnitude to be achieved.

3.1.3 *Deflection 2 mm with $\sigma = 252$ MPa*

Figure 8. Dial gauge needle for 2mm deflection

Giving numbers to specimens applies to the three specimens with a defect of 2 mm namely Specimen 7, Specimen 8, and Specimen 9 as shown in the Figure 8 above. The appearance of the Dial Gauge small needle shows the number 2 which means 2 mm, where the long needle on the three specimens has reached number 0 which further clarifies the absolute magnitude to be achieved.

The Figure 9 below is a comparison of cracks in specimens of 4 mm, 3 mm and 2 mm deflection.

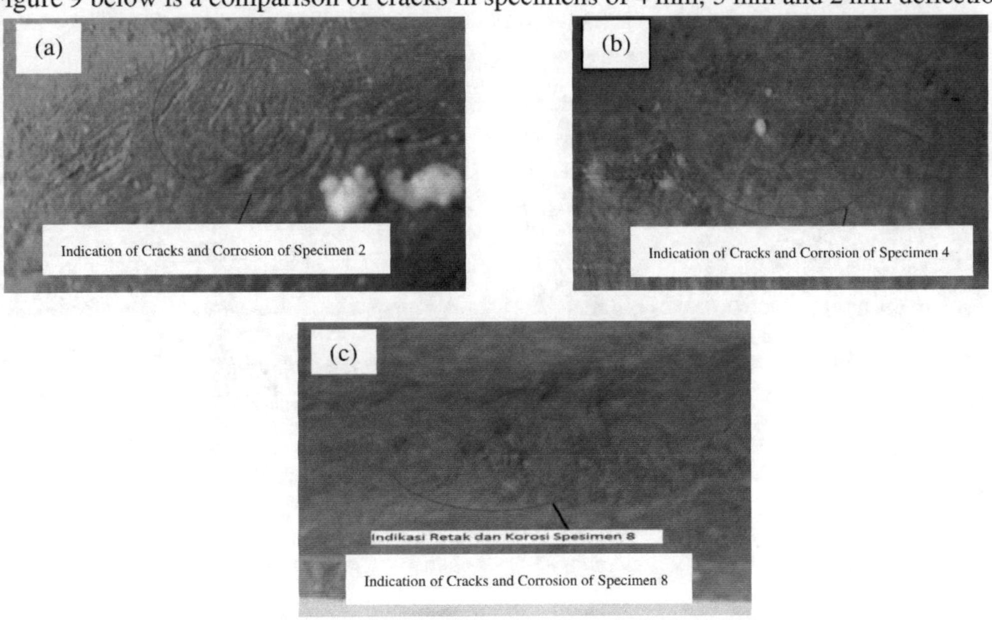

Figure 9. Crack and corrosion results with a 4mm deflection (a), 3mm deflection (b), 2mm deflection (c)

3.2. Hardness Testing Results

The Brinell Hardness value produced from Al-10Si-5Mg is 63.39 kg/mm^2 as shown in Figure 10. This value is above the standard number of AA6063-T5 (Al-Mg-Si) alloy which has value 60 kg/mm^2 BHN. So it is certain that the value of this hardness enhances toughness to the alloy for its resistance to being loaded.

(a)

No	F (kg)	D (mm)	d (mm)	$\sqrt{D^2-d^2}$	BHN (kg/mm^2)	∑BHN (kg/mm^2)
1	60	1.588	1.1	1.15	54.36	
2	60	1.588	1	1,23	67.90	63.39
3	60	1.588	1	1,23	67.90	

(b)

Figure 10. Results of untreated Al-10Si-5Mg Hardness Testing. Visual Appearance (a) and Calculation Table (b)

3.3. Tensile Test Results

- Average Stress (σ) for reflected as 4 mm specimens
  $\bar{\sigma} = (\sigma 1 + \sigma 2) / 2 = (0.325 + 0.343) / 2 = 0.334 kgf / mm^2$
- Average Stress (σ) for reflected as 3 mm specimens
  $\bar{\sigma} = (\sigma 4 + \sigma 5 + \sigma 6) / 3 = (0.149 + 4.726 + 2.953) / 3 = 2.609 kgf / mm^2$
- Average Stress (σ) for reflected as 2 mm specimens
  $\bar{\sigma} = (\sigma 7 + \sigma 8 + \sigma 9) / 3 = (5.895 + 7.838 + 3.180) / 3 = 5.638 kgf / mm^2$

It can be concluded that specimens that have a larger deflection have less power.

3.4. Macro Observation

Figure 11. Photograph of a fracture of a specimen with a deflection of 4 mm (a) deflection of 3 mm (b) deflection of 2 mm (c)

In Figure 11 above, it can be seen the appearance of all fractures of the specimen is brittle fracture which features a surface that is almost even (not concave / convex). The macro structural shows that the spreading evenly of Mg_2Si's precipitation on the matrixes that has formed and gives an effect of maximally strength and hardness properties unto the materials.

4. Conclusion

- The stages of specimens in achieving corrosion are :
  a. Occurrence of the Attached Brown Blob Area
  b. Crack Indication
  c. Color Indication Changes
  d. Crack and Corrosion Indications
- The strength of the aluminum material decreases for specimens which has applied 3 point bending test treatment. The strongest material is the speciment that got the lowest deflection, and vice versa. (σ y = 4 mm <σ y = 3 mm <σ y = 2 mm).
- No fracture while the stress had been applied on the appearing corrosive material.
- So with the result of resistance to SCC which does not cause a break in the test material, the Al – 10Si – 5Mg alloy can be recommended as an ocean current turbine blade material.

Acknowledgements

The focus of learning gained in the Energy Security Graduate Program of the Faculty of Defense Management, the Indonesia Defense University is a gratitude for the writer to be guided during the Renewable Energy lesson. This paper is prepared to support the plan to build a marine current energy generator that will be designed by the Government of Indonesia in the success of the target of the application of renewable energy in Indonesia by 2025 by 23% of the energy mix that contained in the Electricity Provision General Plan as known as Rencana Umum Penyediaan Tenaga Listrik (RUPTL) 2019-2028 literature.

The author hopes that there will be an increase in the use of Aluminum material for ocean current turbine blade applications given the large availability in Indonesia and the stability of the moment when turbines spin in producing electricity production that is effective in meeting the power requirements of military and general facilities in the form of an Off Grid System in preventing the failure of the main electricity network to support Indonesia's energy security.

Reference

[1] Ai Yuningsih, A. Masduki 2011 *Potensi Energi Arus Laut Untuk Pembangkit Tenaga Listrik Di Kawasan Pesisir Flores Timur NTT* (Pusat Penelitian dan Pengembangan Geologi Kelautan, Bandung)
[2] Information on http://www.washingtonpost.com/wp-dyn/content/article/2008/09/19/AR200809 1903729.html
[3] Dr. R. A. Cottis 2000 *Guides to Good Practice in Corrosion Control: Stress Corrosion Cracking* (National Physical Laboratory, NPL, Serco Group. UK)
[4] Corren, D., Colby, J., & Adonizio, M. A. 2013 *Improved Structure and Fabrication of Large, High-Power KHPS Rotors-Final Scientific/Technical Report* (No. DOE/GO18168). Verdant Power, Inc.
[5] Davis, J. R. 2001 Aluminum and aluminum alloy, p351-416. *ASM International*, 351-416.
[6] Darowicki, K., Orlikowski, J., Arutunow, A., & Jurczak, W. 2005 *Novel method of the initiation stage of stress corrosion cracking monitoring with respect to marine constructions*. (Polish Journal of Environmental Studies, 14, 161).
[7] ASTM Standard G 39 – 99 1999 *Standard Practice for Preparation and Use of Bent-Beam Stress-Corrosion Test Specimen* (ASTM International, West Conshohocken, PA)

IOP Conf. Series: Materials Science and Engineering **686** (2019) 012017 doi:10.1088/1757-899X/686/1/012017

Optimizing the Process of Multi-wire Sawing Silicon Rods with Consolidated Abrasive Diamonds Based on RSM

Wu Chaohua[1], Wang Ye[1], Guo Ming[2], Liu Shun[2], Lin Jiangjing[2], Fu Chunliang[2] and Zhou Bin[1,*]

[1] Hubei Digital Manufacturing Key Laboratory, School of Mechanical and Electronic Engineering, Wuhan University of Technology Wuhan 430070, China
[2] Wuhan Second Ship Design and Research Institute, Wuhan, 430205, China

[*]The corresponding author: Zhou Bin, zhb_8250@163.com

Abstract. In order to improve the processing efficiency and processing quality of silicon rods, and to optimize the process parameters during the cutting process of silicon rods, the sawing force of wire saw is studied from the aspects of cutting deformation and friction. Meanwhile, the mathematic model is established between the sawing force and some cutting process parameters. Besides, the quadratic regression model is also established by Box-Behnken center combination experiment and the response surface methodology(RSM), which can reflect the wire saw speed, workpiece feed rate and wire saw radius of the interaction of three process parameters on the wire saw sawing force. The results show that the model can accurately predict the sawing force and optimize the process parameters during the cutting process as well.

1. Introduction

In the many slicing methods, the consolidated abrasive diamond wire saw is widely used in the cutting process of silicon materials in solar cells due to its high cutting speed, low cost of consumables and environment protection [1]. However, the quality of multi-wire saw products is affected by parameters such as wire saw speed. Therefore, it is necessary to study and analyze the sawing force during the wire saw cutting process. The influence of process parameters on the cutting force of wire saw is analyzed, which lays a foundation for optimizing process parameters.

Chaohua Wu et al. [2] established a single abrasive grain sawing force model during the wire saw cutting process, and studied the influence of parameters such as wire saw speed on the deformation of a single abrasive grain by finite element analysis; Hao Wu et al. [3] studied the influence of diamond abrasive shape on the cutting depth and cutting force of SiC by theoretical calculation and finite element analysis. Li Shujuan et al. [4] established the cutting force model of wire saw during the wire saw cutting process. Zhang Liaoyuan [5], Cui Dan et al. [6] studied the effects of different process parameters on the sawing force. Meng Jianfeng et al. [7] analyzed the sawing force from the angle of vibration, and established the transfer function of the wire saw cutting operation equation. Gao Yufei et al. [8] studied the influence of wire speed and feed speed on the surface roughness of silicon wafer. Firstly, the sawing force of wire saw is studied from the aspects of cutting deformation and friction, and the mathematical model between wire saw sawing force and cutting process parameters is established. Then the response surface methodlogy is used to analyze the influence of each parameter on the sawing force of the wire saw, and the optimal parameters are obtained. The accuracy of the model is verified by experiments.

IOP Conf. Series: Materials Science and Engineering **686** (2019) 012017 doi:10.1088/1757-899X/686/1/012017

2. Mathematical model of sawing force

2.1. Analysis of sawing force of single abrasive particle

Due to its many excellent properties, diamond is especially suitable for processing hard and brittle non-metallic materials such as silicon [9]. Therefore, diamond is usually electroplated on a fine wire to form a wire saw for processing photovoltaic silicon rods. Figure 1 shows a SEM photograph of a wire saw with a diameter of 100um. Therefore, the idealized model is established, that is, the abrasive diamond grains have the same shape and are conical. At the same time, the abrasive diamond particles are evenly distributed on the surface of fine steel wire. The side view of the idealized saw is shown in Figure 2. As shown in the figure, Silicon rod move up with speed v_w, and diamond wire move outside (the direction perpendicular to the X axis and Y axis) with speed v_s.

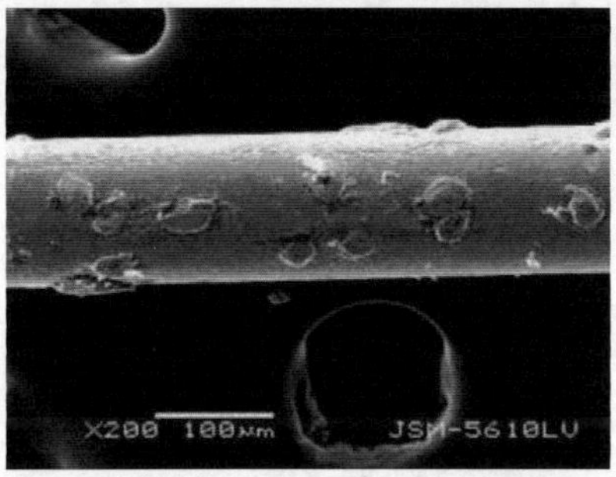

Figure 1 Scanning electron micrograph of wire saw with diameter of 100um

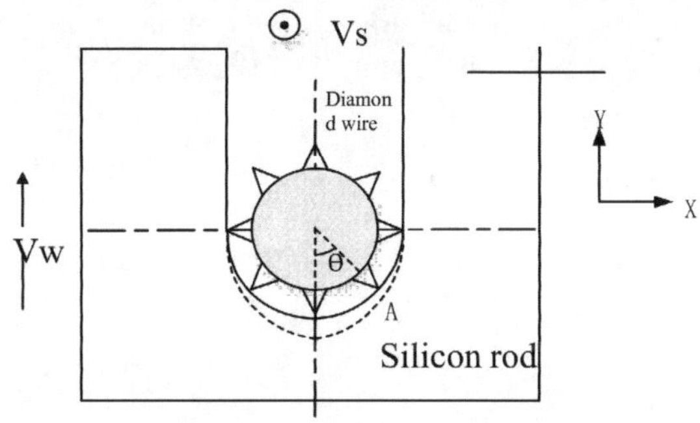

Figure 2 Side view of idealized saw

In the process of abrasive diamond sawing silicon rods, the wire saw process can be considered as the extrusion deformation and friction between many diamond abrasive particles and the workpiece to remove the material [10]. The force analysis diagram of a single diamond abrasive at point A is shown in Figure 3. Among them, Fnc is the normal sawing force, Fns is the normal friction , Ftc is the tangential cutting force, Fts is the tangential friction . The sawing force acting on a single abrasive particle is as follows:

$$\begin{aligned} Fna &= Fnc + Fns \\ Fta &= Ftc + Fts \end{aligned} \tag{1}$$

In the formula, Fna is the normal sawing force, Fta is the tangential sawing force.

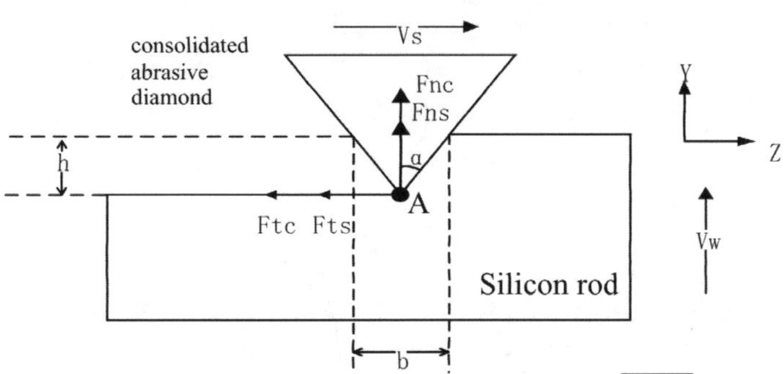

Figure 3 The force analysis diagram of the single diamond abrasive grain

The process of consolidating diamond abrasive wire sawing photovoltaic silicon rods is similar to traditional grinding, so the conclusions of grinding can be used to lead to similar results, as shown in Formula (2) and Formula (3).

$$Fnc = KQ_1$$
$$Q_1 = \frac{1}{2}bh \tag{2}$$
$$Fns = \bar{p}Q_2$$
$$Q_2 = \frac{1}{8}b^2 \tag{3}$$

In the above formula, K is the cutting force per unit area, Q_1 is the average cross-sectional area of a single diamond abrasive, Q_2 is the normal projection area of the contact area between a single diamond abrasive particle and the workpiece, $\bar{p}$ is the average contact pressure between a single diamond abrasive and silicon rod, b is the average cutting width of a single diamond abrasive, h is the average cutting height of a single diamond abrasive.

Figure 4 is a diagram of a consolidated diamond abrasive mitochondrial saw photovoltaic silicon rod, R is the radius of wire saw, If $d\theta$ is a very small angle, then its corresponding arc length dS is:

$$dS = Rd\theta \tag{4}$$

In unit time, the volume Q_3 of the silicon rod for section dS arc saw is:

$$Q_3 = V_s V_w \cos\theta dS \tag{5}$$

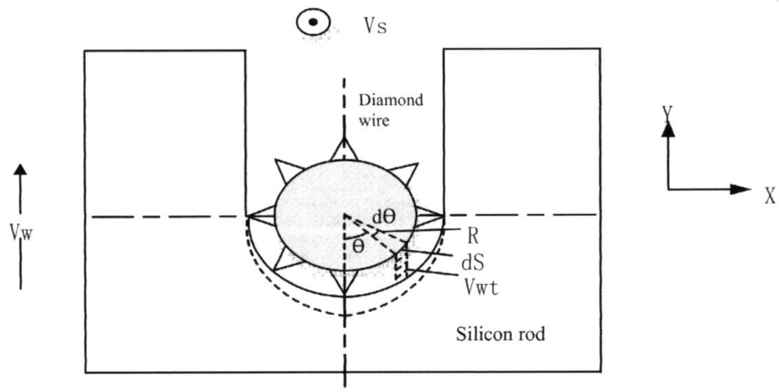

Figure 4 The schematic diagram that consolidated diamond abrasive saws photovoltaic silicon rods

At the same time, the volume Q_3 of the dS arc wire sawing silicon rod can be calculated by formula (6) per unit time.

$$Q_3 = V_s Q_1 C V_s dS \tag{6}$$

In the above formula , $V_s Q_1$ represents the volume of a single diamond abrasive saw per unit time, C is the number of diamond abrasives per unit area, and $V_s dS$ is the area through which the dS arc length passes per unit time. As can be seen from Figure 3, the relationship between b and h can be obtained:

$$h = \frac{b}{2\tan\alpha} \tag{7}$$

United equations (2), (5), (6), and (7), the average cutting width b and the average cutting height h of a single diamond abrasive can be obtained as follows:

$$b=\sqrt{\frac{4V_w\cos\theta\tan\alpha}{CV_s}} \tag{8}$$

$$h=\sqrt{\frac{V_w\cos\theta}{CV_s\tan\alpha}} \tag{9}$$

From formulas (8) and (9), Fns and Fnc can be obtained and substituted into Formula (1) to obtain Fna:

$$Fna=(K+\frac{\bar{p}\tan\alpha}{2})\frac{V_w\cos\theta}{CV_s} \tag{10}$$

The relationship between tangential cutting force and normal cutting force is as follows:

$$Ftc=\frac{\pi}{4\tan\alpha}Fnc \tag{11}$$

The tangential friction produced under normal friction is as follows:

$$Fts=\mu Fns \tag{12}$$

After the above derived formula, the tangential sawing force of a single abrasive grain is obtained as follows:

$$Fta=\left(\frac{\pi K}{2\tan\alpha}+\mu\bar{p}\tan\alpha\right)\frac{V_w\cos\theta}{2CV_s} \tag{13}$$

2.2. Analysis of cutting force of wire saw

In the process of consolidating diamond abrasive miter saw photovoltaic silicon rod, if the length of photovoltaic silicon rod is L, the contact area dA of the wire saw corresponding to angle $d\theta$ is:

$$dA=LdS=LRd\theta \tag{14}$$

The cutting force of a wire saw in the differential region is as follows:

$$\begin{cases} dFn=(K+\frac{\bar{p}\tan\alpha}{2})\frac{V_w\cos\theta}{CV_s}CLRd\theta \\ dFt=\left(\frac{\pi K}{2\tan\alpha}+\mu\bar{p}\tan\alpha\right)\frac{V_w\cos\theta}{2CV_s}CLRd\theta \end{cases} \tag{15}$$

When formula (15) is integrated in the range of $(-\frac{\pi}{2},\frac{\pi}{2})$, the sawing force of wire saw photovoltaic silicon rod is as follows:

$$\begin{cases} Fn=(2K+\bar{p}\tan\alpha)\frac{V_w}{V_s}LR=K_1\frac{V_w}{V_s}LR \\ Ft=\left(\frac{\pi K}{2\tan\alpha}+\mu\bar{p}\tan\alpha\right)\frac{V_w}{V_s}LR=K_2\frac{V_w}{V_s}LR \end{cases} \tag{16}$$

Therefore, the sawing force of the wire saw is a function related to V_s, V_w, R and L, The coefficients K_1 and K_2 mainly depend on the material of the workpiece and the wire saw, and the lubrication between the wire saw and the workpiece.

3. Analysis results of RSM

3.1. Establishment of mathematical model and its significance analysis

The dimension of workpiece L is fixed when cutting silicon rod with ultra-thin diamond wire cutting machine. Box-Behnken is used to carry out 3 factors and 3 levels experiment. The input factors of this experiment are wire saw speed V_s, workpiece feed rate V_w and wire saw radius R, and the output factor is normal sawing force Fn.The experiment factors and levels are shown in Table 1.

Table 1. Test factors and levels.

Factors	Level		
	-1	0	1
V_s/m·min^{-1}	800	1000	1200
V_w/mm·min^{-1}	0.2	0.4	0.6
R/um	50	70	90

Through regression fitting with Design-Expert software, its quadratic regression model is as follows:

$$Fn = 3.69 - 0.75V_s + 1.84V_w + 1.05R - 0.38V_sV_w - 0.22V_sR + 0.52V_wR + 0.11V_s^2 - 0.041V_w^2 - 0.041R^2$$

Table 2. The test design and results of response surface analysis

Serial number	V_s/m·min^{-1}	V_w/mm·min^{-1}	R/um	Fn/N
1	-1	1	0	6.765
2	0	0	0	3.708
3	0	0	0	3.712
4	0	1	1	6.958
5	1	1	0	4.510
6	0	0	0	3.722
7	0	0	0	3.698
8	1	0	1	3.866
9	-1	-1	0	2.255
10	1	0	-1	2.148
11	0	0	0	3.608
12	0	-1	-1	1.289
13	0	-1	1	2.319
14	-1	0	-1	3.221
15	1	-1	0	1.503
16	-1	0	1	5.799
17	0	1	-1	3.866

Design-Expert software was used to analyze variance of response value Fn in Table 2, and the results were shown in Table 3. Values of "Prob > F"less than 0.0500 indicate model terms are significant. In this case V_s, V_w, R, V_sV_w, V_sR, V_wR, V_s^2 are significant model terms. The "Lack of Fit F-value" of 2.32 implies the Lack of Fit is not significant relative to the pure error,.Non-significant lack of fit is good. So the model has high stability.

Table 3. The variance analysis results of normal force model

Source	Sum of Squares	df	Mean Square	F Value	p-value Prob > F
Model	42.39	9	4.71	1396.44	< 0.0001
V_s	4.52	1	4.52	1339.94	< 0.0001
V_w	27.14	1	27.14	8045.35	< 0.0001
R	8.86	1	8.86	2626.79	< 0.0001
V_sV_w	0.56	1	0.56	167.44	< 0.0001

V_sR	0.18	1	0.18	54.82	0.0001
V_wR	1.06	1	1.06	315.45	< 0.0001
V_s^2	0.050	1	0.050	14.95	0.0062
V_w^2	7.009×10^{-3}	1	7.009×10^{-3}	2.08	0.1926
R^2	7.009×10^{-3}	1	6.923×10^{-3}	2.08	0.1951
Residual	0.024	7	3.373×10^{-3}		
Lack of Fit	0.015	3	4.997×10^{-3}	2.32	0.217
Pure Error	8.619×10^{-3}	4	2.155×10^{-3}		
Cor Total	42.41	16			
Multiple correlation coefficient R^2=0.9994, adjusted coefficient of determination R^2_{Adj}=0.9987					

3.2. Analysis of response surface figure

Furthermore, the multiple correlation coefficient and adjusted coefficient of determination of the model are close to 1, which further verifies the accuracy of the model,Figure 5-7 shows the effect of the interaction of wire saw speed v_s, workpiece feed rate v_w and wire saw radius R on the normal sawing force Fn.

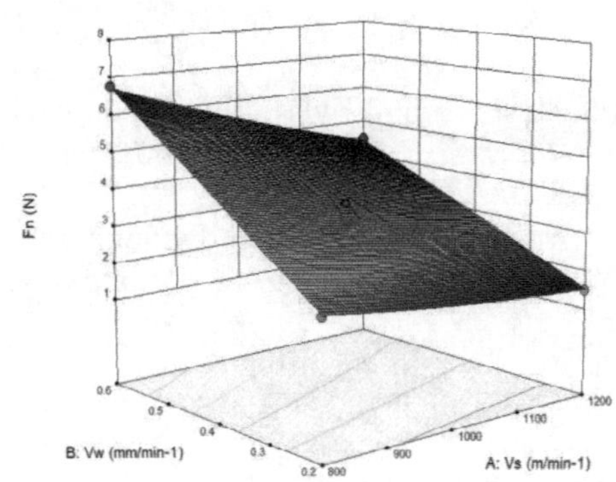

Figure 5 $Fn = f(V_s, V_w)$ response surface

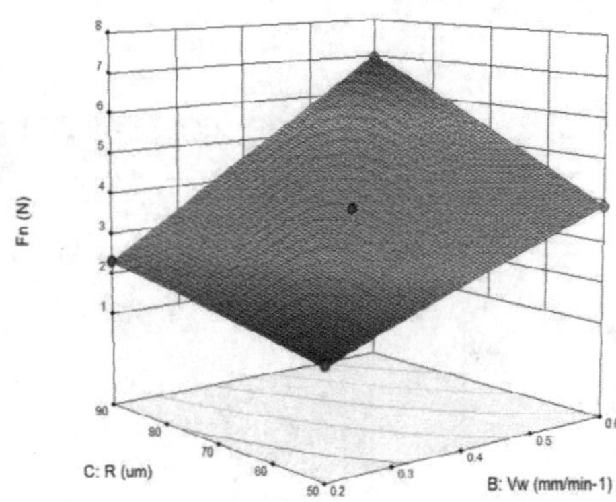

Figure 6 $Fn = f(V_w, \mathrm{R})$ response surface ($Vs = 1000m\cdot min^{-1}$)

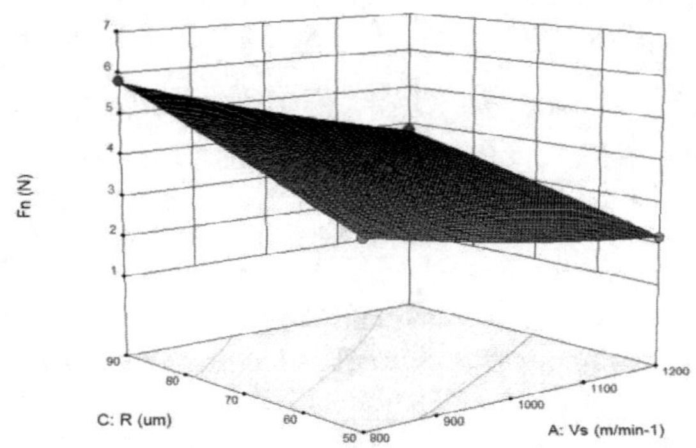

Figure 7 $Fn = f(V_s, \mathrm{R})$ response surface ($Vw = 0.4mm\cdot min^{-1}$)

As can be observed from the above three figures, when the other two process parameters are fixed, a higher workpiece feed rate or a larger wire saw radius will result in a larger normal sawing force. Conversely, the lower the wire saw speed, the larger the normal sawing force. The optimal process parameters are obtained by Design-Expert software, as shown in Table 4, which verified the correctness of Equation (15). However, in the actual processing, the values of these three process parameters must be within a reasonable range, otherwise there will be wire breakage, fragmentation and other phenomena.

Table 4. The optimal process parameters

v_s /m·min^{-1}	v_w /mm·min^{-1}	R /um	Fn /N
800	0.6	90	8.469

4. Conclusions

The model shows that the sawing force of wire saw is a function related to the wire saw speed, the workpiece feed rate, the radius of wire saw and the length of photovoltaic silicon rod. Therefore, within a reasonable range, the mathematical model can accurately predict the sawing force in the cutting process.

Acknowledgements

The research was supported by the Fundamental Research Funds for Central Universities(WUT:2017-IVA-117)and (WUT:2019III138CG).

References

[1] Hongwen D, Zhenqian X 2010 The research of fixed abrasive diamond wire saw technology *J. Modern Manufacturing Engineering*,2010(6):99-101.

[2] Chaohua W and Zhengfeng J et al 2017 Finite element analysis of multi-wire saw silicon rods with consolidated abrasive diamonds *J.The International Journal of Advanced Manufacturing Technology*,90(1):241-248.

[3] Hao W,Chris Y et al 2016 Modeling and analysis of the grit level interaction in diamond wire sawing of silicon *J.The International Journal of Advanced Manufacturing Technology*, 84(5):907-913.

[4] Shujuan L, Yong L, Xiaoli H et. al. 2015, Analysis and modeling cutting force for SiC monocrystal wafer processing *J. Journal of Mechanical Engineering*, 51 (23): 189-204.

[5] Liaoyuan Z, Yuquan G and Yanyan Z, 2008 Research on sawing forces during ultrasonic vibration cutting of hard and brittle materials with electroplated diamond wire saw *J. Diamond & Abrasives Engineering*, 2008(6): 49-54.

[6] Dan C, Shujuan L, Chao H 2014, Modeling and prediction for cutting force in SiC monocrystal wafers processing process *J. Mechanical Science and Technology for Aerospace Engineering,* 33(8):1161-1166.

[7] Jiangfeng M, Jianfeng L, Peiqi G et al 2005 Study on Random Vibration of Electroplate Diamond Wire Saw Subjected to Sawing Force *J.Chinese Mechanical Engineering*,16(24):2231-2234.

[8] Yufei G, Peiqi G and Shaojie L 2009 Study on the machining performance of single crystal silicon wafer cut by using reciprocating electroplated diamond wire saw *J.Journal of Synthetic Crystals*, 38(2):372-377.

[9] Weidong Y and Zhidong L 2014 Situation and development of fixed diamond abrasive wire cutting technology *J. Diamond & Abrasives Engineering*, 34(6):69-75.

[10] Lei M, Shujuan L, Hao L et al 2016 Experimental Research on Monocrystalline Si High-speed Wire Sawing Force Based on RSM *J.Journal of Synthetic Crystals*, 45(8):2067-2073.

Study on the Mechanism of Static Comparison between Vibrating String Strain Gauge and Fiber Grating Strain Gauge for Bridge Structure Monitoring

Lu Peng[1,a], Genqiang Jing[1,b], Zhu Luo[1,c], Yixu Wang[1,d], Zhidan Lin[1,e] and Na Miao[1,*,f]

[1]Highway Research Institute, Ministry of Transport, Beijing 100088, China

Corresponding Email: [*, f]305373174@qq.com;

[a]LuPeng@ncmerb.com, [b]GenqiangJing@ncmerb.com; [c]ZhuLuo@ncmerb.com; [d]YixuWang@ncmerb.com; [e]1345619269@qq.com

Abstract. The key component commonly used in strain monitoring is the strain sensor (strain gauge), which is usually embedded inside the measured object or installed on its surface to monitor the changes of corresponding variables for a long time. Therefore, the reliability, accuracy, consistency and stability of the strain sensor are important guarantees for the accurate measurement of strain. In order to monitor bridge changes more accurately, two strain sensors with a wide range of applications (i.e. vibration string strain sensor and fiber grating strain sensor) were selected for comparative analysis of the calibration model in the laboratory. We'll use national standard of 6061 aluminum alloy cylindrical bar and fix two strain sensors separately in the symmetric position. In order to eliminate the detection error caused by the current sensor detection when the metering clamping device and the strain sensor are under stress, 0-100kn compression test was conducted on the bar by using the superposition force standard machine. The device collects data of the two sensors, conducts data regression analysis, obtains the new model and hence conducts data analysis and comparison. The results show that the accuracy of the model is much higher than that of the strain sensor. The results of this study provide technical support for the development of real-time bridge detection technology and the realization of rapid calibration and detection of strain monitoring sensors used in the transportation industry before installation.

Key Words. traffic measurement, fiber grating, vibrating string strain gauge, Matlab, data model

1. Introduction

In industries and fields such as bridges, tunnels, oil pipelines and aerospace, long-term monitoring of structural health is required, especially the strain monitoring within the structure is very important [1-4]. Common strain sensors include vibrating string strain gauge, resistance strain gauge and fiber grating strain gauge, etc. [5] Their performance testing is very important. With the improvement of the manufacturing quality of strain gauge, its measurement accuracy becomes higher and higher, and the calibration measurement characteristics are gradually improved. In many fields, the need for high-precision strain gauge calibration becomes more and more urgent [6]. In order to monitor bridge changes more accurately, this paper established a more accurate monitoring model by comparing the fiber Bragg grating strain sensor and the vibrating string strain sensor bridge health monitoring calibration model. This provides technical support for the development of real-time bridge detection technology and the realization of rapid calibration and detection of strain monitoring sensors used in the transportation industry before installation.

2. The Working Principle of Vibrating String Strain Gauge and Fiber Grating Strain Gauge

2.1 Theoretical Model of Vibrating String Sensor

The vibrating string sensor works in the way of steel string vibration, and the force is represented by the change of vibration frequency, the actual output frequency signal is delivered. Therefore, there are no problems that strain gauge must be calibrated on the spot, signal drifts and poor durability for long use. Aluminum bar works well. It solves the shortcoming of strain gauge's instability in long-term use. It is currently used on a large scale in bridge monitoring.

The vibration string sensor has good measurement characteristics. It can achieve less than 0.1% in non-linearity, 0.05% in sensitivity and less than 0.1%/10°C error in temperature. After the strain gauge is manufactured, its steel string has a certain initial tension T0, so it has an initial frequency F0. When the strain gauge is installed, the tension of the vibrating string changes with the deformation. The strain can be measured by the tension of the vibrating string.

2.2 The Working Principle of Fiber Grating Strain Gauge

Fiber Bragg Grating Sensor (Fiber Grating Sensor) is a kind of optical Fiber Sensor and it is based on Fiber Bragg Grating sensing process. It is a kind of wavelength modulated fiber optic sensor, which obtains sensing information by modulating the wavelength of fiber Bragg with external physical parameters. Fiber Bragg grating sensors can directly measure physical quantities such as temperature and strain. Since the wavelength of the fiber Bragg grating is sensitive to both temperature and strain, that is, temperature and strain cause the shift of the coupled wavelength of the fiber Bragg grating, it is impossible to distinguish between temperature and strain by measuring the shift of the coupled wavelength of the fiber Bragg grating. Therefore, to solve the problem of cross sensitivity and to realize the discriminating measurement of temperature and stress is the premise of sensor application. The discriminating measurement of temperature and stress is realized by measuring the stress and temperature changes through certain technology . The basic principle of these technologies is to use two fiber Bragg gratings with different temperature and strain response sensitivity to form double grating temperature and strain sensor. By determining the temperature and strain response sensitivity coefficient of two fiber Bragg gratings, the temperature and strain can be solved by using two binary primary equations. There are two kinds of measurement techniques: multi-fiber grating measurement and single-fiber grating measurement.

3. Test Plan

In this test, the superposition force standard machine, which has been approved by the national metrology institute of China, is adopted, and the sinusoidal strain sensor of manufacturer A and fiber grating sensor of manufacturer B are widely used in the domestic testing market.

In the 6061 T6 aluminum rod, a vibrating string-type strain sensor and a fiber Bragg grating sensor are respectively fixed at both ends of the penetrating nut. See figure 1.The superposition force standard machine was used to conduct the pressure test on the aluminum rod. The aluminium index show of the table 1.The test was divided into three groups according to the test pressure, namely, 50KN group, 75KN group and 100KN group.

The test pressure of 50KN group was determined according to the formula $F_{1n} = 5*(n\text{-}1)$ (1 in the formula represents the 50KN test group; N = 1,2, 3,4,5,6,7,8,9,10,11, representing the test level of this group; F stands for test horizontal pressure, unit KN)

In the pressure test at each pressure level, the pressure loading of different types of sensors lasted for 120s, the test was repeated for 10 times, and the test data was collected.

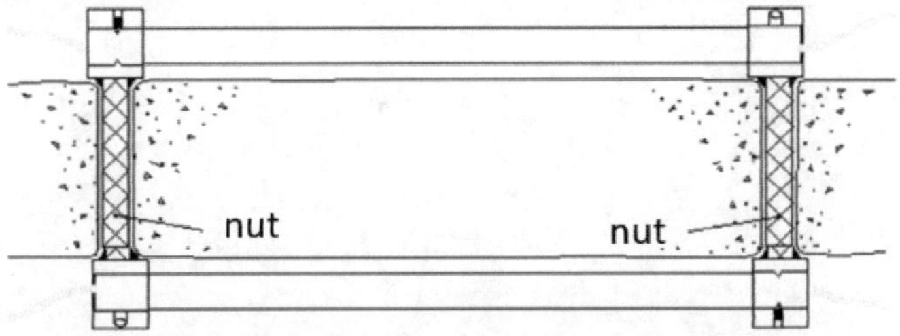

Figure 1. Diagram of experimental equipment

Table 1. The aluminum index of 6061 is:

Tensile strength σb	yield strength σ0.2	Elongation δ5 (%)	elastic coefficient	Ultimate yield strength	Bearing Yield Strength	fatigue strength
≧180MPa	≧110MPa	≧14	68.9GPa	228MPa	103MPa	62.1MPa

Note: longitudinal mechanical properties of bar materials at room temperature

4. Test Results and Analysis

In this study, MATLABR2014 software was used to analyze the test data, and Origin8.5 was used to draw the graph. figure. 2, 3 and 4 are data fitting curves of fiber Bragg grating sensors (50KN group, 75KN group and 100KN group). figure. 5, 6 and 7 respectively represent the data fitting curves of the vibrating chord sensor groups of 50KN, 75KN and 100KN.In this study, regression analysis was conducted on the results of the two sensors to obtain the new models Y1 and Y2.

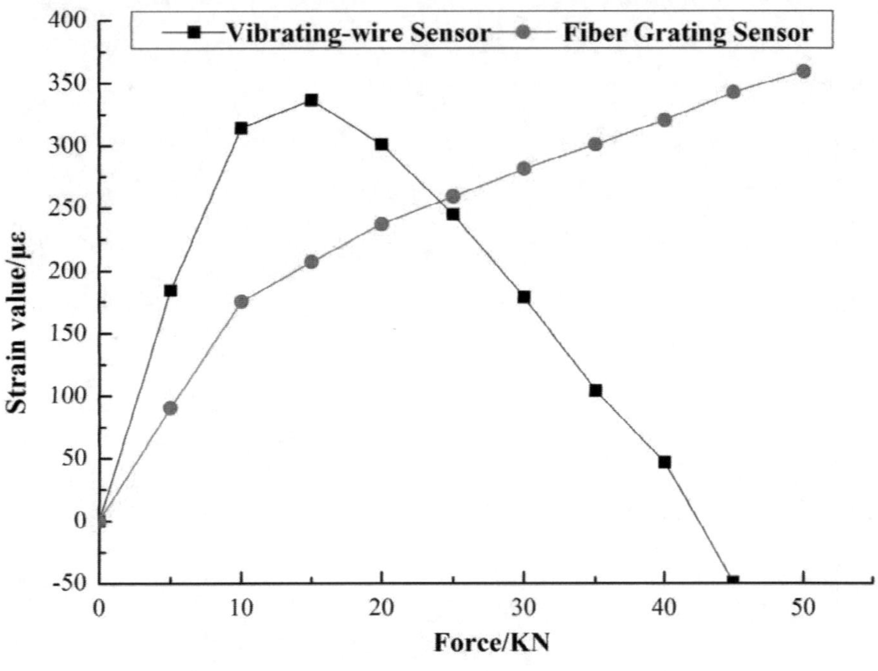

Figure 2. Data graph of 50kn test group

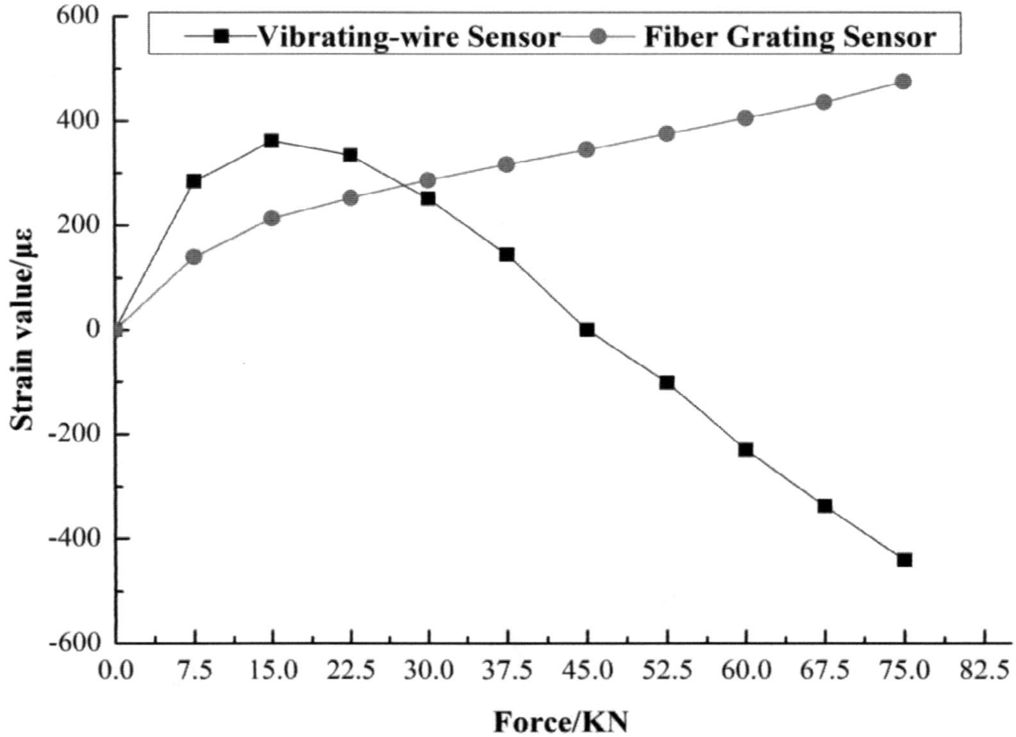

Figure 3. Data graph of 75kn test group

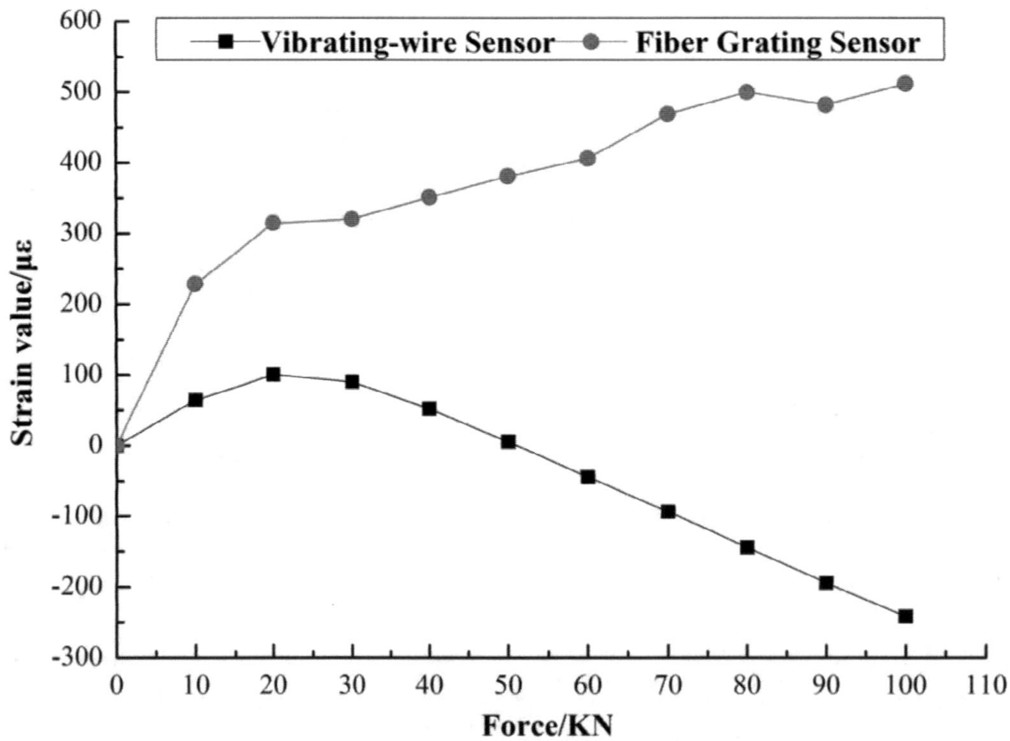

Figure4. Data graph of 100kn test group

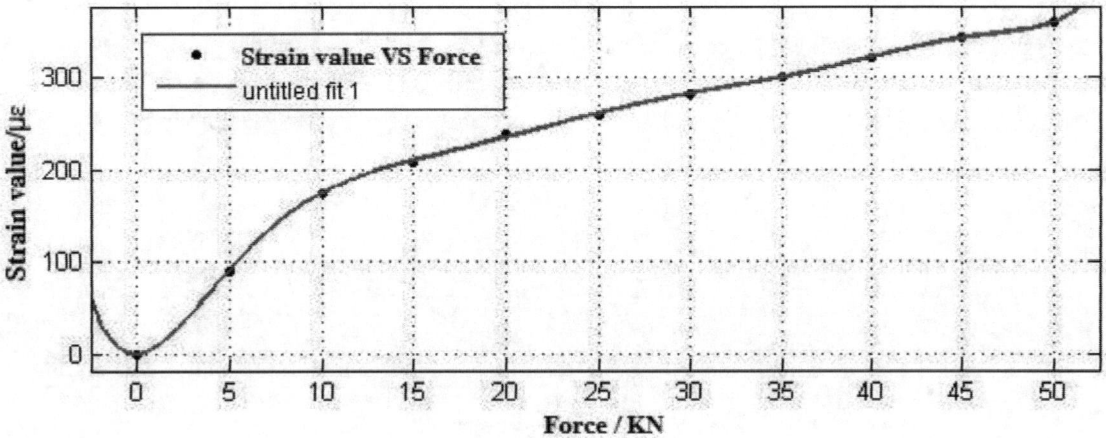

Figure 5. Fitting curve of fiber grating 50kn

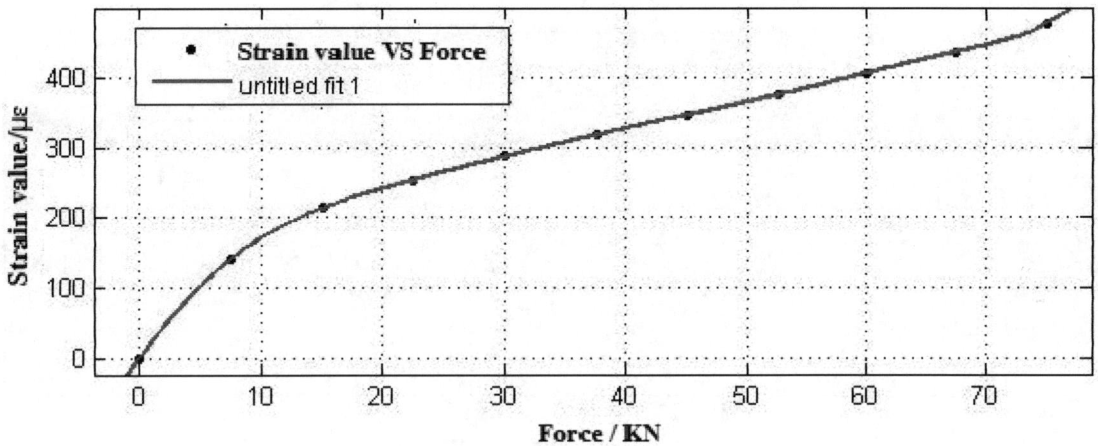

Figure 6. Fitting curve of fiber grating 75kn

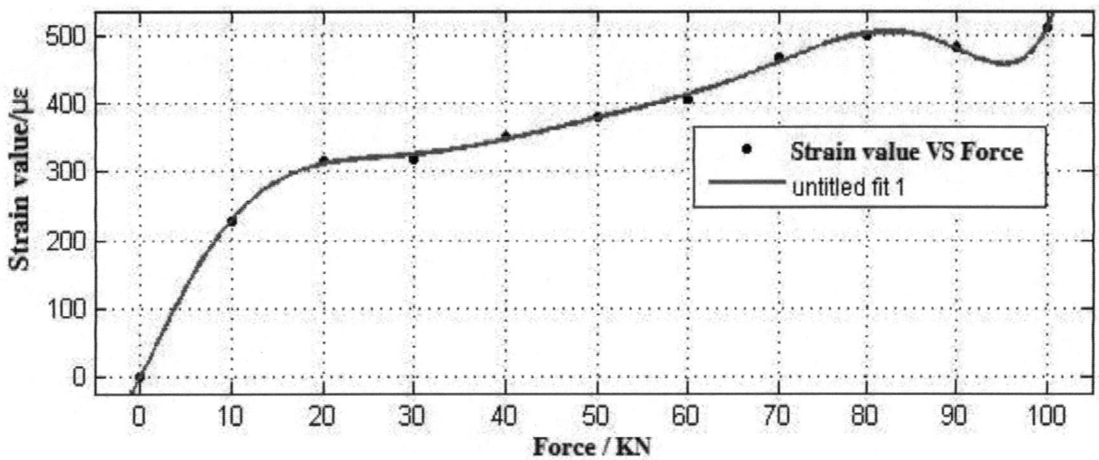

Figure 7. Fitting curve of fiber grating 100kn

IOP Conf. Series: Materials Science and Engineering **686** (2019) 012018 doi:10.1088/1757-899X/686/1/012018

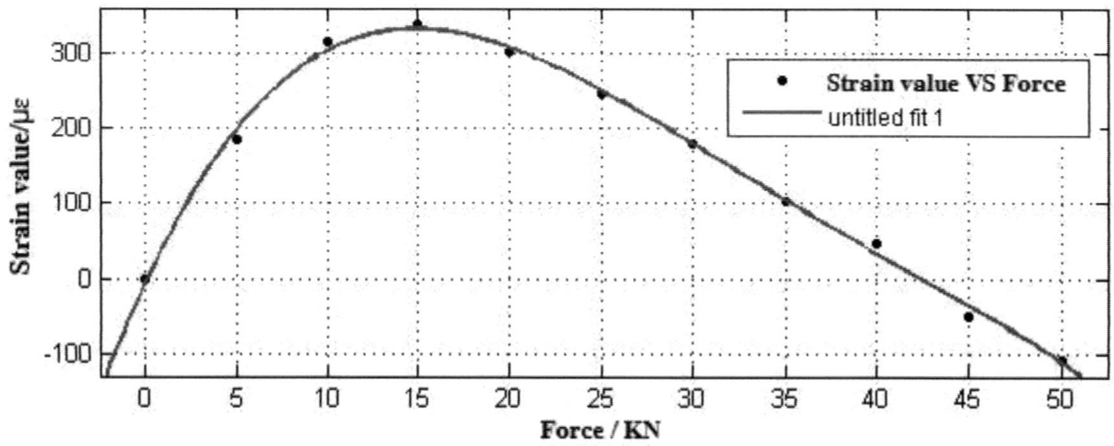

Figure 8. Fitting curve of vibration chord 50kn

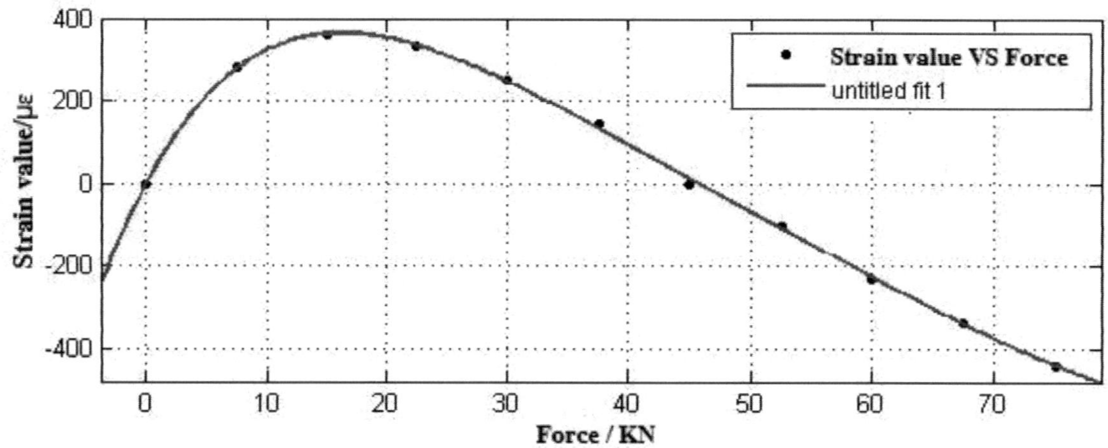

Figure 9. Fitting curve of vibration chord 75kn

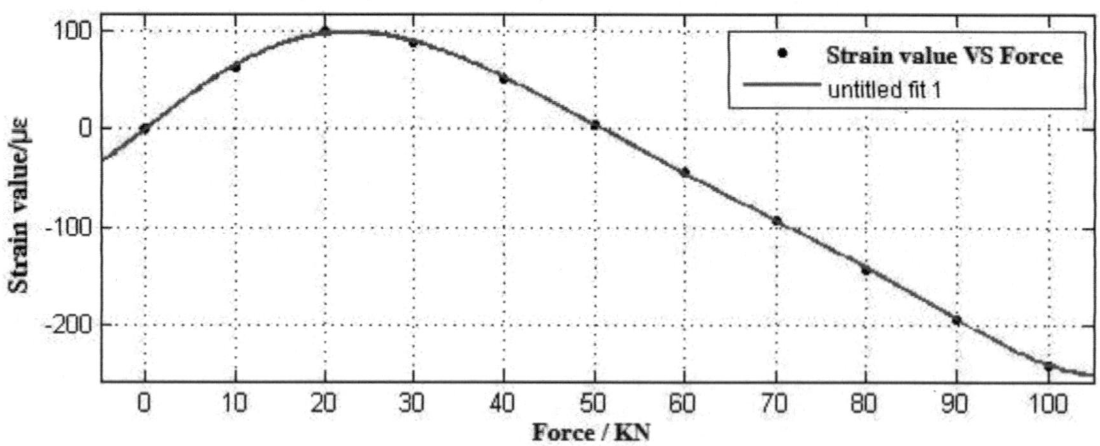

Figure 10. Fitting curve of vibration chord 100kn

In this paper, Origin is used to draw data graphs of two kinds of sensors under three pressure value tests, as shown in figure 2-4. At the same time, the least square method in MATLAB software was used to fit the test data curves of the two sensors under three pressure levels, as shown in figure 5-10.

It can be seen from the comparison of the above figures that the data of the fiber Bragg grating sensor increases with the increase of the force value, which is monotonically proportional. Whereas the numerical value of the vibrating chord sensor and force value do not increase monotonically. Therefore, the fiber Bragg grating sensor can characterize the change of strain variables more accurately than the vibrating chord sensor

The curve fitting indexes of the experimental model were obtained through the least square analysis, as shown in table 2. SSE in the table is the sum of squares due to error; MSE is Mean squared error: mean square deviation, variance; RMSE: Root mean squared error; R-square: Coefficient of determination; Adjusted R-square: degree-of-freedom adjusted coefficient of determination

Table 2. Indicators of the Fitted Curve

	50KN test group		75KN test group		100KN test group	
	Fiber grating sensor	vibrating chord sensor	Fiber grating sensor	vibrating chord sensor	Fiber grating sensor	vibrating chord sensor
SSE	20.79	197.2	6.502	284.5	193.4	14.65
R-square	0.9998	0.9992	1	0.9996	0.9991	0.9999
Adjusted R-square	0.9991	0.9979	0.9999	0.9982	0.9957	0.9997
RMSE	3.224	7.022	1.472	11.93	9.832	1.914

It can be seen from table 1 that the data of the 75kn test group of fiber Bragg grating sensor are best fitted by the least square method.

The fitting curve expression is:

f(x) = -2.461*x8 + 13.79*x7 + 3.198*x6 -37.6*x5 -40.57*x4 +131.9*x3 -52.72*x2 -413.5*x +137.1

5. Conclusion

Aiming at the difficulty of on-line calibration of strain gauge in practical use, a static comparison method based on high-precision force standard machine is proposed. Taking the micro deformation of the structure's characteristic position as the research scene and the uniform metal material as the medium, two kinds of equidistant sensors with different working principles were compared and verified. Conclusions are as follows: 1) The accuracy and real-time of the calibration model of the fiber grating strain sensor is much higher than that of the vibrating string strain sensor. It can also be understood that the vibrating string type sensor needs a long time to stabilize after each load to reduce the data instability caused by the tension deformation of the steel string itself. 2) Fiber Bragg grating technology has a good sense of the environment temperature, and has a good temperature compensation system. Thus, the data are more stable and reliable. Fiber Bragg grating strain gauge is more suitable for dynamic real-time acquisition and can be used as a reference sensor for on-site calibration. This study provides technical support for the development of bridge structure online comparison technology in the future and the rapid calibration and detection of strain monitoring sensors used in the transportation industry before installation.

Acknowledgment

The research was supported by the Project of the National Key R&D Program of China (No. 2017YFF0206305) and the Key Projects of Public Scientific Research Institutes (2018-9031).

IOP Conf. Series: Materials Science and Engineering **686** (2019) 012018 doi:10.1088/1757-899X/686/1/012018

References

[1] Wu B, Lu H, Chen B, *et al.* Study on Finite Element Model Updating in Highway Bridge Static Loading Test Using Spatially-Distributed Optical Fiber Sensors, *Sensors,* 2017, 17(7):1657.

[2] Zhang Wei, Zhang Yue, Zhang Zhimin, *et al.* Dynamic characteristics of strain force sensor and its dynamic compensation, *Acta metrology sinica*, 2012,33(1): 35-38.

[3] Ren L, Jiang T, Jia Z G, *et al.* Pipeline corrosion and leakage monitoring based on the distributed optical fiber sensing technology , *Measurement,* 2018, 122:57–65.

[4] Tsushima N, Su W, Gutierrez H, *et al.* Monitoring multi-axial vibrations of flexible rockets using sensor-instrumented reference strain structures, *Aerospace Science & Technology,* 2017, 71:285–298.

[5] HUANG W, ZHANG W, *et al.* Swept optical SSB-SC modulation technique for high-resolution large-dynamic-range static strain measurement using FBG-FP sensors, *Optics Letters,* 2015, 40(7):1406-9.

[6] Park H , Lee H , Choi S , et al. A Practical Monitoring System for the Structural Safety of Mega-Trusses Using Wireless Vibrating Wire Strain Gauges. *Sensors*, 2013, 13(12):17346-17361.

IOP Conf. Series: Materials Science and Engineering **686** (2019) 012019 doi:10.1088/1757-899X/686/1/012019

Aerodynamic Performance of Variable-Pitch Propellers for High-Altitude UAVs

Jianhua Xu*, Wenping Song, Xudong Yang and Han Nie

National Key Laboratory of Science and Technology on Aerodynamic Design and Research, Northwestern Polytechnical University, Xi'an, 710072, China

*Email: xujh@nwpu.edu.cn

Abstract. As the air density changes dramatically from high to low altitude, the matching of high-altitude propeller and electric motor under low-altitude condition is one of the main challenges that aerodynamic design of high-altitude propeller faces. To solve this problem, variable-pitch propeller is the most commonly used and feasible solution. In this paper, the Reynolds-average Navier-Stokes solver and an efficient surrogate-based optimization toolbox are used for the aerodynamic design of a fixed-pitch propeller. Based on the optimal fixed-pitch propeller, a series of pitch angle are employed to obtain the variable-pitch propeller with maximum thrust. Then, the aerodynamic characteristics of fixed-pitch propeller and variable-pitch propeller under typical working conditions are studied and compared. The results show that under the condition of low-altitude and low wind speed, the fixed-pitch propeller reaches the rated torque of the electric motor at a lower rotational speed, resulting in a lower shaft power and thrust. By adjusting the pitch angle, the variable-pitch propeller can achieve higher rotational speed under the same rated torque constraints, thereby absorbing more shaft power and achieving larger thrust. The lower the altitude, the more obvious the advantage of the variable-pitch propeller. Compared with fixed-pitch propeller, the maximum relative increment of the maximum thrust can reach more than 80% for variable-pitch propeller. It is concluded that variable-pitch propeller can significantly improve the matching of propeller and electric motor, bringing wider flight envelope of UAVs.

1. Introduction

Propellers are widely used in high-altitude vehicles including airships[1][2] and unmanned aerial vehicles[3] (UAVs). One of the challenges in the design of aerodynamic components for high-altitude vehicles is low air density. At the altitude of 20km, the atmosphere density is 7.25% of that on the ground, and the air density at 25km is only 3.27% of that on the ground. Under such atmospheric condition of low density, the propeller propulsion system is currently the only available power system for low speed aircrafts. Due to the low air density, the working Reynolds number of high-altitude propellers is very low[3], usually less than 300,000. In this range of Reynolds number, the aerodynamic performance of the airfoil may change severely with Reynolds number, which results in the deterioration of the airfoil's aerodynamic performance (typically around the Reynolds number of 100,000) and decrease of the propeller efficiency.

Another key problem of the high-altitude propeller design is the matching of propeller and electric motor under low altitude. When the high-altitude propeller is used at a low altitude, where the air density is much higher than that at a high altitude (above 10km), the maximum rotational speed of the propeller is lower due to the restriction of the maximum torque, resulting in a lower shaft power and thrust. A common and available technic to improve the matching of propeller and electric motor is variable-pitch propeller. With the pitch angle of propeller blade decreased, the effective attack angle of propeller section, lift and drag all decrease, and meanwhile the rotational speed, shaft power and thrust increase. The variable-pitch propeller has already been widely used in the large transport airplane. However, the problem of matching the electric motors at high and low altitudes is more urgent in the propeller design of high-altitude aircrafts.

This article studies the aerodynamic performance of the fixed-pitch propeller and the variable-pitch propeller. A baseline propeller (fixed-pitch propeller) is obtained by surrogate-based optimization, whose design point is under high altitude working condition and design target is highest efficiency. Then, the variable pitch propellers are obtained under a series of off-design conditions, with the angle of pitch decided by maximum thrust. The aerodynamic performances of the fixed-pitch propeller and the variable-pitch propeller are compared, followed by an analysis of the aerodynamic advantages of the variable-pitch propeller.

2. RANS solver for propeller

The integral form of Reynolds-averaged Navier-Stokes (RANS) equations expressed in the blade-attached rotational frame can be written as follows:

$$\frac{\partial}{\partial t}\iiint_{\Omega}\mathbf{Q}dV+\iint_{\partial\Omega}\mathbf{F}\bullet\mathbf{n}dS-\iint_{\partial\Omega}\mathbf{F}_v\bullet\mathbf{n}dS+\iiint_{\Omega}\mathbf{G}dV=0 \quad (1)$$

where, t is time, Ω and $\partial\Omega$ stand for control volume and the corresponding boundary, respectively. dS and dV are elemental area and volume of the control volume, respectively. $\mathbf{n}$ is the outward unit normal. $\mathbf{Q}$ is the conservation fluid variables. $\mathbf{F}$ and $\mathbf{F}_v$ are the inviscid and viscous flux term, respectively. $\mathbf{G}$ is the Coriolis force term.

The cell-centered finite-volume method proposed by Jameson[4] is used to solve the above governing equations on the chimera grid, and an in-house RANS solver, named ROTNS is developed. An improved Newton-like LU-SGS method[5] is utilized for time stepping, and a high-efficient FAS multi-grid method[6] on Chimera grid is developed to improve the computational efficiency. Jameson's central scheme (JST)[4] is adopted for the spatial discretization scheme. Spalart-Allmaras (S-A)[7] one-equation model is used for turbulence enclosure.

The 2-bladed propeller with a diameter of 6 inches is simulated to validate the solver. This propeller model is designed by Wichita State University and tested in their 3ft × 4ft Low Speed Wind Tunnel[8]. The blade pitch angle that is represented by the angle at 75% of the blade radius ranges from 15° to 30°. The Reynolds number based on chord at 75% of the blade radius is within the range of 90,000 to 120,000 when the rotational speed is fixed at 6000rpm. Figure 1 shows the propeller shape with a blade pitch angle of 15°. Figure 2 demonstrates that results of the thrust coefficient, power coefficient and efficiency against the advance ratio by ROTNS are in good agreement with the experimental results.

Figure 1. The 2-bladed propeller with a diameter of 6 inches and a blade pitch angle of 15°.

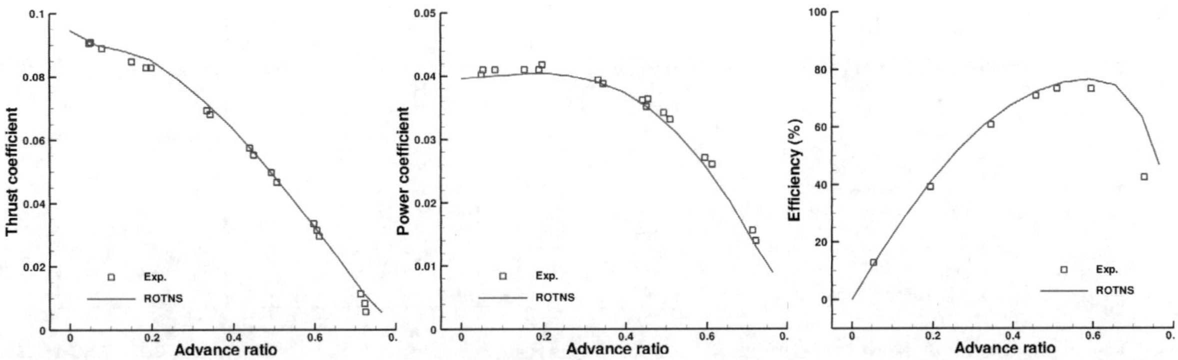

(a) Advance ratio-thrust coefficient (b) Advance ratio-power coefficient (c) Advance ratio-efficiency

Figure 2. Results of the thrust coefficient, power coefficient, and efficiency against the advance ratio by ROTNS are in good agreement with the experimental results[8]

3. Aerodynamic design optimization of propeller

Based on an in-house surrogate-based optimizer, "SurroOpt"[9][10], an aerodynamic design platform for the propeller is established, as shown in Figure 3.

Typical distributions of chord length and twist angle for a propeller are shown in Figure 4. The distribution functions are parameterized by quadratic function. According to the geometrical characteristic of the chord length distribution, the distribution function is defined by two quadratic functions split at the maximum chord length point, which can be written as:

$$C_R = \begin{cases} a_1\left(x - x_{c\max}\right)^2 + C_{R\max} & x \le x_{c\max} \\ a_2\left(x - x_{c\max}\right)^2 + C_{R\max} & x > x_{c\max} \end{cases} \tag{2}$$

in which,

$$a_1 = \frac{C_{Rroot} - C_{R\max}}{\left(x_{root} - x_{c\max}\right)^2}, a_2 = \frac{C_{Rtip} - C_{R\max}}{\left(x_{tip} - x_{c\max}\right)^2}$$

where, C_R is the chord length of blade section, x is the radial coordinate, x_{cmax} is radial coordinate at maximum chord length point, C_{Rmax} is the maximum chord length, C_{Rroot} is the chord length of blade root section, x_{root} is radial coordinate at blade root section, C_{Rtip} is the chord length of blade tip section, x_{tip} is radial coordinate at blade tip section.

The distribution of the twist angle can be written as a general form of following quadratic function:

$$\beta_R = a_\beta x^2 + b_\beta x + c_\beta \tag{3}$$

where, β_R is the twist angle of blade section, a_β, b_β and c_β are coefficients of the function.

Moreover, at the blade tip (x>0.8R) the chord length and twist angle are kept as a constant, and the blade section is translated by ds along the chord direction, which is defined as

$$ds = a_s\left(x - x_s\right)^2 \tag{4}$$

in which, $a_s = \dfrac{ds_{Rtip}}{\left(x_{tip} - x_s\right)^2}$

where, ds_{Rtip} is the translation length at the blade tip . x_s and x_{tip} are the starting radial location and the blade tip radial location. In this article, x_s is fixed to 0.8 and x_{tip} is 1.0.

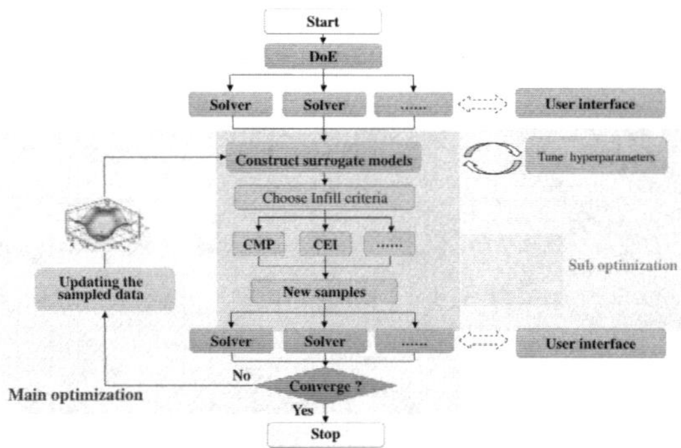

Figure 3. Framework of an efficient surrogate-based optimizer: SurroOpt.

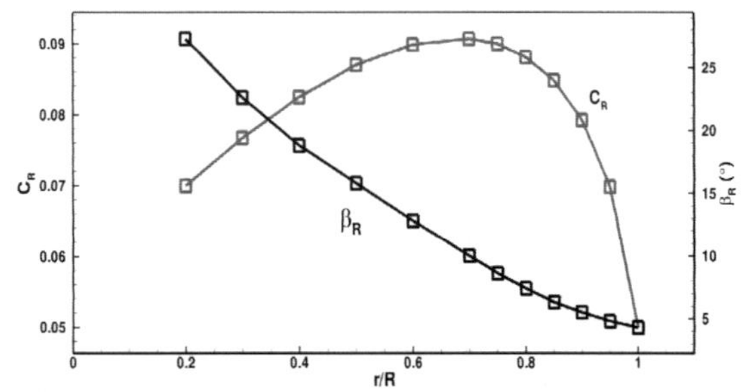

Figure 4. Typical distributions of chord and twist for a propeller.

Therefore, the 8 design variables of the geometry parameterization are $x_{c\,max}$, $C_{R\,max}$, C_{Rroot}, C_{Rtip}, a_β, b_β c_β, and ds_{Rtip}.

4. Variable-pitch propeller

The pitch angle $\Delta\theta$ of the fixed-pitch propeller is defined as 0 (Figure 5), while the pitch angle of variable-pitch propeller can be adjusted according to the working conditions of UAVs. Using the altitude and wind speed of the fixed-pitch propeller at cruise condition as baseline, the principle of the variable-pitch propeller is given as follows.

For higher altitude and wind speed, the rotational speed reaches the rated value first, while the torque and the shaft power are lower than rated value. Therefore, a positive pitch angle ($\Delta\theta > \mathbf{0}$) is employed to increase the torque and shaft power at the rated rotational speed. The maximum thrust is obtained when the torque or shaft power reaches the rated value.

For lower altitude and wind speed, the torque reaches the rated value first, while the rotational speed and the shaft power are lower than rated value. Therefore, a negative pitch angle ($\Delta\theta < \mathbf{0}$) is employed to increase the rotational speed and shaft power at the rated torque. The maximum thrust is obtained when the rotational speed or shaft power reaches the rated value.

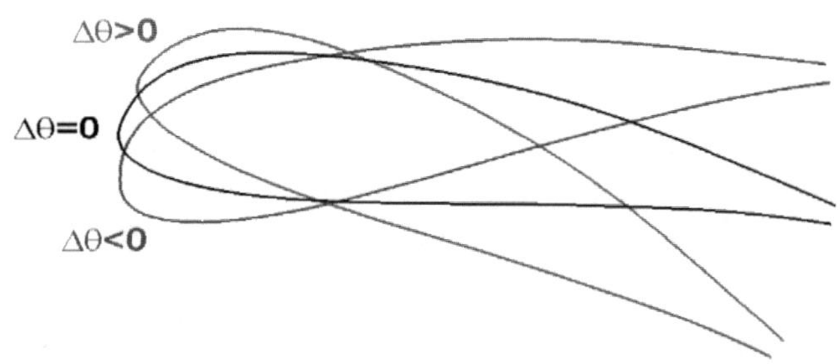

Figure 5. Blade section of fixed-pitch ($\Delta\theta = 0$) and variable-pitch propellers.

5. Results and discussion

5.1. Design optimization of baseline propeller (fixed-pitch)

Take 2 meters and 2-bladed high-altitude propeller for example. The classic low Reynolds number airfoil E387 is used for each section. The design working condition is at the altitude of 25km, wind speed of 40m/s and rotational speed of 1200r/min. The optimization target is max thrust with restriction to the shaft power less than 1.0 kW. The optimization model is described as

$$\begin{array}{ll} Maximize: & T_s \\ s.t. & P_{s0} - P_s > 0 \end{array} \tag{5}$$

in which, T_s is the thrust of the propeller. P_s is the shaft power and the subscript "s0" means maximum shaft power. The optimized geometry is shown in Figure 6. Figure 7 demonstrates the pressure contours and streamlines of flow around the propeller at the section of 75%R.

Figure 6. Geometry of the baseline (fixed-pitch) propeller (25km, 40m/s).

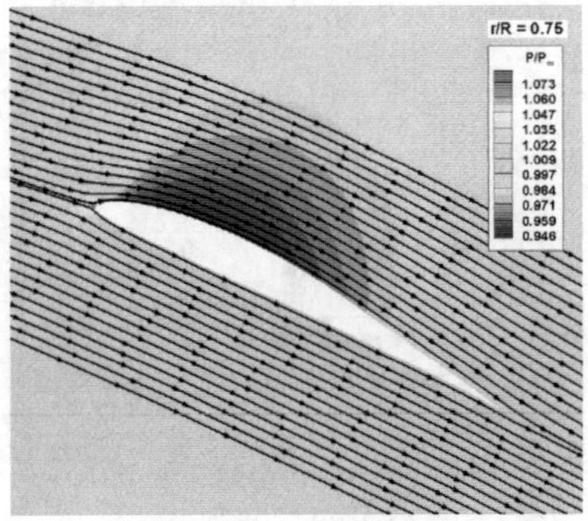

Figure 7. Pressure contours and streamlines at 75%R (25km, 40m/s, 1200rpm)

5.2. Comparison of aerodynamic performance between fixed-pitch and variable-pitch propeller

At the altitude of 10km and wind speed of 10m/s, the aerodynamic performances of fixed-pitch propeller and variable-pitch propeller are calculated under the constraints of rated torque 8N.m, shaft power 1Kw and rotational speed 1200r/min. Figure 8 shows that the rotational speed and shaft power go up with the decrease of pitch angle. Due to the restriction of rated torque, the rotational speed and shaft power are unable to reach their rated value. Figure 9 shows that a peak of maximum thrust is located at a pitch angle of -19°. When the pitch angle is less than -19°, the maximum thrust decreases in spite that the shaft power is higher, for the reason of a negative effective attack angle and large flow separation.

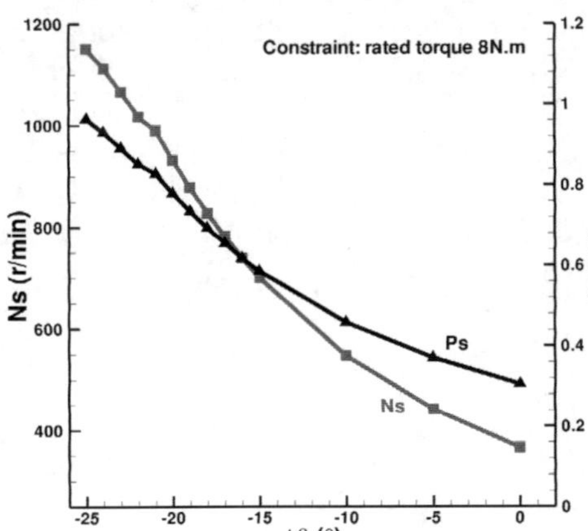

Figure 8. Rotational speed and shaft power of variable-pitch propeller under the constraints of rated torque, shaft power and rotational speed.

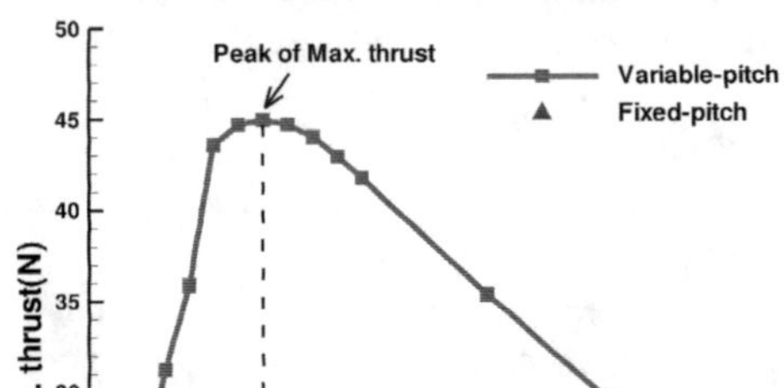

Figure 9. Maximum thrusts of fixed-pitch propeller and variable-pitch propeller with different pitch angles.

Table 1 shows the comparison of aerodynamic performance between fixed-pitch and variable-pitch propeller, at different altitude and wind speed. With the constraint of rated torque, the rotational speed of variable-pitch propeller is significantly higher than that of fixed-pitch propeller. The lower the altitude, the greater the relative increase in rotational speed. Under the same torque constraints, the relative increase of shaft power is nearly the same as that of rotational speed. Particularly, the relative increase of rotational speed and shaft power can reach as much as 1.5 times for variable-pitch propellers, under the ground conditions. Although the efficiency of variable-pitch propeller is lower than that of the fixed-pitch propeller, its maximum thrust is greatly increased for the reason the shaft power of the variable-pitch propeller is significantly higher than that of the fixed-pitch propeller. The maximum relative increment of the maximum thrust can reach more than 80%. In addition, the rotational speed and torque of the variable-pitch propeller are closer to their rated value of electric motor. Thus the matching of the variable-pitch propeller and the electric motor is better, resulting in a more efficient working condition of electric motor.

Table 1. Comparison of aerodynamic performance between fixed-pitch and variable-pitch propeller.

Altitude (km)	Wind speed (m/s)		Pitch angle (°)	Rotational speed (r/min)	Torque (N.m)	Shaft power (kW)	Max.thrust (N)	Efficiency (%)
		Fixed-pitch	0	836	7.986	0.699	22.89	81.84
20	25	Variable-pitch	-9	1200	7.915	0.995	31.24	78.52
		Relative Δ %	/	+43.54	/	+42.35	+36.47	-4.06
		Fixed-pitch	0	542	7.997	0.454	24.30	80.32
15	15	Variable-pitch	-17	1150	7.995	0.963	42.54	66.28
		Relative Δ %	/	+112.17	/	+112.11	+75.06	-17.48
		Fixed-pitch	0	366	7.967	0.305	24.56	80.43
10	10	Variable-pitch	-19	878	7.987	0.734	44.98	61.25
		Relative Δ %	/	+139.89	/	+140.66	+83.14	-23.85
		Fixed-pitch	0	278	7.961	0.232	23.95	86.68
5	8	Variable-pitch	-19	671	7.973	0.560	42.98	61.38
		Relative Δ %	/	+141.37	/	+141.38	+79.46	-29.19
		Fixed-pitch	0	213	7.900	0.176	24.16	82.27
0	6	Variable-pitch	-20	547	7.958	0.456	43.21	56.88
		Relative Δ %	/	+156.81	/	+159.09	+78.85	-30.86

6. Conclusions

The aerodynamic performance of variable-pitch propellers for high-altitude UAVs are studied. The maximum thrust of fixed-pitch propeller is limited by the rated torque of electric motor at lower altitude, resulting in a poor matching of propeller and electric motor. Using variable-pitch propeller can obviously increase the rotational speed and shaft power under the constraint of rated torque, leading to a larger maximum thrust. The rotational speed and torque of the variable-pitch propeller are closer to their rated value of electric motor, resulting in a better matching of propeller and electric motor, and electric motor operates more efficient. Compared with fixed-pitch propeller, the maximum relative increment of the maximum thrust can reach more than 80% for variable-pitch propeller.

References

[1] Colozza A J 1998 NASA/CR-1998-208520
[2] Colozza A J 2002 NASA-CR-2002-211866
[3] Mueller T J 2001 *Fixed and Flapping Wing Aerodynamics for Micro Air Vehicle Applications* (Virginia, USA: Smerican Institute of Aeronautics and Astronautics) p1
[4] Jameson A, Schmidt W, Turkel E 1981 AIAA-1981-1259
[5] Xu J H, Song W P, Han Z H 2009 Calculation of Aerodynamic Performance of Propellers at Low Reynolds Number Based on Reynolds-Averaged Navier-Stokes Equations Simulation *Computational Fluid Dynamics 2008* ed Choi H, Choi H G, and Yoo J Y (Berlin Springer Press) pp 283-288.
[6] Jameson A and Baker T J 1984 AIAA-84-0093
[7] Spalart P R and Allmaras S R 1992 AIAA-92-0439
[8] Ghoddoussi A and Miller L S 2016 AIAA-2016-3422.
[9] Han Z H 2016 30th Congress of the International Council of the Aeronautical Sciences
[10] Liu J, Song W P, Han Z H, et al. 2017 Structural and Multidisciplinary Optimization 55 925

IOP Conf. Series: Materials Science and Engineering **686** (2019) 012020 doi:10.1088/1757-899X/686/1/012020

Optimal Sensor Placement of Health Management for Thermal Protection System on RLV

JIANLIN XUAN[1,2], WENHUI ZHOU[1,*] and YUE SONG[1]

[1]The School of Aeronautics, Northwestern Polytechnical University, Xi'an, Shaanxi 710072, China
[2]Yangtze River Delta Research Institute of Northwestern Polytechnical University, Taicang, Jiangsu 215400, China

[*]zhouwenhui@mail.nwpu.edu.cn

Abstract: When a Reusable Launch Vehicle (RLV) travels through the atmosphere there should be strong aero heating around it, so a Thermal Protection System (TPS) would be used to protect the vehicle structures from aero thermal ablation. However, the TPS structure may fail more frequently than normal vehicle structures because of the severe environment and load conditions, while a TPS failure can even lead to a catastrophic breakdown of vehicle structures. To keep the RLV safe, the TPS must be monitored real-time. To learn the heat transfer inside a ceramic TPS tile finite element models are built, and computation made with them indicates that temperature monitoring and frequency monitoring of the TPS structure should be effectual for fault detection. Health management for TPS is one of the key technologies for space vehicle. The paper puts forward a method that analyze and valuation the system states by monitoring the key function index sign of TPS. The concept of health management for Thermal Protection System of space shuttle and the principles of health monitoring are introduced. The principles of sensor selection and layout optimization method for TPS Health Management are discussed.

1. INTRODUCTION

In addition to aerodynamic thermal load primarily undertaken by thermal protection system (TPS) of a reusable launch vehicle (RLV) during flight, it must bear various non-heating loads including alternative temperature between -150℃ and +55℃, aerodynamic noise and deceleration overload, etc.[1]. It is much likely for Insulation tiles that have been struck by thermal insulation materials splitting away off an external tank or space debris to break or fall of. If the number of Insulation tiles falling off has reached the limit, RLV may be broken into pieces due to giant pressure and super high temperature at the time of atmospheric re-entry, which further gives rise severe accidents.

Over the past several decades, heat transfer of the thermal protection structure has been theoretically analyzed and experimentally investigated by multiple scholars. Cunning ton et al. made surveys about effective thermal conductivity of 7 kinds of multilayer insulation structures in 1967 together with analyzing heat transfer in theory; in addition, they also took advantage of an approximation method for radiant heat absorption analysis [2]. According to Keller et al., the approximation method described above and the modified diffusion approximation method are both utilized to analyze radiation heat transfer proposing an internal heat conduction analysis model for different multilayer insulation structures [3]. The two-flux theory is adopted by Kamran et al. who has discussed about optimal multilayer structure design to analyze radiative transfer on one hand; on the other hand, the finite-difference numerical analysis method is further used to obtain the corresponding solution and the finite volume method to analyze energy conservation control equation [4,5]. Moreover, countries of great power in aerospace such as America have been also dedicated to improving aerodynamic heating prediction accuracy based on computational fluid mechanics, perfecting transient response analysis approaches for TPS temperature field and combining such two analysis processes together to enhance mass estimation precision of TPS [6,7,8].Among them, some investigations on health management of the structure have been involved with TPS or taken it into account [9,10]. Regarding health monitoring technique test targeted at nose cone TPS performed by overseas scholars who resort to X-38, X-38 is a Lifting body re-entry vehicle designed as emergency return spacecraft for International Space Station crew developed by NASA, they lay emphasis on studying several sensor technologies available for TPS health monitoring, including acoustic emission, micrometer and infrared imaging[11]. As for researches on thermal protection structure surface damages and strain isolation layer degumming faults, rapid non-destructive evaluations on these failures in the process of ground maintenance were stressed [12,13,14].

Experts in ground adopt electric equipment-based monitoring to comprehend conditions of Insulation tiles in a better manner and further carry out fault analysis and judgment. However, photos and images that are not clear enough give rise to information incompletion and non-determinacy. Consequently, it becomes less likely to estimate causes to such faults and the overall health condition of a space shuttle. Moreover, neither instantaneity of fault processing nor high accuracy and reliability of fault diagnosis is achieved as fault analysis consumes a long time, which also signifies that safety of RLV return pro-cess cannot be exactly assessed in advance.

Therefore, greater importance should be attached to studies on health management of RLV TPS for the purpose of effectively preventing catastrophic failures incurred by phenomena described above. We have studied the composition, failure mode and influence analysis of Thermal Protection System on RLV, on the basis of which, this paper mainly the selection of sensors and the optimal layout method.

2. SELECTION OF TPS HEALTH MONITORING SENSOR

Referring to the experience of the US NASA in X-33 and X-38 using Fiber Bragg Grating （FBG） high temperature sensor temperature sensor for health monitoring, TPS health management has the following three requirements for sensors.

(1) The sensor (with its signal transmission line) must be adaptable to high-low temperature environment where it is located at and guaranteed to be in a position to run reliably and stably in limiting temperature.

(2) Q-percentile life of the sensor (with its signal transmission line) must be longer than that of thermal protection materials in the same position; otherwise, monitoring system maintenance cannot make up for thermal protection structure disassembly.

(3) The sensor must be sufficiently consistent and reliable.

By taking accuracy, cost and quality, etc. into consideration, Fiber Bragg Grating (FBG) high temperature sensor was selected in this paper. The corresponding installation mode was designed to be internally embedded.

Generally speaking, the FBG Intelligent Health Monitoring system mainly consists of the following three parts: FBG sensor system, signal transmission and collection system, data processing and monitoring system. Among them, FBG sensor system, including the choice of FBG sensors, FBG sensor network in the structure of the layout scheme, select specific modulation and performance requirements of FBG sensor, and consider the topology of FBG sensor. The signal transmission and acquisition system include the calibration of FBG sensors, the sampling module and the storage structure and mode of the massive real-time data. Data processing and monitoring system is the core of health monitoring system, including the analysis of the validity of a large number of data, the selection of parameters of the structure health performance, the visualization system of structure running state and the corresponding disaster early warning function.

3. OPTIMAL SENSOR PLACEMENT OF TPS

Optimal sensor network design was conducted for TPS by focusing on sensor placement and employing temperature as the principal monitoring object.

Fiber optic sensor (FOS) was selected as the research object of this study to perform sensor placement optimization. As assumed, the mass of a single sensor was denoted to be *m* and the corresponding detection probability to be *g*; in zones of high, medium and low temperature, sensors with a number of $x, y, z \in N$ were respectively placed to constitute an extensive health monitoring sensor network. The goal of such optimization is to achieve a detection rate up to the standard by virtue of the least quantity of sensors. As long as the type of sensor has been determined, total mass of the entire sensor network can be deemed as a function of the sensor quantity, that is, $M = M(n_1, n_2, n_3)$; besides, TPS fault detection rate *P* acquired by the whole sensor network is a function of both the sensor quantity and the position matrix *J* of sensor placement. To be specific, J is constituted by a coordinate vector *(x, y, z)* of each sensor placement point and $P = P(n_1, n_2, n_3, J)$. Therefore, sensor network placement optimization should be conducted on the premise of maximizing *P*, but minimizing *M*. In the case that it is aimed at *M* minimization, the relevant constraint condition can be described as fault detection probability no less than the required value; by contrast, if the objective of it is to maximize fault detection probability *f(M)*, such a condition turns into the total mass of this sensor network no more than the specified value.

Provided that the system detection rate is not below 90%, the number of sensors can be equal to a variety of values. Nevertheless, RLV has a very strict requirement for system mass control. Considering this, the sensor quantity should be proper, not too much. As a result, a scheme utilizing the minimized quantity (i.e., mass minimization) should be selected from various solutions up to relevant requirement for placement. In this condition, the mathematical model of such an optimization problem is simplified as follows,

$$\begin{aligned} & Min\ (M(n_1, n_2, n_3)) \\ & s.t. \quad P \geq 0.9, x, y, z \in [u, v, w] \text{ and } n_1, n_2, n_3 \in N \end{aligned} \tag{1}$$

where $P \geq 0.9$ is referred to as inequality constraint acting as a constraint condition of this optimization model, signifying TPS fault detection probability achieved by the sensor network; n_1, n_2, n_3 are optimization or decision variables respectively, that is the number of sensors; $[u, v, w]$ is space R^n subjected to a non-void condition, if solved in a set *D* and $D = \{(n_1, n_2, n_3) | n_1, n_2, n_3 \in N\}$, *D* stands for the feasible region of this optimization problem. After the optimal number of sensors (i.e., optimal sensor network mass) has been achieved by solving this problem, system detection rate in the corresponding condition can be figured out.

Mass distribution of the entire TPS should be under control. Besides, the proper sensor quantity and placement position be defined, so as to gain a maximum TPS fault detection rate. Contrary to the optimization model described above, such a model is expressed as follows,

$$\begin{aligned} & Min\ (-P(n_1, n_2, n_3, J)) \\ & s.t. \quad M \leq M_{\max}, x, y, z \in [u, v, w] \text{ and } n_1, n_2, n_3 \in N \end{aligned} \tag{2}$$

where $M_{\max}$ is the upper limit of sensor network mass, considered to be the constraint condition of this optimization model. Likewise, it is solved in a set *D*, $D = \{(n_1, n_2, n_3) | n_1, n_2, n_3 \in N\}$ and *D* serves as the feasible region of the related optimization problem. In order to obtain the optimal system detection rate, this problem should be resolved to further figure out the number of sensors and their placement positions (i.e., the optimal sensor network mass distribution) under the corresponding circumstance.

Optimal RLV sensor network placement design is a multidisciplinary and multi-objective optimization problem involving subject knowledge of flight vehicle design, sensor selection, test system design and fault detection, etc. As for sensor placement of TPS, it is a broad problem of massive sensors and should be oriented by local optimization and minimum fault isolation in line with real needs. We discuss the detection rate based optimal sensor placement for thermal protection structure.

When the finite element mode has been established for the thermal protection structure together with the determination of corresponding parameters, each node of this model represents one possible sensor placement position, the measured value (or statistical or sensitive information of distortion) of which can be obtained through finite element analysis. As for sensitive data of other parameters, such as model form and stiffness degradation, it is probable for us to acquire them by virtue of the FEM model that takes material performance reduction into account.

Sensor placement is meant to find a set of sensor placement positions to guarantee that sensors distributed on the thermal protection structure are able to accurately identify state of the structure on the premise of satisfying identification probability required by analysis and evaluation. In general cases, the approach to optimal sensor placement is described as follows. "n position(s) to be selected and a sensor(s) to be utilized are given on the premise of $a = n$ to ensure the highest detection capability."

Optimal sensor placement realized by Snob fit is an optimization scheme applicable to boundary restrictions of the target function subjected to interference. Major advantages of adopting such optimization software lie in the fact that only the following optimization problem needs considering without determining positions of sensors to be selected in advance.

$$\begin{aligned} &\min f(x) \\ &s.t. x \in [u, v] \end{aligned} \tag{3}$$

where x is continuous and $[u,v]$ constrained in non-void space R^n.

The above equation of optimization problem plays a role in defining the position of one sensor. x refers to coordinate of a sensor placed to recognize configuration state at the maximum probability when the thermal protection structure has been sound or damaged. Px denoting correct configuration state recognition probability is acquired to realize application of optimization function in Equation (3) applied, in which case, $f(x) = -Px$ and restricted space $[u,v]$ of x depends on physical dimension of the structure. So optimal sensor placement can be determined by solving this equation.

Simulation analysis on local thermal protection structures is implemented by Ansys. Principle of transient analysis is to motivate the structure by a sinusoidal alternating dynamic load, so that its vibration frequency changes to 600Hz in 0.2s. As auxiliary input of an active detection algorithm, such a excitation is a normal random variable with a mean value of 0.254mm and a variable coefficient of 0.5. Considering that the excitation only acts on one position, the piezoelectric actuator, there is no need to regard this parameter as a random process or field. Finite element analysis was repeatedly loaded for the model to provide sample sources for further statistical analysis. Sound operating state and damaged operating state were respectively simulated with 30 excitations each, which could be realized by identical random excitation amplitude, so as to set up mean eigenvector and feature covariance matrix required by damage recognition and detection algorithm. After that, 30 random excitations were utilized to simulate 30 sound and 30 damaged operating states separately to estimate advantages and disadvantages of sensor placement.

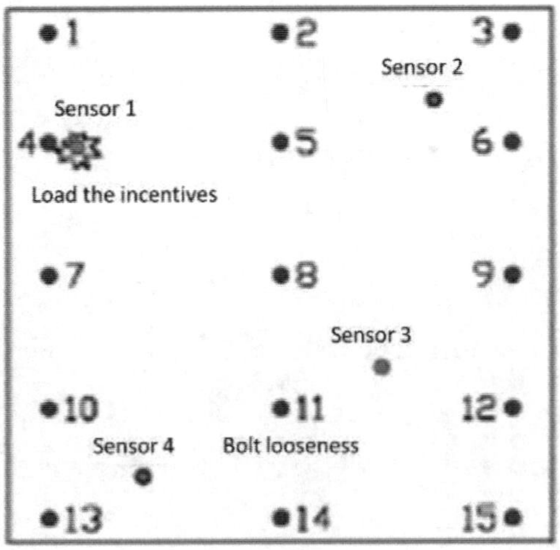

Figure 1. Typical Sensor Placement of Insulation Tile

Figure 1 presents the typical sensor placement. That located at sensor 1 is a fixed loading excitation point, while those at sensors 2, 3 and 4 are measuring points whose positions keep changing. To connect the insulation tile and the fuselage, 15 bolts were put into use; under this circumstance, bolt looseness and damage are incurred at point 11.

A series of features were extracted from time displacement data of analog input and they corresponded to signals of given parameters. Features that have been selected in this paper consisted of time and frequency domains. In detail, the former is generated by response autocorrelation and cross correlation of 3 different sensors. Time data of such 3 sensors were adopted to gain three functions of autocorrelation and three functions of cross correlation, where all these functions have definite shapes; meanwhile, the first two order responses of correlation functions could be also achieved by 4th central moment. As a result, 18 features in time domain were produced. Likewise, autocorrelation and cross correlation functions of the sensors' frequency domains were obtained dependent on frequency domain features; with the first order response of the correlation function acquired by 4th central moment as well, 12 frequency domain features are gained. In total, there is a 30-dimension eigenvector used for state classification. Furthermore, a state classifier was acquired by Bayes' decision theory and detection rate maximization. Classification of each variety of configuration states should be based on mahalanobis distance shown in equation (4).

$$d_j(x) = (x-\mu_j)^t \sum_j^{-1} (x-\mu_j) \tag{4}$$

where j serves as an index of configuration state, x as eigenvector for classification, μ_j and $\sum_j$ respectively as mean vector and feature covariance matrixes for "learning" data of the given configuration state. The feature covariance matrix must be nonsingular, because inverse matrix of $\sum_j$ is demanded by the mahalanobis distance. To ensure it, some features should be removed from such series. Finally, relevant states were classified by a 12-dimension eigenvector. Based on outcomes of the first round of sound and damaged operating state simulation with 30 excitations, output feature vector recognition and classification foundations were established for sound and damaged states.

Then, evaluations based on classification equation for sound and damaged operating state simulation with 30 excitations each were performed to continue state classification. Additionally, relevant states were judged by the minimum value. A set of sensor placement generated a configuration state detection classification matrix correspondingly.

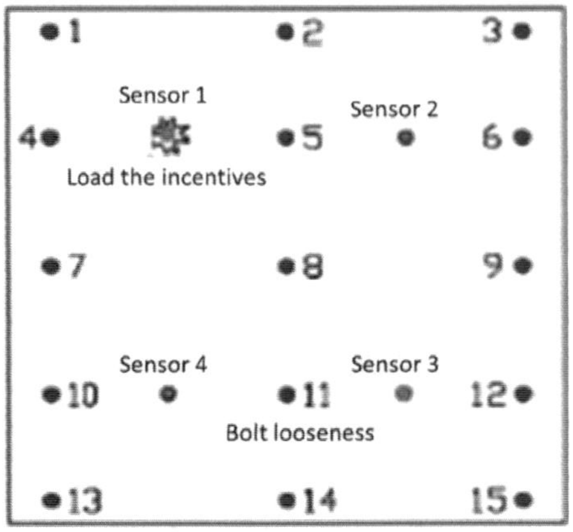

Figure 2. Schematic Diagram of Sensor Placement Optimized

According to this classification matrix, both fault detection rate and false alarm rate of a corresponding sensor array can be figured out, which should be repeated targeted at different kinds of sensor placement to realize optimal sensor placement by Snob fit as an optimization toolkit of MATLAB. Figure 2 is a schematic diagram of sensor placement after the optimization. In this case, positions of four sensors are in the middle of 4 and 5, 5 and 6, 10 and 11, and 11 and 12 respectively. Through calculations, fault detection rate turned out to be 97.8% and the false alarm rate to be 3.0%.

4. SUMMARY

(1) According to the special requirements of RLV heat-resisting system, the fiber Bragg grating high temperature sensor is selected as the temperature measurement sensor.
(2) The optimum design criterion for the sensor layout of the heat-resistant system is minimization criterion of identification error, model reduction criterion and interpolation fitting criterion.
(3) The sensor arrangement of the heat-shield system is based on the actual requirements of local optimization and minimum fault isolation as the main guiding ideology. In this paper, detection rate based optimal sensor placement for thermal protection structure is studied.

REFERENCES

[1] Hu Changshou 1999 *Handbook of aerospace Reliability Design* (China:China Machine Press).
[2] Streed, E. R. , Cunningtont, G. R. , & Zierman, C. A. . 1966 Performance of multilayer insulation systems for the 300° to 800° k temperature range. *Thermophysics & Temperature Control of Spacecraft & Entry Vehicles*, 735–71.
[3] Keller, K. , Hoffmann, M. , W. Zörner, & Blumenberg, J. . 1992 Application of high temperature multilayer insulations. *Acta Astronautica*, **26(6)** 451–8..
[4] Daryabeigi, & Kamran. 2002 Thermal analysis and design optimization of multilayer insulation for reentry aerodynamic heating. *Journal of Spacecraft and Rockets*, **39(4)** 509–14.
[5] Darya, B. K. 2002 Heat transfer in high-temperature fibrous insulation. *Journal of Thermophysics & Heat Transfer*, **17(1)** 10-20.
[6] Blosser, M. L. , Martin, C. J. , & Myers, D. E. 2013 *Parametric Weight Comparison of Current and Proposed Thermal Protection System (TPS) Concepts*. NASA Langley Technical Report Server.
[7] Shih, P. K. , Zwan, A. D. , & Kelley, M. N. 2013 Thermal protection system optimization for a hypersonic aerospace vehicle. *Thermophysics, Plasmadynamics & Lasers Conference*.
[8] Chen, Y. K. , & Milos, F. 2006. Solution strategy for thermal response of nonablating thermal protection systems at hypersonic speeds. *Aerospace Sciences Meeting & Exhibit*.
[9] Prosser, W. H. . 2003 Development of structural health management technology for aerospace vehicles. *Nasa Larc Jannaf Cs/ Aps/ Pshs/ Mss Joint Subcom Meeting*, **31(30)** 1–5.

[10] Milos, F. S. , Watters, D. G. , Pallix, J. B. , Bahr, A. J. , Huestis, D. L. , & Kundu, T. . 2001 Spie proceedings [spie 6th annual international symposium on nde for health monitoring and diagnostics - newport beach, ca (sunday 4 march 2001)] advanced nondestructive evaluation for structural and biological health monitoring - . *Proceedings of SPIE - The International Society for Optical Engineering*, **4335** 74-82.

[11] Graue, R. , Reutlinger, A. , & Werner, J. 1999 Tps health monitoring on X-38. *Proceedings of SPIE - The International Society for Optical Engineering*, 3668..

[12] Luo, H., & Hanagud, S. . 2006. Detection of debondings in thermal protective system tiles by using nonlinear structural dynamic response. *Aiaa/asme/asce/ahs/asc Structures, Structural Dynamics, & Materials Conference & Exhibit.*

[13] Trivedi, A., Mitra, K., & Subramanian, C. . 2013 Numerical and Experimental Study of Non-Destructive Detection of Thermal Tile System Debonding Using Short-Pulse Laser. *Aiaa Thermophysics Conference*.

[14] Cooper, E. G. , Jones, S. M. , Goode, P. W., & Vazquez, S. L. 1992 Automated anomaly detection for orbiter high temperature reusable surface insulation. *Proceedings of SPIE - The International Society for Optical Engineering*, **1829**, 330–40.

IOP Conf. Series: Materials Science and Engineering **686** (2019) 012021 doi:10.1088/1757-899X/686/1/012021

Tool for Teaching Physical Model Making in Product Design

Supradip Das* **and Amarendra Kumar Das**

Department of Design, Indian Institute of Technology Guwahati

*supradip.das@iitg.ac.in

Abstract. As physical prototypes and models can solve many design issues during the New Product Development (NPD) process, it has been introduced as an essential course in many design schools. With the advancement of 3D CAD technologies students find it easier to make digital models, but they find the physical prototyping and model making difficult. This difficulty arises due to lack of understanding of materials, processes and product form correlation required for physical prototyping. In order to address this issue, a tool has been developed and introduced to design students. The tool would guide them to select materials and processes, based on the product form. The tool is a matrix of variables viz. product form, processes and materials. To formulate this matrix, altogether three hundred fifty numbers of products were considered, and clustered them according to their form and processes. On this basis, possible materials have been identified for fitment and suitability of the processes. The tool has been tested with two groups of design students in the course of 'modelling and simulation'. The effectiveness of the tool has been analysed by conducting an independent t-test and also the Felder-Silverman Index of learning. It has been observed that the tool is highly effective. This paper discusses this effort in details.

1. Introduction

Physical model making is used for different purposes in different disciplines. With the technological advancement, the model making process has changed over time, but physical model making is the process, which is still used by designers in professional work and continues to be embraced [1]. Physical model making is extensively and continuously used by product designers because models are not merely a representation method, but it also makes complex realities comprehensible, operational and workable. Models can also act as exploratory devices by working as self-communication tools and allowing new questions to be raised, alternative solutions to be explored, hypotheses to be formulated [2]. The construction of 3D physical models has some advantages for the product design engineer, such as: visualizing design concepts, communicating ideas, stimulating creativity, aiding understanding, identifying real fabrication difficulties, allowing form modifications and form generation. In product design students learn through 'learning by doing', and model making helps them to take the idea "out of the paper" for visualizing it better and especially for making it tangible [3]. Despite that, novice product designers are not using models or prototypes always in all design phases [4]. Novice product designers are considering prototypes to be made towards the end of the design process to test and evaluate only, rather exploiting the advantages of prototypes in the whole design process [5]. This is true also for the Department of Design, IIT Guwahati. This paper illustrates the research work to overcome issues regarding physical model making for students in design.

2. Problem in physical model making

Prior research explicates that novice student has a narrow perception of a prototype [6], but it is also important to check the understanding of prototype making among novice designers. To identify the problems faced by design students for physical model making a study was conducted with students of the second year second semester Bachelor of Design students of the Department of Design, IIT Guwahati. For the rest of the paper, these students will be considered as Batch1. There were a total of 36 students in the Batch1. These students were admitted through the Joint Entrance Examination of India. They all had taken the same basic courses offered by the institute in the first year. In the second year odd semester also they all had taken the same core courses offered by the Department of Design. It is important to mention that they all had a compulsory course Material and Processing. The study was conducted in four steps. These are:

- Step 1. Students of B.Des II year were asked to select iconic furniture design to make a model of it.
- Step 2. Students were allowed to use workshop facility (machine tools, hand tools, equipment and instruments required) to prototype, but to take suggestions or help from workshop instructors.
- Step 3. Qualitative data of the output of the individual students were recorded.
- Step 4. Students were asked to write about their experiences (good/bad) of model making.

After the study, all the students' voices were converted into technical voices. These are represented in Table 1:

Table 1. Problems in the model making among novice designers and frequencies

Problems	Frequencies
1. Problem in the selection of material	10
2. Problem in the selection of processes	8
3. Problem in the selection of both material and processes	9
4. Problem in the selection of hand tools	2
5. Problem in the selection of finishing material	2
6. Problem in the selection of finishing techniques	1
7. Others	4

To overcome come such issues and facilitate students in physical model making, a structured tool is required. Aiming to make a structured tool to help beginning design students, a systematic literature review was carried out.

3. Background study

Systematic background study had been done as defined by Okoli and Schabram [7].

3.1 Purpose of the Background Study

The intended objective of conducting the systematic literature review was to identify the drivers, principles or guidelines and support tools for model making. For this, the following definitions were considered:

- Drivers in the model making are those dimensions, which are intrinsic or extrinsic to prototype and prototyping process [8].
- Design principles, guidelines or heuristics are criteria, method or decision factors, which give direction to a particular process [9].
- Tools [10] are an implement that you employ to increase efficiency in one or several phases of the process in the form of manual, framework, card or computer-based application. In this context, background study had been done for universal principles or guidelines for design tools also, which will help in making a structured tool for model making.

3.2 Practical Screen
Total one hundred and thirteen references were searched and sixty research publications were selected after filtering. The considered literature spans from 1983 to 2019, where 52% (31 items) of the considered publications are from journals, 25% are from books or book chapters, 20% are from Conference Proceedings and only 3% are from Gray literature.

3.3 Data Extraction
Data had been extracted from the considered literature based on three aspects: 1. Drivers, 2. Principles/ Guidelines, 3. Tools.

3.3.1 Drivers: To understand the drivers in model making, it is very important to understand the characteristics of the model/prototype of model making/prototyping process.

Houde and Hill [11] presented a model which represents what prototypes model/prototype. The model represents a three-dimensional space, which corresponds to important aspects of the design of an interactive artefact. He defined three dimensions of the artefacts as 1.role; 2.look and feel; and 3.implementation. "Role" refers to the function, with which it is useful to the user. "Look and feel" denotes the sensory experience of using an artefact. "Implementation" refers to the techniques and components through which an artefact performs its function. Moreover, Lim [8] describes the anatomy of model/prototype with two different dimensions such as filtering dimensions and manifestation dimensions. Filtering dimensions give an understanding of what to filter and consider while crafting a prototype. But it is also important to consider how to make a model/prototype. Material, resolution and scope determine the specifics of how to make a model/prototype.

3.3.2 Principles/ Guidelines/ Criteria in Model Making/ Prototyping Otto and Wood [12] suggested a list of criteria while selecting materials for prototyping. These are 1. Cost, 2. Availability, 3. Ability to accept changes, 4. Ease of use and forming capability, 5. Scalable geometry, Scalable properties.

3.3.3 Tools (Support) for Model Making/Prototyping. In the recent past, Jessica Menold proposed a framework for prototyping to support the product design process, but it does not have any support tool or method [13]. There are other frameworks available, but Posada [3] is the only design educator, who proposed a tool where material selection is based on the process of addition of material or subtraction of material. This tool does not allow students to select material and processes based on a wide variety of forms.

3.4 Synthesis of Extracted Data
The insight from the background study is illustrated in Figure 1. The figure shows the relationship between filtering dimension, Manifestation dimension, Principles, Processes and related tool.

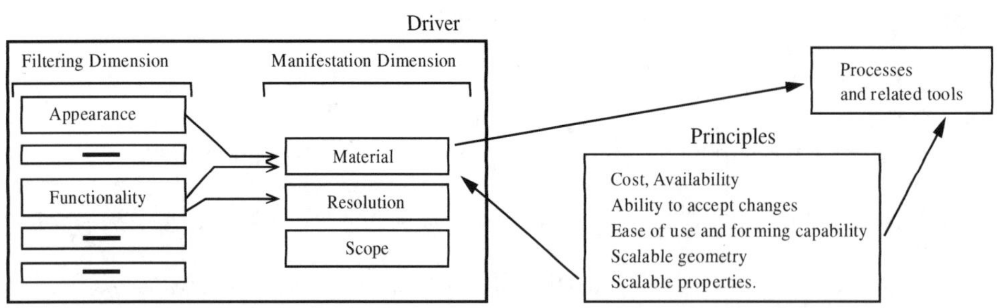

Figure 1. Relationship between filtering dimension, Manifestation dimension, Principles, Processes and related tool

Background study reveals that the tool must explain the material and processes based on the appearance (form) and function, considering the prototyping principles.

4. Development of the tool

With the help of extracted data, a tool was developed to facilitate model making in furniture design (TfMM 0.1-Tool for Model Making 0.1), where material selection can be done based on dominating form and primary elements such as bar, body and surface. Followed by material selection, related process selection, hand tool and machine tool selection is also possible in the developed tool. The tool was developed systematically. Various steps for developing the tool were as under:

- Step 1. Total 350 number of iconic chairs were grouped based on forms (Figure 2). Chairs, which were not possible to group on the basis of forms, they were grouped on the basis of the processes or the common material used. A total of 40 number of groups were formed.

Figure 2. Grouping of chairs based on forms, materials or processes

- Step 2. The furniture of each group was further grouped based on the dominant element [14] between bar, body and surface [15] as shown in Figure 3.

(a) (b) (c)

Figure 3. Furniture with (a) Surface dominant, (b) Bar dominant, (c) Body dominant

- Step 3. Based on the groups, appropriate materials, processes and hand tools were assigned.

The entire process of coding was done by three coders, who are having more than eight years of experience in product design teaching. The inter-rater reliability was checked by Cohen's Kappa (k= 0.85).

The final card-based tool has three columns such as form, material & processes and hand tools. The column form further divided into three rows such as bar, body and surface. Based on the types of furniture, novice designer can select material & processes and related hand tools (Figure 4).

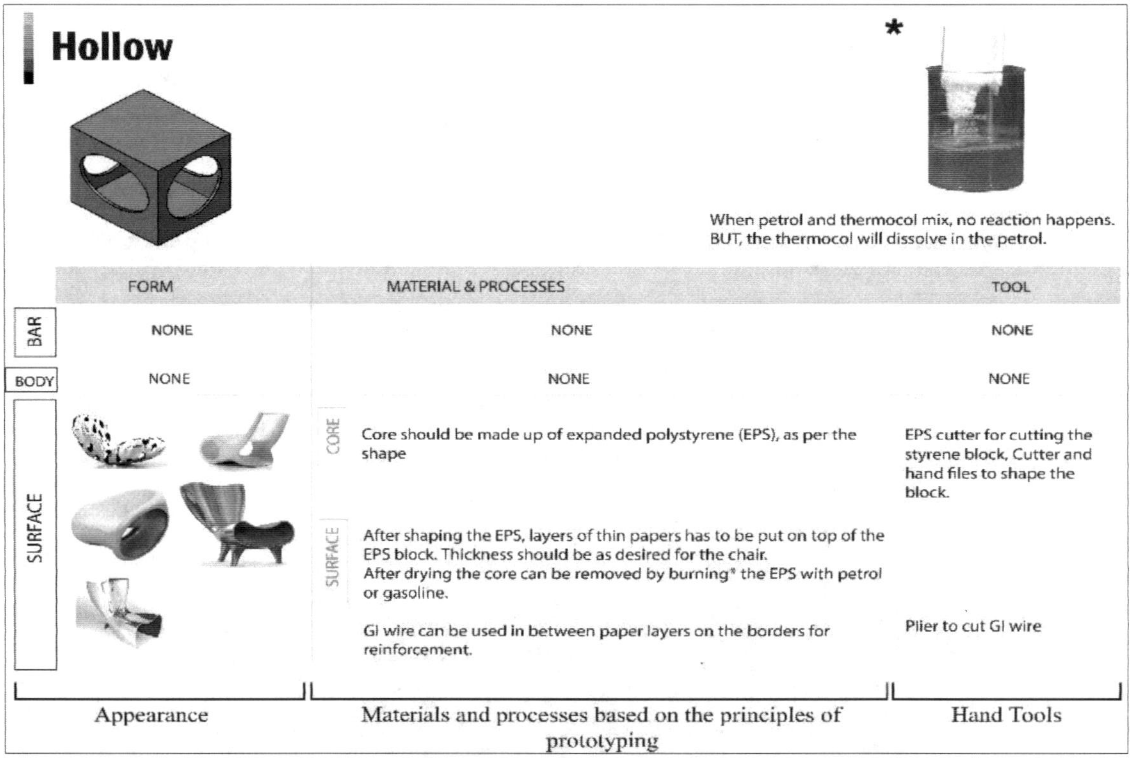

Figure 4. A card-based tool on 'Hollow' form with assigned materials, processes and tools required

5. Teaching with the tool

Students were asked to make a prototype of any iconic chair as per their choice, with the help of the developed tool. The developed tool was introduced to the Batch 2 of 50 students of the second year, second semester and the qualitative outcome was recorded (Figure 5). It was observed that the developed tool facilitate students in prototyping.

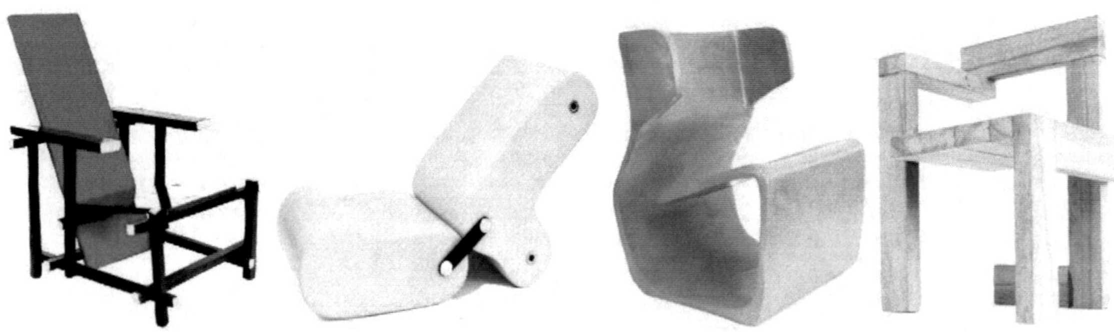

Figure 5. The outcome of the students after the introduction of TfMM

6. Testing of the tool

An independent t-test was conducted to check the impact of the tool. Independent t-test was done between two groups Batch1 and Batch 2 [16]. There was a total of 36 students in the Batch1 and 50 students in Batch2. The tool was not introduced to Batch1, but the tool was introduced to Batch 2 students. The outcomes of both batches students were scored on a scale of 1 to 10 (Table 2). A panel of three external faculties did the scoring process. To test the assumption of normality, we did the Shapiro-Wilks test (Table 3).

IOP Conf. Series: Materials Science and Engineering **686** (2019) 012021 doi:10.1088/1757-899X/686/1/012021

Table 2. Scoring Scale

10	9	8	7	6	5	4	3	2	1
Exceptional	Excellent	Very good	Appropriate	Clear	Satisfactory	Marginal Rudimentary	Insubstantial	Insufficient	Deficient

Table 3. Tests of normality

	Group	Kolmogorov-Smirnov[a]			Shapiro-Wilk		
		Statistic	df	Sig.	Statistic	df	Sig.
Grade	Batch1	.205	36	.001	.900	36	.003
	Batch2	.175	50	.001	.931	50	.006

For the above case, where a = .001, given that p = .003 for the Batch1 and p = .006 for the Batch2 – we would conclude that each of the levels of the Independent Variable (Stress Condition) are normally distributed.

To check the homogeneity of variances a Levene's F Test was conducted (Table 4). It was found that the F value for Levene's test is 3.058 with a Sig. (p) value of .084. As the Sig. value is more than our alpha of .05 ($p > .05$), it was concluded that there is no significant difference between the two group's variances.

Table 4. Independent Sample Test

		Levene's Test for Equality of Variances		t-test for Equality of Means					
		F	Sig.	t	df	Sig. (2-tailed)	Mean Difference	Std. Error Difference	95% Confidence Interval of the Difference Lower
GRADE	Equal variances assumed	3.058	.084	-4.096	84	.000	-1.27556	.31138	-1.89478
	Equal variances not assumed			-4.299	83.922	.000	-1.27556	.29671	-1.86561

The results of the independent t-test indicated that there is a significant difference in performance score between the two groups, $t(84)=-4.096$, $p =.000096$. These results suggest that individuals in the Batch2 ($M = 7.22$; $SD = 1.58166$) scored better than individuals in the control group Batch1 ($M = 5.9444$; $SD = 1.16972$) after the introduction of the tool [17]. The size of this effect ($d = -0.9170$), as indexed by Cohen's (1988) coefficient d was found to exceed the convention for large effect size ($d = .80$).

7. Learning from the tool

The Tool for Model Making (TfMM) was designed to facilitate novice designers and it has some explicit and implicit learning component.

IOP Conf. Series: Materials Science and Engineering **686** (2019) 012021 doi:10.1088/1757-899X/686/1/012021

7.1 Explicit learning from the tool

- Learning about the physical prototyping
- Understanding of required material and processes according to the form

7.2 Implicit learning from the tool

- Types and constructs of furniture

To check the effectiveness of the tool in a course, students' response to Felder-Silverman index of learning [18] was recorded based on the four domains such as processing information, perceiving information, receiving information and understanding information. The average score is illustrated in Figure 6, Figure 7, Figure 8 and Figure 9.

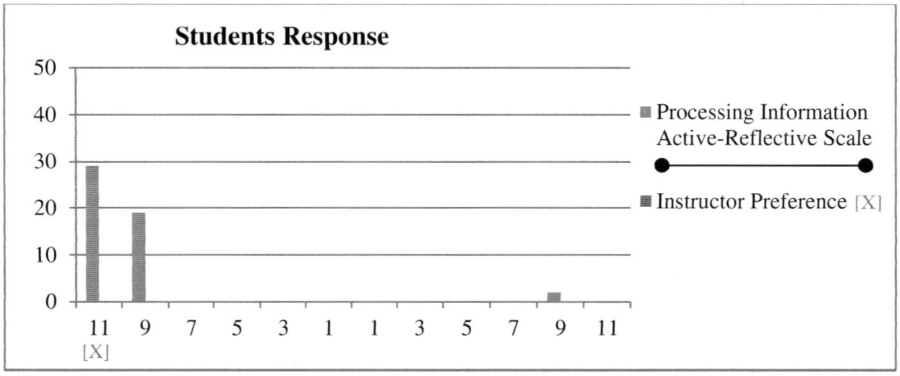

Figure 6. Felder-Silverman Index of learning for Processing Information in Active-Reflective Scale

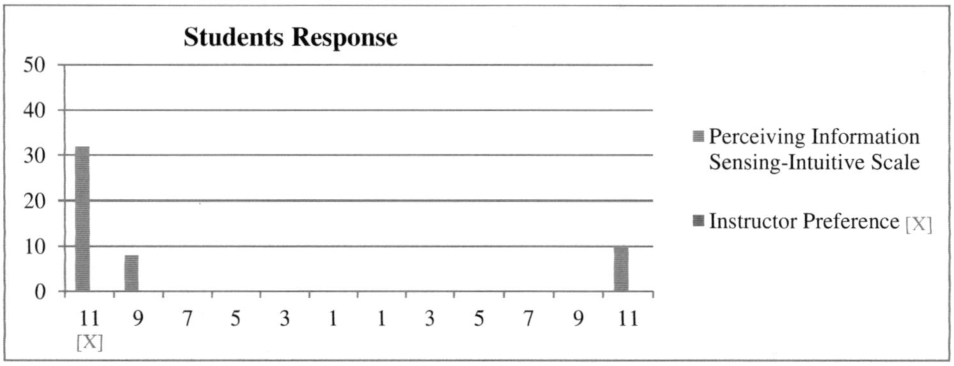

Figure 7. Felder-Silverman Index of learning for Perceiving Information in Sensing-Intuitive Scale

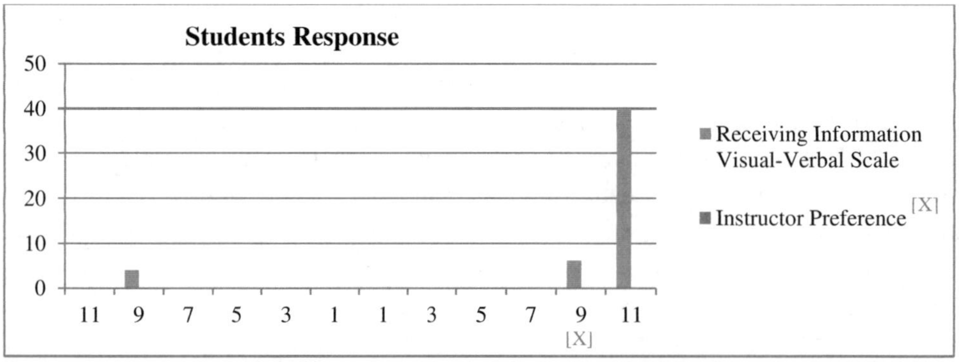

Figure 8. Felder-Silverman Index of learning for Receiving Information in Visual-Verbal Scale

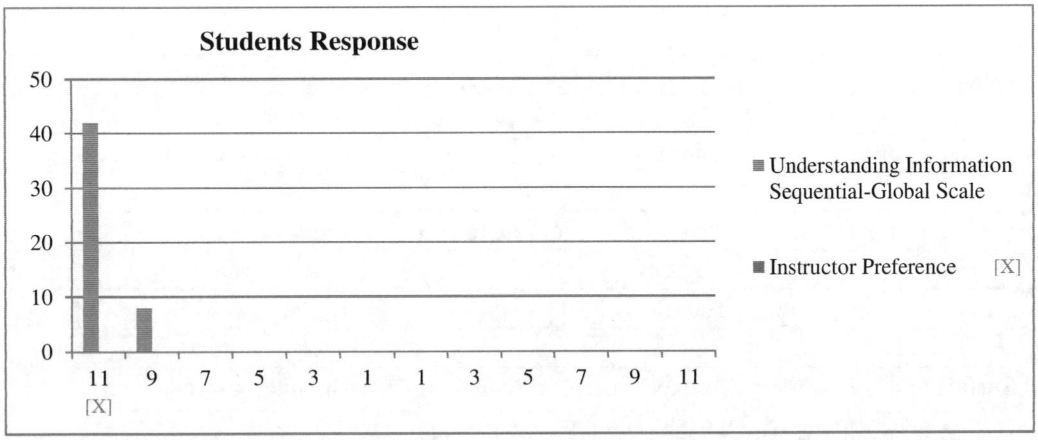

Figure 9. Felder-Silverman Index of learning for Understanding Information in Sequential-Global Scale

The tool was developed to learn prototyping techniques by active self-engagement and students' response (Figure 6) shows that they processed information by making prototypes. The developed tool has structured information with examples and the majority of the students perceived information through that (Figure 7), however, a group of students (Total No: 10) found that the grouping of the furniture also triggers ideas to develop new furniture. Students also received information from the verbal information given in the tool, but few students received information from the visual information (Figure 8). The tool also helped them making prototype sequentially, by selecting material and process (Figure 9).

8. Future scope

It was observed that the tool may be used as furniture design heuristics and it will be tested for its effectiveness as design heuristics for new furniture development. Future work will identify the aspects of design heuristics and refine the tool for a new version as heuristics.

9. Discussion and conclusion

When we are discussing 4th industrial revolution and prototype driven innovation [19], it is very important to empower beginning designers with the knowledge of manufacturing and processing for prototyping. The TfMM is developed towards prototype driven innovation for the beginning designers to provide the manufacturing and process know-how. The tool TfMM was developed from an analysis of existing chair geometries and the tool is then used to prototype existing chair again because the derivation was done by product design expert and hence it is important to check whether students understood the tool or not.

The tool has three-segment 'form', 'material and processes' and 'hand tool', where form guides the material and process selection. The effectiveness of the tool was tested through a t-test and it was found that the developed tool is fulfilling its objective to guide students for prototyping.

In the first version of the tool, a set of 40 cards were developed (Figure 10) and these are mentioned in Table 5. This is a unique tool in the field of product design (specifically furniture design) and will be helpful for product designers.

As the tool was developed by grouping furniture based on the forms and geometries, it also triggers ideas for a new design, which indicates that the tool may be used as heuristics [20] for developing new forms of furniture. The effectiveness of the tool as design heuristics has to be tested scientifically and qualitatively in future.

Table 5. List of the cards

Polyhedron	Space Truss	Wave Surface
Origami	Shaft to Blade Loop	Reinforcement
Street Pattern Tower	Cantilever	Continuous Block
Hollow	Transformed Uniform Grid	Transformed Non-Uniform Grid
3D Print	Tensile Surface Structure	Möbius Loop
Stack Body	Stack Bar	Stack Surface
Tensegrity	Hand Fan	Bulge/ Swelling
Wiggle	Bent Surface	Peeled Surface
Stud Surface	Locked String or Rope (Knot/ Knit/ Weave)	Brush Stroke
Nervous System	Revolving Bundle	Comet Structure
Locked Cushion (Knot/ Knit/ Weave)	Swollen Street	Bulge Surface
Genealogical Tower	Bars & Panels	Collapsible
Inflatable	Flat-Pack	Log-Joint
Transformed Body		

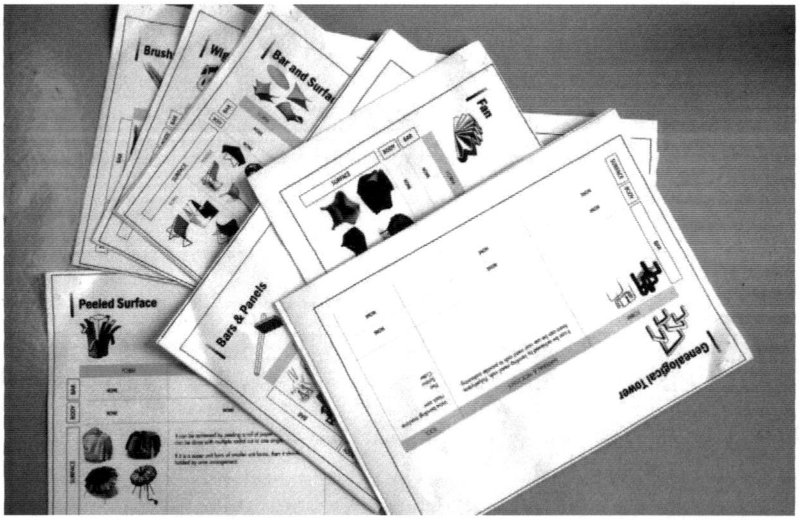

Figure 10. Tool for Model Making Cards

References

[1] Hallgrimsson B 2012 *Prototyping and Model Making for Product Design* (London: Laurence King Publishing Ltd)
[2] Cannaerts C 2009 Models of / Models for Architecture: Physical and Digital Modelling in Early Design Stages *27th eCAADe Conference Proceedings* pp 781–6
[3] Velásquez-Posada A 2005 MODEL MAKING TECHNIQUES AS A TEACHING TOOL IN PRODUCT *Crossing Design Boundaries* (Taylor & Francis) pp 445–51
[4] Deininger M, Daly S R, Sienko K H and Lee J C 2017 Novice designers' use of prototypes in engineering design *Des. Stud.* **51** 25–65
[5] Hamon C L, Green M G, Dunlap B, Camburn B A, Crawford R H and Jensen D D 2014 Virtual or physical prototypes? Development and testing of a prototyping planning tool *ASEE*

Annual Conference and Exposition, Conference Proceedings (Indianapolis)

[6] Lauff, Carlye, Daria Kotys-Schwartz and M E R 2017 Perceptions of prototypes: pilot study comparing students and professionals *International Design Engineering Technical Conferences and Computers and Information in Engineering Conference* (American Society of Mechanical Engineers) p V003T04A011-V003T04A011

[7] Okoli C and Schabram K 2010 A Guide to Conducting a Systematic Literature Review of Information Systems Research *Sprouts Work. Pap. Inf. Syst.* **10**

[8] Lim Y-K, Stolterman E and Tenenberg J 2008 The Anatomy of Prototypes: Prototypes as Filters, Prototypes as Manifestations of Design Ideas *ACM Trans. Comput. Interact.* **15** 1–27

[9] Katherine K, Yang M C, Design K L W, Fu K K, Yang M C and Wood K L 2015 Design Principles : The Foundation of Design *International Design Engineering Technical Conferences and Computers and Information in Engineering Conference* p V007T06A034-V007T06A034

[10] Lutters E, Van Houten F J A M, Bernard A, Mermoz E and Schutte C S L 2014 Tools and techniques for product design *CIRP Ann. - Manuf. Technol.* **63** 607–30

[11] Houde S and Hill C 1997 What do prototypes prototype? *Handb. Hum. Comput. Interact.* 1–16

[12] Otto K N and Wood K L 2003 *Product Design: Techniques in Reverse Engineering and New Product Development* (Pearson Education India)

[13] Menold J, Simpson T W and Jablokow K W 2016 The Prototype for X Framework: Assessing the Impact on Desirability, Feasibility, and Viability of End Designs *ASME 2016 International Design Engineering Technical Conferences & Computers and Infomation in Engineering Conference* (Charlotte, North Carolina)

[14] Hannah G G 2002 Elements of Design: Rowena Reed Kostellow and the structure of visual relationships

[15] Roland Knauer 2008 *Transformation : Basic Principles and Methodology of Design* (Switzerland: Birkhauser)

[16] Viswanathan V K C 2012 *Cognitive Effects of Physical Models in Engineering Idea Generation* (Texas A&M University)

[17] Hinton P R 2014 *Statistics Explained* (London and New York: Routledge)

[18] Willems J 2011 Using learning styles data to inform e-learning design : A study comparing undergraduates, postgraduates and e-educators *Australas. J. Educ. Technol.* **27** 863–80

[19] Schrage M 1996 The Culture(s) of Prototyping *Bringing Design to Software* (New York: ACM Press) pp 191–205

[20] Yilmaz S, Daly S R, Seifert C M and Gonzalez R 2016 Evidence-based design heuristics for idea generation *Des. Stud.* **46** 95–124

IOP Conf. Series: Materials Science and Engineering **686** (2019) 012022 doi:10.1088/1757-899X/686/1/012022

CNC Planing Process for Turbine Twisted Blade Profile

Tao Ding[1], Jia Li[1,*], Yongquan Jin[1] and Juncheng Zhang[2]

[1]Key Laboratory of Machinery Theory and Equipment Design, Ministry of Education, Tianjin University, No.135, Yaguan Road, Jinnan District, Tianjin, China
[2]Harbin Turbine Ltd., No. 345, Three Power Roads, Xiangfang District, Harbin, China

[*]Corresponding author: jli@tju.edu.cn

Abstract. In this paper, a multi-axis follow-up CNC planing process for twisted blade profile is proposed to reduce the machining cost and achieve high machining efficiency. Firstly, the planing process of the turbine twisted blade is analysed to determine the planing shaping movements; then the machine layout and CNC planing system needed to realize the shaping movements are constructed; then the implementation method of the multi-axis follow-up function which is unique to the CNC planing system is studied; Finally, a proof experiment of this machining process has been done to verify its feasibility.

1. Introduction

Turbine twisted blade is a kind of complex curved surface parts that usually machined on the five-axis machining center. This will lead to lower efficiency and higher cost. However, it is not only the cost of equipment and tools can be reduced, but also the impact and vibration caused by cutter teeth cut-in and cut-out in milling can be avoided while planing process is carried out. Also the efficiency of material removal will be higher than that of conventional milling, and the economic benefits of turbine manufacturing enterprises will be improved.

Up to now, the planing of surface has been realized for several kinds of simple surface, including cylindrical, conical, helical surface and tooth surface of spur bevel gears. For cylindrical surface machining, Yuan and Wang adopt the method of fixing the X and Y axes in cutting stroke and feeding the X and Y axes in non-cutting stroke [1,2]; Wang and Shen adopt the method of feeding the Y axis in cutting stroke and feeding the X axis in non-cutting stroke [3,4]. For the machining of conical surface and cylindrical surface, the planing tool moves in a straight line relative to the workpiece [5,6]. For helical surface machining, Ueno and Wang adopt the method of Z-axis and C-axis two coordinates linkage, and the tool makes helical motion relative to workpiece [7,8]. For the machining of spur bevel gears, Li adopts the method of planar envelope planing and the Z axis of the planer's main motion participates in the multi-axis linkage [9]. Except for reference [6] whose main motion of planer is driven by servo motor with ball screw, the main motions of planer used in all the references are driven by hydraulic cylinder, just like that of common planer.

The surface of twisted blade is much more complex than those surface. This is not only the shape and area of its cross-sections are different along the length of the blade, and there is also a relative torsion between two adjacent cross-sections [10]. In reference [1,2,5,6], the cutting motion is linear, and the curved surface machined by the cutter tip along the trajectory is straight line. So the machined surface is a ruled surface. In reference [3,4], 2-axis linkage planing was adopted, but only reference [3] discusses the generation of planing tool path for curved surfaces other than cylindrical surfaces, whose

IOP Conf. Series: Materials Science and Engineering **686** (2019) 012022 doi:10.1088/1757-899X/686/1/012022

tool path is the intersection lines of a series of planes and surface. It's hard to achieve high machining accuracy in this way because it didn't take the tool radius compensation into consideration. The workpiece will be over-cut when this method is applied to surfaces other than cylindrical surfaces. The CNC planing of the helical surface cannot be solved by using the method of two-axis linkage planing in reference [7,8] because of the continuous change of the section shape of the twisted blade profile. The planar envelope planing method in reference [9] is suitable for the machining of convex surfaces, but not for the machining of twisted blades profile which consist of both concave and convex surfaces. Viewing from current situation, there is no available solution for CNC planing of twisted blade profile, which is worth exploring to improve machining efficiency for turbine blade manufacturing.

2. Planing Process analysis

As shown in figure 1, the twisted blade of a steam turbine consists of three parts: the blade tip, the blade profile and the blade root. Among them, the blade profile is the blade surface to be machined, including the convex surface part called the back arc, the concave surface part called the inner arc, and the transition arc surface between them, called the leading edge and the trailing edge.

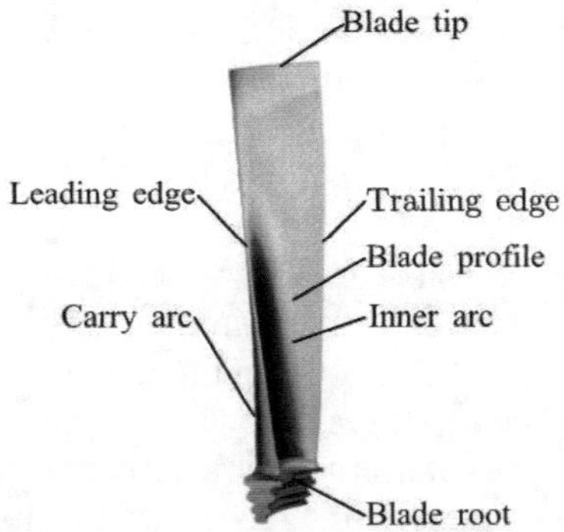

Figure 1. Basic structure of twisted blade in steam turbine.

Obviously, the curvature distribution of the blade profile is very uneven. The curvature distribution from leading edge to the trailing edge varies greatly, while the curvature changes slightly from the blade tip towards the blade root. For the profile with this curvature distribution, if the horizontal planing mode shown in figure 2 is adopted, the problem of negative rake angle cutting and interference of flank surface shown in figure 3 cannot be avoided even by adjusting the workpiece swing angle. If the longitudinal planing mode shown in figure 4 is adopted, not only the problems of negative rake angle cutting and interference of flank surface can be avoided, but also the efficiency advantage of planer in machining long and narrow workpiece can be fully exerted. Therefore, it is more reasonable to adopt the longitudinal planing mode.

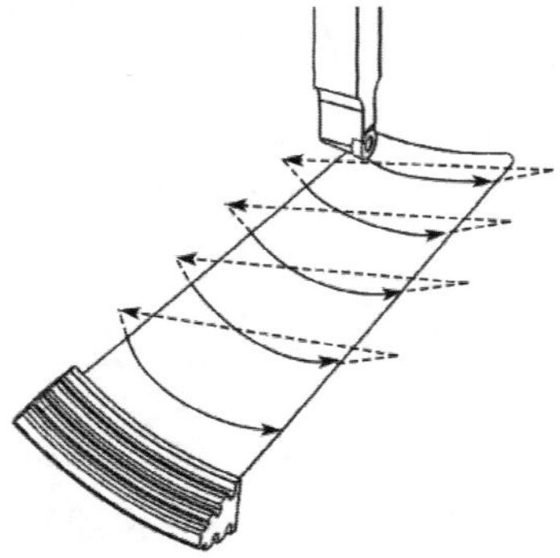

Figure 2. Horizontal planing mode.

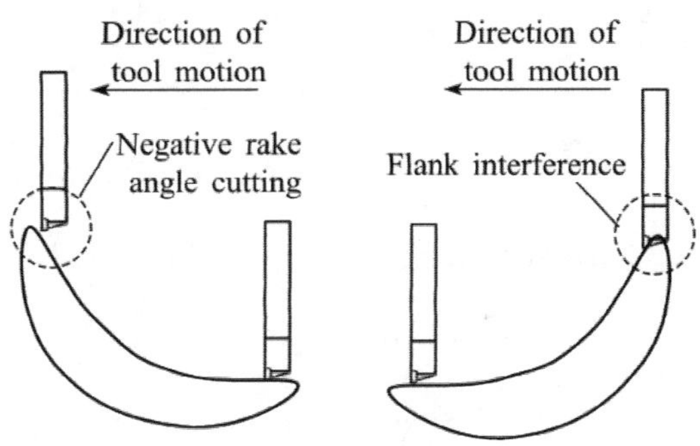

Figure 3. Tool interference in horizontal planing mode.

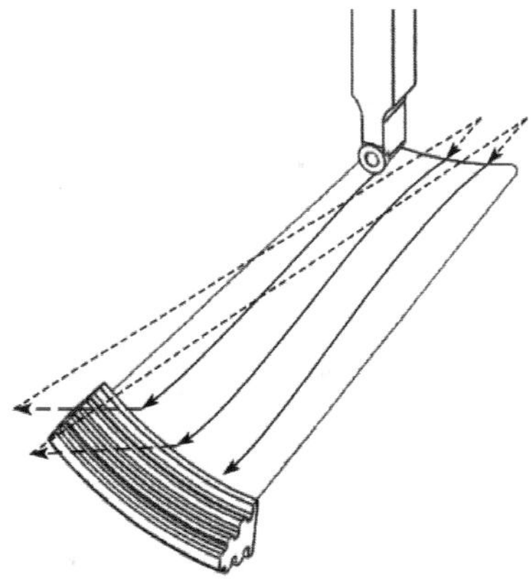

Figure 4. Longitudinal planing mode.

Considering it is difficult to meet the requirement of surface machining accuracy by using the re-grinded tool, the non-re-grinding tool is more suitable. Among various kinds shapes of non-regrinding tool blades, round blade is better for heavy cutting because of its sufficient strength. This feature is just suitable for the need of CNC planing of twisted blade profiles, and the cutting residual height of round blades is smaller. Therefore, round blade non-regrinding tool is selected as the cutting tool for the profile planing.

Because the width of the blade profile is narrowing from the tip to the root of the blade, it couldn't make lateral movement in the cutting stroke to adapt to the change of the width if using the method of two-axis linkage mentioned in reference [3]. By using the method of non-swing 3-axis linkage planing, the interference between workpiece and tool rest shown in figure 5 will occur when the workpiece is planed near the blade tip. However, the interference phenomenon can be avoided if the workpiece can swing, Therefore, four shaping movements are needed in the machining of twisted blade profile, namely, the main movement of the tool, the lateral movement of the tool relative to the workpiece, the longitudinal movement of the tool relative to the workpiece, and the swing of either the workpiece or the tool.

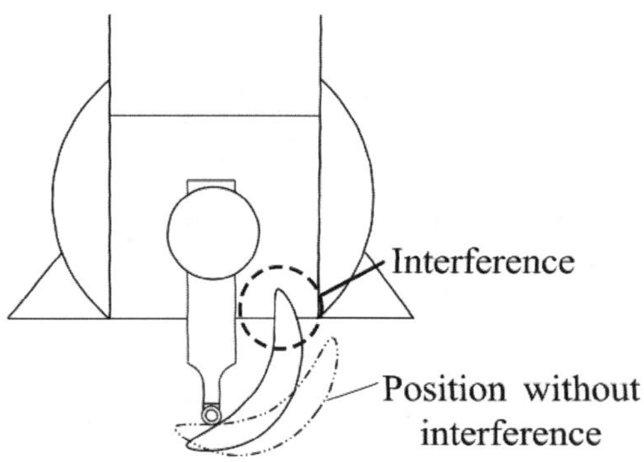

Figure 5. Interference between blade surface and tool rest.

3. Construction of Planing Machine Tool

3.1. Machine tool layout

According to the four shaping movements and their driving force, stroke, as well as the characteristics of the planing process, the layout of the shaping machine for the twisted blade profile is constructed as shown in figure 6. In which, Z-axis motion is defined as the main motion, X-axis realizes the lateral motion of the workpiece, Y-axis realizes the longitudinal motion of the cutter, and C-axis realizes the swing of the workpiece.

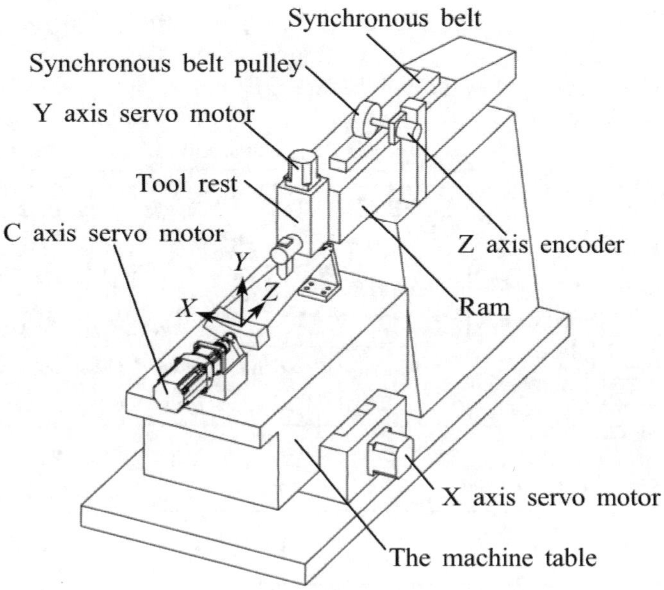

Figure 6. Machine tool layout.

As the main motion, the Z-axis should maintain the linkage with the other three movements, and it also needs to provide enough cutting force and power because the cutting force of planer is very large. To satisfy these demand, there are three choices for Z-axis driving. The first choice is to drive the ram via drive ball screw with servo motor, but the cost of heavy-duty ball screw is too high at present. The second choice is hydraulic servo drive, which not only cost a high price but also hard to achieve high accuracy. The third is the traditional hydraulic cylinder drive. The disadvantage of this driving method is that it can't accurately control the Z-axis position., but the position sensor can be installed on the ram to measure the Z-axis position. The CNC system can calculate and control the other axes movements to obtain the accuracy of the linkage motion according to the measure result. Considering the third choice, the most widely used one, provide large and stable forces with cost low, it is chosen as the Z-axis driving method. And all the axes of X, Y and C are driven by PMSM servo motor.

In the machine layout shown in figure 6, the synchronous belt, the synchronous pulley and the encoder are the elements for measuring the position of Z axis. The synchronous belt is fixed on the ram, and the synchronous pulley is rigidly connected with the encoder, which is mounted on the machine tool bed. When the ram is moving, synchronous pulley and encoder are driven to rotate synchronised with the ram's movement. The encoder generates pulse signals and sends them to the CNC system. The CNC system calculates the corresponding axes of X, Y and C according to the encoder signals, and sends them to the servo driver.

3.2. CNC system

The core function of the CNC planing system is to control servo motors of X, Y and C axes according to Z-axis position measured by encoder. Whether the servo motors can respond to the change of Z axis position in real-time will affect the accuracy of shaping movements and the machining accuracy. Therefore, the real-time performance of the field bus used in the servo driver is of vital importance. Among the mainstream field bus protocol, EtherCAT has the highest real-time synchronization performance [11]. In the CNC system products supporting EtherCAT bus, the products of Beckhoff Automation Company can make multiple slave axes synchronized to one main axis [12], which conforms to characteristics of the CNC planing system. The hardware of the CNC system is selected and constructed as shown in figure 7. The system uses EtherCAT bus to connect Beckhoff C61400 IPC, Beckhoff IO-Link terminals and Tsino-Dynatron Cool Drive A8 servo driver to form the CNC system. The pulse signals from Z-axis encoder is processed by EL5151 Beckhoff IO-Link terminal module and sent to C61400 IPC through EtherCAT bus. C61400 IPC calculates corresponding axes movements of X, Y and C according to Z-axis position and sends the control signal to servo motors of X, Y and C Cool Drive A8 servo driver through EtherCAT bus.

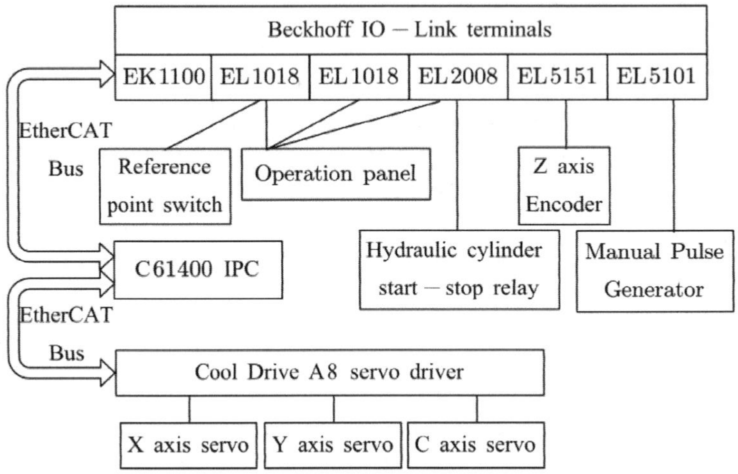

Figure 7. Hardware structure of CNC system.

For this hardware structure, the software of CNC system is developed by using Beckhoff TwinCAT development platform. The composition of software function modules is shown in figure 8. Among these functions, multi-axis follow-up function is the unique and core one, which needs further study to make a realization.

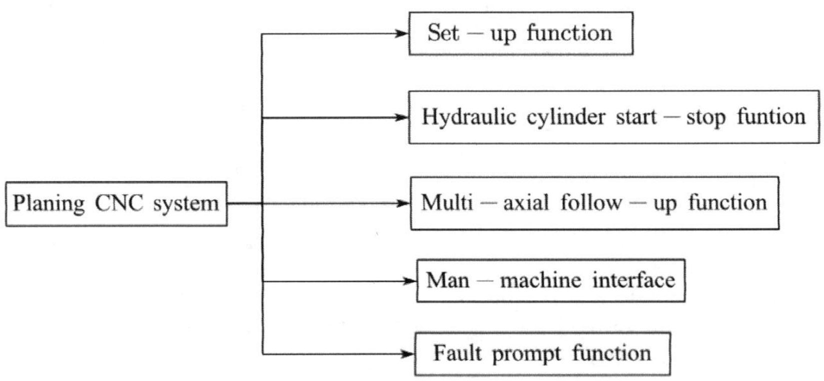

Figure 8. Software function modules.

4. Realization of Multi-Axis Follow-up Function

As mentioned earlier, the multi-axis follow-up function calculates the corresponding X, Y and C axis coordinates according to the Z axis coordinates obtained by the encoder, and sends them to the servo driver. The corresponding relationship of coordinate values of each axis depends on the cutter location and cutter axis vector, but the cutter location and cutter axis vector of CNC planer are different from that of CNC milling, which need to be calculated in a unique way.

4.1. Cutter Location and Cutter Axis Vector Calculation

For the machined surface, the cutter contact should be the contact point, so the cutter location corresponding to the contact point can be calculated according to the characteristics of the machined surface. The machined surface, that is the twisted blade profile, is expressed by bicubic B-spline surface as follow.

$$\boldsymbol{P}(u,v)=\sum_{j=0}^{n}\sum_{k=0}^{m}\boldsymbol{d}_{j,k}N_{j,3}(u)N_{k,3}(v) \tag{1}$$

In this equation, $u,v\in[0,1]$; $\boldsymbol{d}_{j,k}$ ($j=0,1,\mathrm{K},n$, $k=0,1,\mathrm{K},m$) are the control points of bicubic B-spline surface obtained from the shape points data on each section of blade; $N_{j,3}(u)$ is u directional cubic B-spline basis functions; $N_{k,3}(v)$ is v directional cubic B-spline basis functions.

By using the iso-parametric algorithms, the cutter contact trajectory can be generated on the surface, as shown in figure 9.

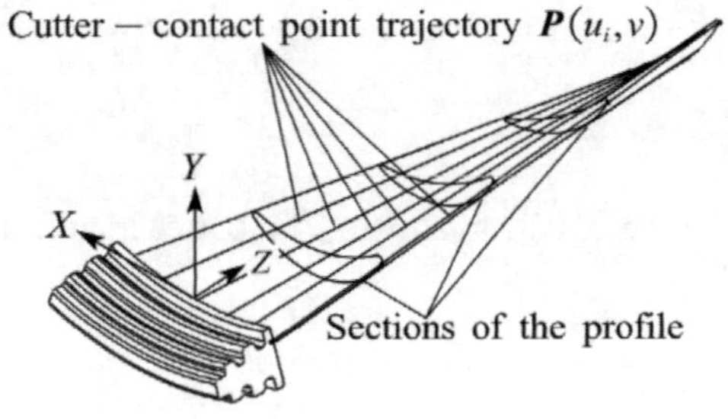

Figure 9. Cutter-contact point trajectory.

In this case, the cutter location can be obtained by offsetting a tool radius along the normal direction of the workpiece surface at the cutter contact point if the ball-end milling cutter is selected. However, the cutter location is not in the normal direction of the surface, but in the rake face of the cutter parallel to the XOY plane according to the aforementioned round blade planing cutter, so the offset direction of the tool radius needs to be calculated by the following methods.

The local coordinate system shown in figure 10 is established at the cutter contact point. The unit normal vector $\boldsymbol{n}$ of the cutter contact is rotated around the X-axis to the XOY plane to be $\boldsymbol{n}_{XOY}$. The second rotation shown in figure 11 is used to obtain the vector $\boldsymbol{n}'$, so as to avoid the local gouge phenomenon shown in figure 12. At this point, the cutter location can be obtained by offsetting the tool radius along the vector $\boldsymbol{n}'$ from the cutter contact point.

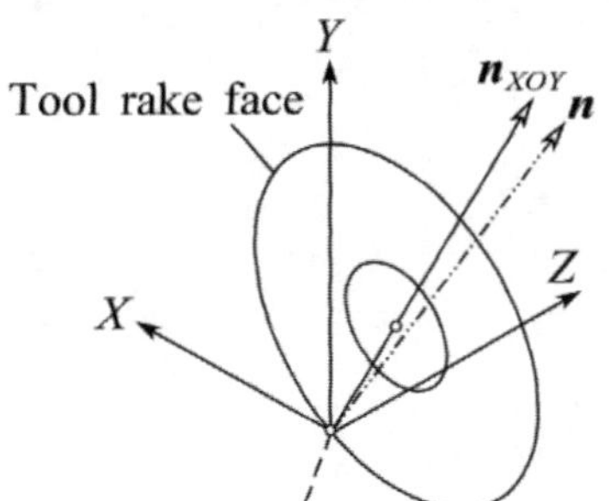

Figure 10. Local coordinate system.

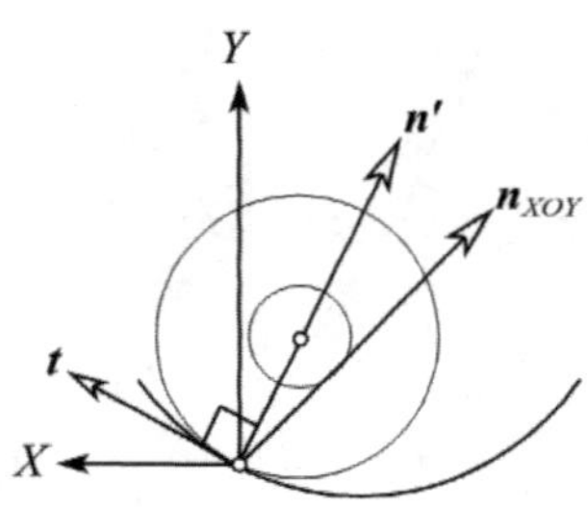

Figure 11. The second rotation of unit normal vector.

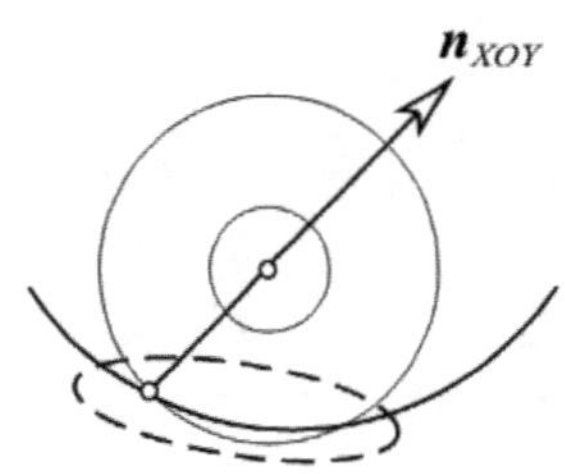

Figure 12. Local gouge.

According to the above method, the first step is calculating the unit normal vectors of surface as

$$\boldsymbol{n}=\frac{\partial\boldsymbol{P}(u,v)}{\partial u}\times\frac{\partial\boldsymbol{P}(u,v)}{\partial v}\left(\left|\frac{\partial\boldsymbol{P}(u,v)}{\partial u}\times\frac{\partial\boldsymbol{P}(u,v)}{\partial v}\right|\right)^{-1} \tag{2}$$

In the local coordinate system shown in figure 10, $\boldsymbol{n}$ is rotated around the X-axis to the XOY plane, and the vector $\boldsymbol{n}_{XOY}$ on the XOY plane is obtained by

$$\boldsymbol{n}_{XOY} = \left(n_X, \sqrt{n_Y^{\,2} + n_Z^{\,2}}, 0\right) \tag{3}$$

In the equation, n_X , n_Y and n_Y are coordinate components of $\boldsymbol{n}$.

It is noted that the axial section of the blade is parallel to the XOY plane. On this section, the unit tangent vector of the blade section curve is

$$\boldsymbol{t} = \frac{\partial \boldsymbol{P}(u,v)}{\partial u}\left(\left|\frac{\partial \boldsymbol{P}(u,v)}{\partial u}\right|\right)^{-1} \tag{4}$$

From the perpendicular relation between $\boldsymbol{n}_{XOY}$ and $\boldsymbol{t}$, as shown in figure 11, an equation can be derived as

$$\boldsymbol{t} \cdot \boldsymbol{n}' = 0 \tag{5}$$

Where $\boldsymbol{n}'$ can be solved from the equation above. The cutter location can be obtained by offsetting the tool radius along $\boldsymbol{n}'$ as

$$\begin{cases} X_{CL} = P_X(u_i, v) + R \cdot n'_X \\ Y_{CL} = P_Y(u_i, v) + R \cdot n'_Y \\ Z_{CL} = P_Z(u_i, v) \end{cases} \tag{6}$$

In the equation, $P_X(u_i,v)$, $P_Y(u_i,v)$ and $P_Z(u_i,v)$ are coordinate components of $\boldsymbol{P}(u_i,v)$. n'_X and n'_Y are coordinate components of $\boldsymbol{n}'$.

It is noticed that the cutter edge of the round blade is just tangent to the section curve of the profile at the cutter contact point as figure 13 if the cutter location sets on $\boldsymbol{n}'$, then the interference shown in figure 5 can be avoided. Therefore, $\boldsymbol{n}'$ is chosen as the cutter axis vector.

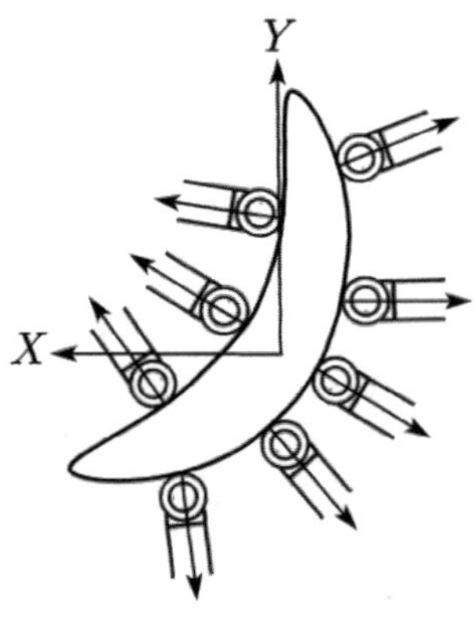

Figure 13. Tool orientation.

4.2. Post-processing of cutter location data

According to the characteristics of planer machining, the cutter location data are post-processed as follow.

As mentioned previously, the machine tool has no tool swing axis. The vector direction of tool axis is adjusted by workpiece swing, that is, the vector $\boldsymbol{n}'$ of tool axis is swing to Y direction through C-axis swing. The swing angle of C axis is the angle from the direction of $\boldsymbol{n}'$ to +Y, which is equal to the angle θ between $\boldsymbol{n}'$ and +Y direction unit vector $\boldsymbol{e}_Y$ as (shown in figure 14):

$$C_{CL}' = \begin{cases} \theta, n'_X \ge 0 \\ -\theta, n'_X < 0 \end{cases} = \begin{cases} \operatorname{acos}\left[\boldsymbol{n}' \cdot \boldsymbol{e}_Y \left(|\boldsymbol{n}'||\boldsymbol{e}_Y|\right)^{-1}\right], n'_X \ge 0 \\ -\operatorname{acos}\left[\boldsymbol{n}' \cdot \boldsymbol{e}_Y \left(|\boldsymbol{n}'||\boldsymbol{e}_Y|\right)^{-1}\right], n'_X < 0 \end{cases} = \begin{cases} \operatorname{acos}(n'_Y), n'_X \ge 0 \\ -\operatorname{acos}(n'_Y), n'_X < 0 \end{cases} \tag{7}$$

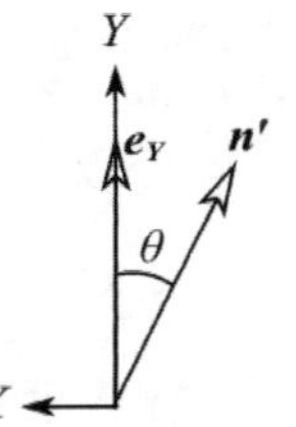

Figure 14. Tool orientation turned to +Y direction.

Combined with the C axis swings, the cutter location becomes

$$\begin{cases} X_{CL}' = X_{CL} \cdot \cos\left(C_{CL}'\right) - Y_{CL} \cdot \sin\left(C_{CL}'\right) \\ Y_{CL}' = Y_{CL} \cdot \cos\left(C_{CL}'\right) + X_{CL} \cdot \sin\left(C_{CL}'\right) \\ Z_{CL}' = Z_{CL} \end{cases} \tag{8}$$

The CNC machining data of blade profile can be obtained by equation (7) and (8) as

$$\begin{cases} C_{CL}' = \begin{cases} \text{acos}\left(n_Y'\right), n_X' \geq 0 \\ -\text{acos}\left(n_Y'\right), n_X' < 0 \end{cases} \\ X_{CL}' = \left[P_X\left(u_i,v\right) + R \cdot n_X'\right] \cdot \cos\left(C_{CL}'\right) - \left[P_Y\left(u_i,v\right) + R \cdot n_Y'\right] \cdot \sin\left(C_{CL}'\right) \\ Y_{CL}' = \left[P_Y\left(u_i,v\right) + R \cdot n_Y'\right] \cdot \cos\left(C_{CL}'\right) + \left[P_X\left(u_i,v\right) + R \cdot n_X'\right] \cdot \sin\left(C_{CL}'\right) \\ Z_{CL}' = P_Z\left(u_i,v\right) \end{cases} \tag{9}$$

4.3. Realization of Multi-Axis Following Function

According to equation (9), the parametric variable v of tool contact trajectory $\boldsymbol{P}(u_i,v)$ satisfies the following equation.

$$P_Z\left(u_i,v\right) - Z_{CL}' = 0 \tag{10}$$

The multi-axis follow-up function will be realized by the encoder collecting Z-axis coordinate values Z_{CL}', substituting it into equation (10) to solve for v, and then substituting v into equation (9) to calculate X, Y and C axes CNC machining data.

It is noted that equation (10) is a non-linear equation, which usually needs to be solved iteratively. If the solution method is unsuitable, the convergence rate of iteration will decrease, which will affect the real-time performance of the CNC system.

The widely used algorithms for solving nonlinear equations are dichotomy, secant method and Newton method. The convergence rates of the three methods are different. The secant method and Newton method are both locally convergent, so distance between the initial iteration value and the exact solution has a great influence on the convergence rate. In order to improve the convergence rate and make the initial value of iteration closer to the exact solution, a set of initial solutions of equation (10) are calculated by uniformly changing the coordinate value of Z-axis at a certain interval in advance, which is used as the initial value of iterative solution of multi-axis follow-up function. In order to find the appropriate interval of initial iteration values, a comparison of the time-consuming of iteration under different intervals has been made by experiments, and there is also a comparison of the time-consuming of the three methods when the initial solutions of equation are the same though experiments. The comparison results are shown in figure 15. It can be seen that the secant method is the fastest in most cases when the interval is 20 mm, but it is slower than Newton method in some cases. When the interval is 10 mm and 5 mm, all three methods has become faster, and secant method is the fastest in all cases. When the interval is reduced to 2.5 mm, the change of time consuming of the three methods are not obvious. Therefore, the secant method is used to solve the equation (10). The realization process of multi-axis follow-up function is shown in figure 16.

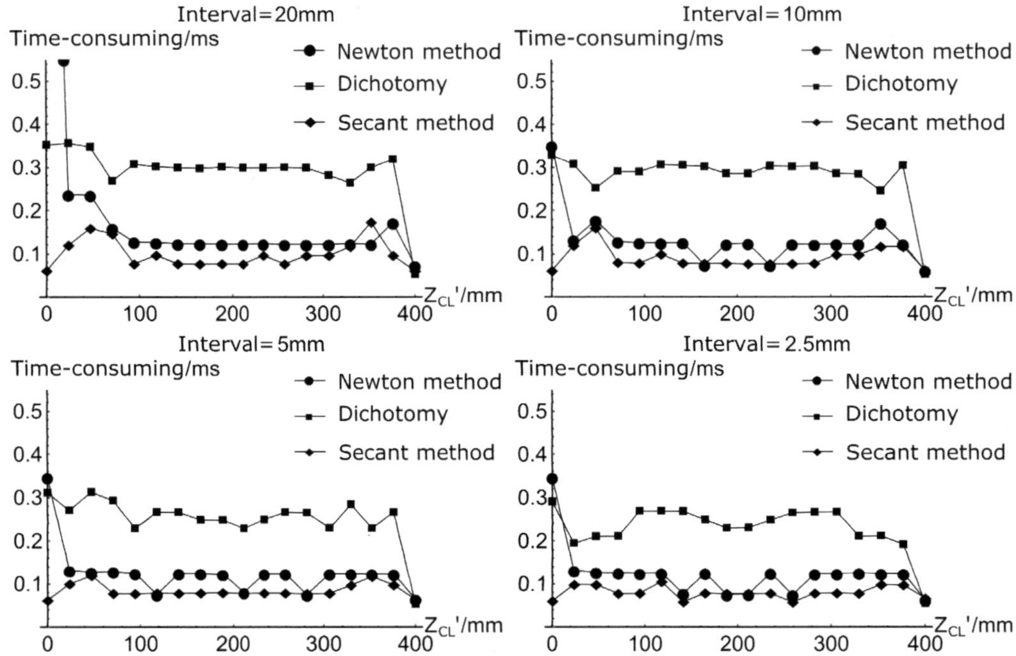

Figure 15. Comparison results of time-consuming of iteration.

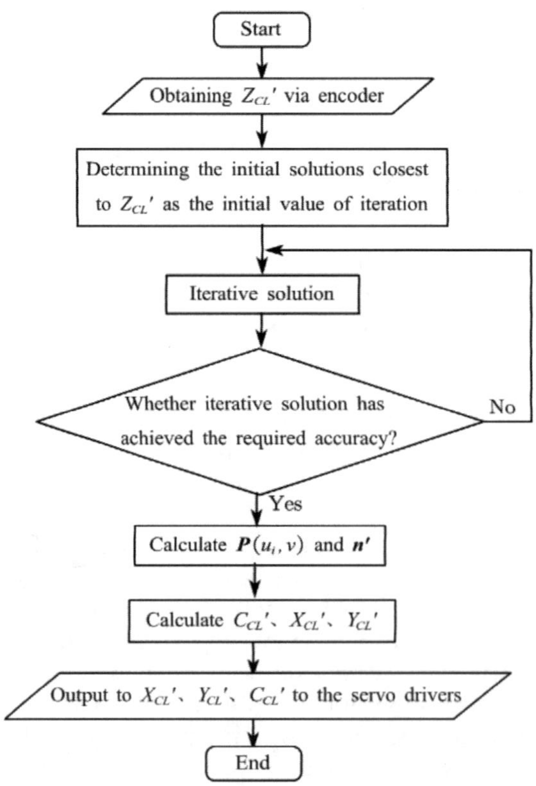

Figure 16. The realization process of multi-axis follow-up function.

5. Machining experiment

According to the method mentioned above, a BY60100B hydraulic planer is reformed to the experimental system of twist blade profile CNC planing (figure 17).

Figure 17. The experimental system of twist blade profile CNC planing.

The twisted blade used for machining experiment is 400 mm long and 110 mm wide, with the twist angle of 15 degrees. The blade profile has 11 cross-section original data, each cross-section has 200 shape points, as shown in figure 18. The experimental cutting parameters are in Table 1.

Table 1. Experimental cutting parameters.

Cutting parameters	Value
Cutting speed (m/min)	10.0
Maximum residual height in roughing (mm)	1.00
Quantity of roughing cutting trajectories	1850
The cutting depth of roughing (mm)	1.50
Maximum residual height in finishing (μm)	12.5
Quantity of finishing cutting trajectories	999
The cutting depth of finishing (mm)	1.00

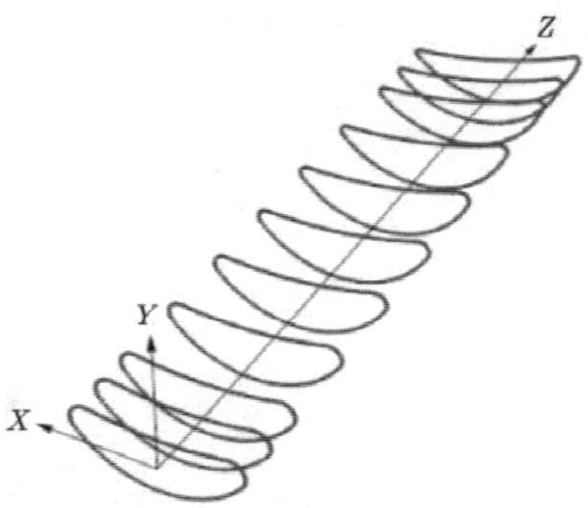

Figure 18. Original data of blade profile.

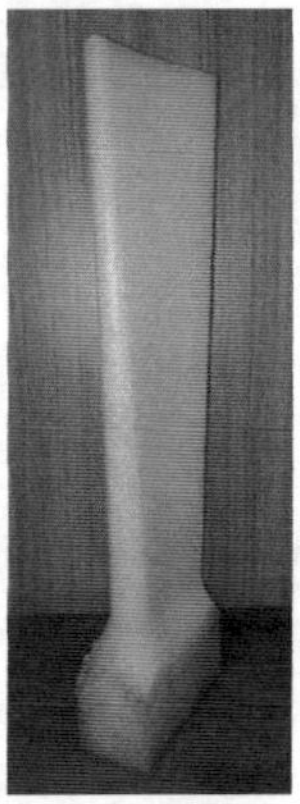

Figure 19. Twisted blade profile obtained by machining experiments.

The twisted blade profile obtained by machining experiment is shown in figure 19, which proves the method is feasible.

6. Conclusion

A multi-axis follow-up planing method is proposed for the planing of twisted blade profile. The corresponding machine tool and CNC system are constructed. The multi-axis follow-up function required for the CNC planing is realized in the CNC system, and the machining experiment is completed. Experimental result show that the proposed method is feasible.

Turbine blade belongs to thin-walled workpiece, which is easy to be deformed during machining. How to select reasonable cutting parameters to improve planing accuracy still needs further studied.

References

[1] Yuan L and Zhao Y 2000 *C. Machine Tool & Hydraulics* **2000(03)** 93-4
[2] Wang N 2006 *C. J. Turbomachinery* **2006(01)** 22-4
[3] Wang Q 1997 *C. Manufacturing Technology & Machine Tool* **1997(03)** 35-7
[4] Shen B and Gao J 2014 *C. Mechanical Research & Application* **2014(03)** 84-6
[5] Yi Y, He D and Deng H 2000 *C. Machine Tool & Hydraulics* **2004(04)** 27-8
[6] Liu H 1995 *C. Manufacturing Technology & Machine Tool* **1995(12)** 39-40
[7] Ueno K 1998 *Gear Technology* **15(2)** 17-9
[8] Wang S, Wang K, Sun X W and Liu X M 2012 *C. Machine Tool & Hydraulics* **2012(09)** 42-4
[9] Li J, Li X and Wang P 2015 *Patent CN* 104722851A
[10] Yue Y 2010 *C. Turbine Technology* **52(03)** 235-7
[11] Qu F, Li Z, Hu H and Liu W 2017 *2017 Chinese Automation Congress (CAC)* (Jinan: China/IEEE) pp 6494-6498
[12] Bartos F J 2008 *Control Engineering C.* **2008(02)** 24-7

IOP Conf. Series: Materials Science and Engineering **686** (2019) 012023 doi:10.1088/1757-899X/686/1/012023

A Portable Fluorescent Microfluidic Detection Device Based on FPGA

Gangshan Wu[1], Chiyuan Chen[2], Ning Yang[2,*] and Peifeng Xu[1]

[1]Jiangsu Polytechnic College of Agriculture and Forestry, Jurong 211121, China
[2]Jiangsu University, Zhenjiang 212013, China

[*]yangning7410@163.com

Abstract. Fluorescence analysis method has high sensitivity, but the difference between photoelectric converters will lead to inaccurate data. To solve the problem of the separation of effective and unwanted light in the process of fluorescence conversion and the noise of small-scale variable amplification. This paper designs a method of modulating the light source makes the fluorescent light be a kind of differential signal that is easily extracted by the photoelectric conversion circuit and amplified, and then uses digital devices FPGA and ADC to achieve signal acquisition and processing, and finally it is presented in the form of a dark room box. In this paper, a novel detection structure is used as a discussion point, and a low-power, high-precision, low-cost signal processing system has been introduced in detail, and mainly used in 500 The fluorescence in the wavelength range of -600 nm was tested. The mathematical relationship between different fluorescence intensities and the post-conversion variables was established. In the subsequent product development, the detection of different wavelengths of light can be achieved by replacing the excitation light source and the corresponding filter, providing a simple circuit processing method for designing photoelectric probes, biomedical detection, and chemical applications.

1. Introduction

There are many electron energy levels in the molecule that are close to each other. When the detection sample is irradiated with light of a specific wavelength, it will cause the electrons in the sample molecule to transition to a high energy level, become extremely unstable, and spontaneously lose energy and return to the original. The level of electrons in which fluorescence or phosphorescence occurs. Sensitivity is an important parameter of fluorescence analysis. Optical experiments often use spectrophotometry and colorimetry, while fluorescence analysis is 2 to 3 orders of magnitude higher [1]. FPGA (field programmable gate array) is very different from traditional PC and microcontroller development. It mainly focuses on parallel computing. The driver uses hardware description language (HDL), which is fast and parallel in the field of signal processing and communication. The advantage is ahead of other processor chips [2]. FPGA design covers communications, automotive electronics, avionics, defense, consumer electronics, military and medical electronics. The scale of contemporary FPGAs has also reached tens of millions of gates, which is suitable for implementing system-on-chip SOC. The design is flexible, the development investment is small, and the degree of fault tolerance is large. The FPGA is less risky than the ASIC, and the stealth cost is lower [3].

In recent years, there has been a unique favor for fluorescence in testing at home and abroad, and many talents have made relevant attempts, and many innovative methods have been proposed. Kulp et

IOP Conf. Series: Materials Science and Engineering **686** (2019) 012023 doi:10.1088/1757-899X/686/1/012023

al. designed a multi-color fluorescence detection system based on high-power white LED module and light rod homogenizing structure for orthogonal microfluidic PCR chip [4], for simultaneous detection of multiple gene targets of unknown pathogens. Immunosorbent assay for fluorescent ligation using an automated microfluidic photonic crystal enhanced fluorescence analysis platform for cancer biomarker detection in human serum [5]. A portable, low-cost, user-friendly and ultra-sensitive microcapsule array based on functional nucleic acid strategy was designed for the quantitative detection of copper ions [6].

The main research of this paper is the signal processing based on FPGA fluorescence detection, which is used for the collection of fluorescent signals in microfluidic chips. The FPGA uses a 12-bit parallel ADC with a sampling frequency of 65 MHz to acquire the fluorescent signal. Photoelectric conversion uses square wave modulation filtering to simplify the signal conditioning circuit.

2. Materials and Methods

2.1. Principle of Fluorescence Quantitative Detection

The reason why the fluorescence method can analyze and measure the sample is because there is a certain quantitative relationship between the fluorescence intensity of the substance and the solubility of the sample solution. According to the white law, the relationship between the transmitted light intensity and the solution concentration is:

$$\frac{I}{I_0} = 10^{\varepsilon bc} \tag{1}$$

The fluorescence intensity F is proportional to the difference between the incident light intensity and the transmitted light intensity (I-I0) and the fluorescence quantum yield φ. When the influence of other factors is small, the relationship between the fluorescence intensity and the incident light intensity is:

$$F = \mathrm{k}\phi\, I_0 2.303\varepsilon bc \approx KI_0 \tag{2}$$

Where ε is the molar coefficient, b is the thickness of the liquid layer, c is the concentration of the fluorescent substance in the solution, and k and K are different proportional constants. It can be seen from equation (2) that there is a linear relationship between the fluorescence intensity of the sample and the concentration of the fluorescent substance, and the basis of fluorescence quantification is this linear relationship. Therefore, increasing the detection sensitivity can be improved by photoelectric conversion and signal amplification. Electronic equipment is all right.

2.2 Theoretical analysis of photoelectric signal conversion

The modulation circuit is used to realize the regular illumination of the excitation light source, and the AC change of the fluorescence signal is indirectly controlled. After the conversion of the silicon photodiode, in the actual test process, the obtained electrical signal is a current signal of several hundred nanoamperes, which is presented. The waveform of a distorted sine wave. The main error of fluorescence measurement lies in the stray light and high-intensity excitation light in the photoelectric conversion environment. These error sources can be eliminated by using a monochromator or a filter. The filter has the advantages of being cheap and simple, and is divided into glass, film and interference. Three kinds of filters; secondly, the influence of various signals on the analog circuit after the conversion, such as the parameters and PCB topology of the op amp itself, etc. Selection, schematic optimal design, signal integrity analysis, and optimized PCB design. What the paper measures is not the precise amount of fluorescence produced, but the number of samples to be measured that is proportional to the amount of fluorescence. Therefore, in the obtained AC waveform, the required amount is the peak-to-peak value or effective value of the AC signal. It represents the difference information of the fluorescence that occurs after modulation. Fluorescence distance is different from that of the photoelectric converter, and the required amplification gain will be different.

Therefore, the INA128 is used to adjust the gain of the adjustable gain, so that the signal obtained by amplifying and filtering the AC signal has a level of one hundred millivolts, reaching the ADC sampling. The magnitude requirement.

2.3 Hardware Design

The hardware uses 1W blue LED chip (withstand voltage of 3.0-3.2V, maximum current is 300mA, emission wavelength is 460nm, luminous flux is 20-30LM), EP4ce5e22c8 main control chip, SG-S1223G silicon photodiode, OPA380, INA128, OPA227 , LTC1966, MAX660, AD9226 and display. The fluorescence reagent used in the paper produces a fluorescence wavelength of about 500-600 nm in the photoluminescence phenomenon. The 530 nm narrow-band filter can be used to filter out excitation light and retain fluorescence.

The system adopts a projection optical path, and the modulation light source is embedded in a conical cavity, leaving a hole to illuminate the microfluidic chip bearing area. Below the carrying area is a signal conditioning circuit board, the circuit board is parallel to the carrying area, and the photodiode and the light emitting The holes are on the same vertical line. The specific 3D structure is shown in Figure 1(a)(b). The box casing has a gap in the microfluidic chip bearing area for replacing the microfluidic chip. In order to avoid the strong excitation light in the transmission process and the stray light around the photodetector affecting the fluorescence detection, a small black cap is made by the filter to cover the photodetector, so that only the light from the vertical direction can enter the detector, and the incident light path is The microfluidic chip is 45 degrees, making the possibility of strong excitation light entering the detector being reduced.

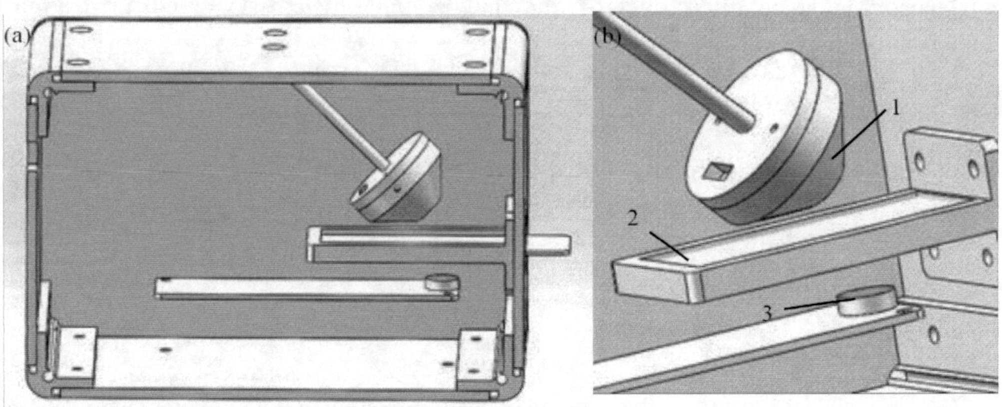

1. Modulated light source; 2. Microfluidic Chip Bearing Zone; 3. Photoelectric Detector

Figure 1. Device Structural Chart

Considering that the photocurrent generated by the photodiode is only 2.6~3.6uA, the transimpedance amplifier formed by the general operational amplifier is affected by the parameters of the operational amplifier itself, such as offset voltage and bias current. 100,000 times magnification can be achieved with the OPA380 and photodiode. The schematic is shown in Figure 2(a).

When designing the intermediate stage amplifying circuit, considering that the photodiode output current is affected by the fluorescence intensity and the distance between the photodiode and the sample to be tested, the size difference is very large, so the INA128 design gain adjustable low noise amplifier circuit is adopted. . The schematic diagram is shown in Figure 2(b). The magnification is controlled by a sliding rheostat with a resistance of 50KΩ, which can achieve 2~1000 times magnification.

In a circuit with two-stage amplification and a magnification of 1000000 times or more, there must be noise interference. Gaussian white noise exists in the theoretical calculation. Therefore, a low-pass center frequency 1000 Hz, 500 Hz bandwidth band-pass filter is designed to preserve the 1 kHz signal waveform. After testing, the waveform has no other interference. The filter is designed using OPA227, and the circuit schematic is shown in Figure 2(c).

Considering that there is error in the signal processing of a digital device and the external device can be used for calibration or measurement when the digital device program has problems, the paper uses LTC1966 to design the RMS detection circuit to realize the RMS detection of the AC signal and the AC signal. The effective value is converted to a voltage value for easy detection. Since the output voltage of the chip is less than 500mV, the voltage amplification is realized by the OPA227, and the sampling accuracy is improved. The schematic diagram is shown in Figure 2(d).

When designing the PCB, considering the problem of weak signal amplification processing, there is theoretically a phenomenon of noise drowning signals. The copper photodiode output pin and the OPA380 input pin are connected as short as possible, and the feedback resistor must be connected between the photodiode and the chip pin near the input to prevent signal feedback current from entering the chip. Copper is placed around the input copper wire and the feedback copper wire and the hole is grounded to the ground to form a low-impedance environment, which avoids parasitic parameters and leakage current of the circuit board and reduces the possibility of signal attenuation. The position of the filter capacitor resistor is placed as much as possible in the schematic topology to reduce the chip workload and result errors. The power supply pins of all chips must be connected to the 100nf capacitor to ground, and this capacitor is as close as possible to the chip to filter out high frequency interference in the supply voltage. The main component of the negative power supply module is copper-cutted, leaving the input and output ports, because the MAX660 is a switching power supply chip to prevent interference with the signal chain. Ensure that the signal chain of the entire signal conditioning circuit is in a straight line, eliminating crosstalk and attenuation. The specific circuit PCB is shown in Figure 2(e).

The AD9226 chip is used as the main chip of the system ADC, and the connection between the pin of the AD9226 and the FPGA is shown in Table 1. The overall circuit schematic is shown in Figure 3.

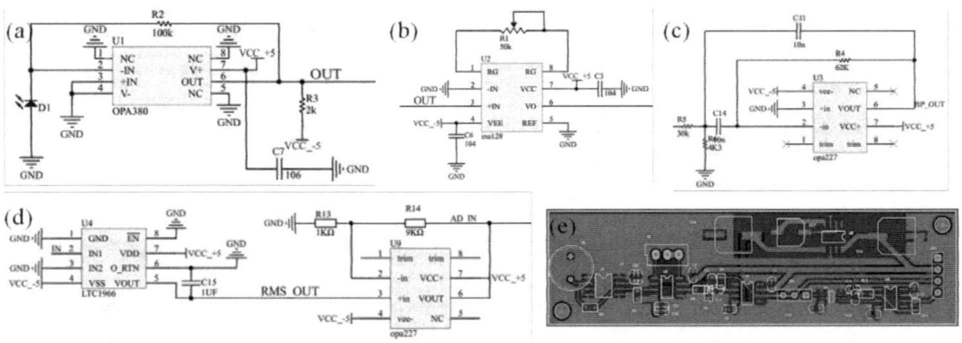

Figure 2. (a) OPA380 transimpedance amplifier circuit; (b) INA128 low noise amplifier circuit; (c) OPA227 bandpass filter; (d) rms detection circuit; (e) signal conditioning circuit PCB.

Table 1. ADC chip and FPGA pin connection.

AD9226	Description	FPGA	AD9226	Description	FPGA
PIN_1	Sample CLK	PIN_72	PIN_8	Data 6	PIN_65
PIN_2	Data12(LSB)	PIN_71	PIN_9	Data 5	PIN_64
PIN_3	Data 11	PIN _70	PIN_10	Data 4	PIN_60
PIN_4	Data 10	PIN_69	PIN_11	Data 3	PIN_59
PIN_5	Data 9	PIN_68	PIN_12	Data 2	PIN_58
PIN_6	Data 8	PIN_67	PIN_13	Data 1(MSB)	PIN_55
PIN_7	Data 7	PIN_66			

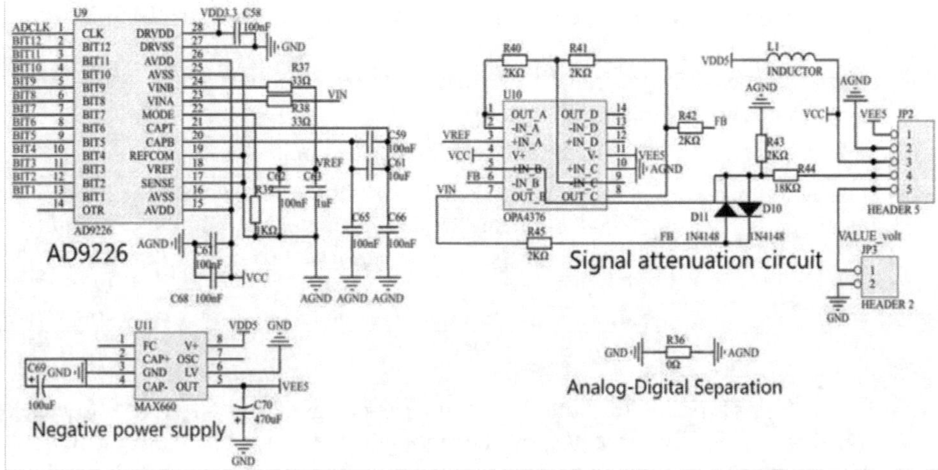

Figure 3. ADC circuit schematic.

2.4. *Internal signal processing*

The signal processing module mainly stores DATA_in into a RAM. The RAM is a dual-port RAM with a bit width of 12 and a depth of 4096 of the M9K of the FPGA, which is convenient for the display module to call display. Because the known signal frequency is 1KHz, only the peak-to-peak information of the signal needs to be known. Since the signal is not a standard sine wave, the system also performs the extraction of the RMS detection when setting the peak-to-peak value, setting two variables. Complement each other and eliminate errors.

The peak-to-peak detection mainly finds the maximum and minimum values of a set of DATA_in in 1 millisecond, finds the average value of the upper and lower values of the previous set, and stores the result as the maximum and minimum registers. The added number is averaged as the maximum and minimum values of the next millisecond.

The RMS detection is mainly obtained by squaring all the DATA_ins in one group and averaging them in 1 millisecond. By using the multiplier core embedded in the FPGA, the maximum 18-bit multiplication can be realized, because of the multiplication in timing. The time required is relatively long, and enough time is reserved in this process to ensure that there is no error, but the structure of the FPGA itself is difficult to perform logarithmic operations, so the data is directly output to the display.

3. Results and Discussions

Signal conditioning circuit when no filter is directly modulated light source brachytherapy photodiode, the distance between 1 to 3 cm, acquired through the use of an oscilloscope waveform of the band pass filter shown in Figure 4. It can be seen that the waveform frequency is about 1KHz, the amplitude is about 3.96V, and the waveform is slightly distorted. From the oscilloscope, the peak-to-peak value is 3.96V, which is the waveform after the magnification is up to 6000000 times.

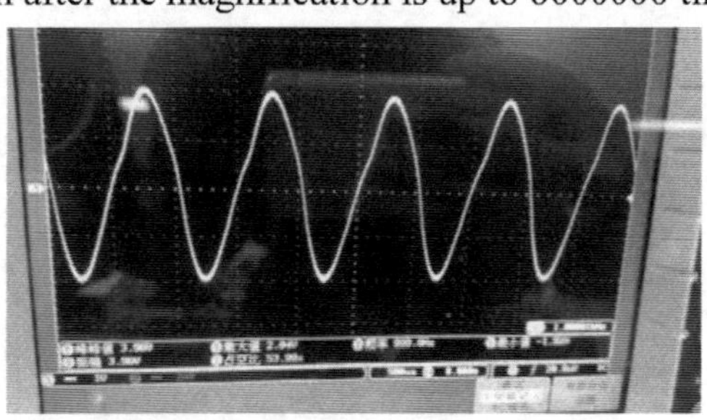

Figure 4. Filter output waveform.

The solution containing the fluorescent reagent powder is arranged on the microfluidic chip, and the test results of the system are turned on as shown in Table 2. It can be found that the concentration of the fluorescent substance is linear with the measured peak-to-peak value and the effective value, and the waveform is complete without distortion.

Table 2. Fluorescence test results.

Fluorescent concentration	10g	20g	30g
Waveform integrity	Complete without distortion	Complete without distortion	Complete without distortion
Peak to peak (mV)	488	1034	1334
Valid value (mV)	172	363	468

4. Conclusions

In this paper, a method of modulating the light source is designed to make the fluorescence become a differential signal that is easily extracted by the photoelectric conversion circuit and amplify it, and then use FPGA and ADC to realize signal acquisition and processing. The paper uses silicon photodiodes to extract optical signals into electrical signals, then performs signal amplification and filtering on the electrical signals, and finally inputs them into the FPGA for digital signal processing to achieve concentration detection of fluorescent reagents, including excitation light source driving design, optical path structure. , Photodetector design, signal conditioning circuit design, and FPGA signal acquisition and processing design. In this paper, a low-power, high-precision, low-cost signal fast processing system is introduced in detail. Experiments are carried out mainly on fluorescence in the wavelength range of 500-600 nm, and the mathematical relationship between different fluorescence intensity and post-conversion variables is established. And designed a simple embedded system based on FPGA, using TFT screen display results. In the subsequent product development, the detection of different light wavelength requirements can be realized by replacing the excitation light source and the corresponding filter, which provides a simple circuit processing method for designing photoelectric probes, biomedical detection, and chemical applications.

Acknowledgements

This research was financially supported by the Project of Jiangsu Polytechnic College Agriculture and Forestry (2018kj11, 2018kj12), Jiangsu Agricultural Science and Technology Innovation Fund (CX(18)3043), Zhenjiang Dantu Key R&D Program (SH2018003), Zhenjiang key R&D plan(NY2019013, SH2019012), the Priority Academic Program Development of Jiangsu Higher Education Institutions (PAPD).

References

[1] Smith L M, Sanders J Z, Kaiser R J 1986 Fluorescence detection in automated DNA sequence analysis, J. Nature. 321(6071) 674.

[2] Monmasson E, Cirstea M N 2007 FPGA Design Methodology for Industrial Control Systems—A Review, J. IEEE Transactions on Industrial Electronics. 54(4) 1824-1842.

[3] Strukov D B, Likharev K K 2005 CMOL FPGA: a reconfigurable architecture for hybrid digital circuits with two-terminal nanodevices, J. Nanotechnology. 16(6) 888.

[4] Kulp M, Bragina O, Kogerman P 2011 Capillary electrophoresis with led-induced native fluorescence detection for determination of isoquinoline alkaloids and their cytotoxicity in extracts of Chelidonium majus L, J. Journal of Chromatography A. 1218(31) 5298-5304.

[5] Metildi C A, Kaushal S, Lee C 2012 An LED light source and novel fluorophore combinations improve fluorescence laparoscopic detection of metastatic pancreatic cancer in orthotopic mouse models, J. Journal of the American College of Surgeons. 214(6) 997-1007.

[6] Dong Y, Wang R, Li G 2012 Polyamine-functionalized carbon quantum dots as fluorescent probes for selective and sensitive detection of copper ions, J. Analytical Chemistry. 84(14) 6220-6224.

IOP Conf. Series: Materials Science and Engineering **686** (2019) 012024 doi:10.1088/1757-899X/686/1/012024

Research on Turbine Rotor Fault Diagnosis Based on CPSO-BBO Optimization SVM

Zhibiao Shi and Wenzhuo Dong*

School of Mechanical Engineering, Northeast Electric Power University, Jilin, China, 132012

*2274971760@qq.com

Abstract: In order to improve the recognition accuracy and efficiency of the turbine rotor fault diagnosis, a fault diagnosis method based on CPSO-BBO (chaos particle swarm optimization biogeography based optimization) algorithm was proposed to optimize SVM (support vector machine). Firstly, four common faults states of turbine rotors were simulated by ZT-3 rotor test-bench to obtain fault data. Secondly, CEEMD (complementary ensemble empirical mode decomposition) is used to decompose the rotor vibration signal, and the more effective intrinsic mode function (IMF) is screened by combining the variance contribution rate, and the corresponding PE (permutation entropy) is calculated as the fault characteristic value. After that, chaos theory and PSO (particle swarm optimization) algorithm are combined into the theory of BBO (biogeography based optimization) to obtain CPSO-BBO algorithm, which is used to optimize SVM to obtain the optimal parameters of the diagnostic model. Finally, the fault identification is studied by using the acquired fault data.

1. Introduction

Since the turbine rotor running environment is complex and the fault features are easy to be submerged in the noise, the fault diagnosis results are affected and the fault occurrence is difficult to be predicted. Therefore, it is very important to study the fault diagnosis of turbine rotor accurately and quickly [1].

SVM by using structural risk minimization criterion instead of empirical risk minimization criterion, has the ideal generalization ability, can solve high dimensional and nonlinear problems [2]. In the process of SVM application, the SVM parameters optimization is important, which will directly affect the learning and generalization ability [3]. At present, some commonly used algorithms are deficient to varying degrees.The ant colony algorithm has slow convergence speed, easy to fall into local optimal and easy to appear stagnation phenomenon. The grid search method needs to be set according to human experience, while the simulated annealing algorithm is easy to fall into local optimal [4]. PSO algorithm has outstanding advantages such as fast convergence and adaptability to dynamic environment, and can effectively solve global optimization problems [5]. The BBO algorithm has strong adaptive ability, high search accuracy, few parameters and easy implementation.

In order to avoid the defects of single algorithm in search ability and robustness, and improve the fault recognition rate and efficiency of SVM, CPSO-BBO algorithm is proposed by combining chaos theory with PSO algorithm and BBO algorithm. CPSO-BBO algorithm is used to optimize the parameters of SVM, penalty factor c and kernel function σ to obtain the optimal diagnosis model, meanwhile, compared with the SVM model optimized by BBO algorithm.

2. Basic principle

2.1 CPSO algorithm principle

Firstly, particle swarm is used to initialize a group of particles, $X = \{x_1, x_2, \Lambda, x_n\}$, each particle has a random position, and the position in the *d* dimensional space, which is $x_i = (x_{i1}, x_{i2}, \Lambda, x_{id})$, and the velocity is $v_i = (v_{i1}, v_{i2}, \Lambda, v_{id})$. Each particle is evaluated by a uniform fitness function $f(x_i)$. The optimal position experienced in the successive evolution is called the individual extreme value P_{best_i}, and the optimal position experienced by all particles in the population is called the global extreme value g_{best}. In the process of each iterative evolution, the particle adjusts its speed and position according to the individual optimal position P_{best_i} and the global optimal position g_{best} [6,7],

$$\begin{cases} v_i(t+1) = \omega v_i(t) + c_1 r_1 (p_{best\_i} - x_i(t)) + c_2 r_2 (g_{best} - x_i(t)) \\ x_i(t+1) = x_i(t) + v_i(t+1) \end{cases} \tag{1}$$

where ω is the weight of inertia, t is the number of iterations, r_1 and r_2 are random numbers between [0,1], and c_1 and c_2 are the learning factors.

In order to avoid falling into local optimum in particle search, logistic mapping was introduced into the population to form chaotic PSO. The global optimal value in the population g_{best} is mapped to the chaotic variable of [0,1] according to the following formula [8,9],

$$c_x^{(t)} = \frac{g_{best}(t) - x_{\min}}{x_{\max} - x_{\min}} \tag{2}$$

where $x_{\min}$ and $x_{\max}$ represent upper and lower bounds of position vectors respectively. Substitute the obtained chaotic variables $c_x^{(t)}$ into the logistic mapping formula.

$$c_x^{(t+1)} = 4c_x^{(t)}\left(1 - c_x^{(t)}\right) \tag{3}$$

M iterations were carried out by logistic mapping formula and the new optimal particle sequence was obtained through the mapping of formula (3).

$$g_{best}(t+1) = x_{\min} + c_x^{t+1}(x_{\max} - x_{\min}) \tag{4}$$

That is the new optimal population X_{best}, which is merged and updated with the original population X to get the new population X'.

2.2 Principle of biogeography algorithm

2.2.1 Migration operation. In this paper, the cosine mobility model closest to the natural law is used to simulate the migration process of species in the biogeographic environment, as shown in figure 1.

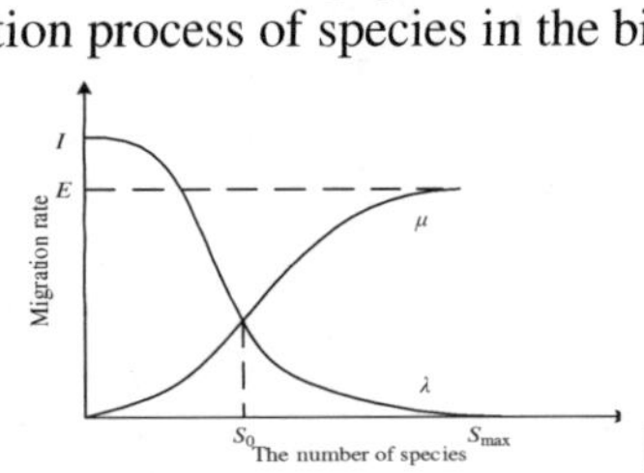

Figure 1 Cosine migration rate model

In the figure, I is the maximum migration rate; E is the maximum removal rate; S_{max} is the maximum number of species in the habitat; λ is the rate of immigration; μ is the removal rate; and S_0 is the equilibrium point. When $S_{max}=n$, the number of species in the habitat is $S=k$, the migration rate $\lambda_k = \frac{I}{2}\left[\cos\left(\frac{k\pi}{n}\right)+1\right]$ and migration rate $\mu_k = \frac{E}{2}\left[\cos\left(\frac{k\pi}{n}+1\right)\right]$.

IOP Conf. Series: Materials Science and Engineering **686** (2019) 012024 doi:10.1088/1757-899X/686/1/012024

BBO with suitable habitat index (HSI) is used to evaluate habitat suitability function, where the characteristic factors related to HSI are expressed by the suitable index vector (SIV), so the optimal solution problem is transformed into the fitness optimization problem.

When carrying out the migration operation, the HSI value of the habitat should be calculated first and arranged in order from large to small. λ is then determined by the rate of immigration to the habitat, then according to the removal rate μ to determine how to exchange with the adjacent habitat, and a SIV was randomly selected from the adjacent habitat to replace one SIV in the habitat. Finally, the HSI value of each habitat was calculated in order, where the highest HSI point corresponds to the optimal solution.

2.2.2 Mutation operation. BBO algorithm improves species diversity through mutation operation. The rate of habitat variation is inversely proportional to the number of species,

$$m_s = m_{\max}\left[\frac{1-P_s}{P_{\max}}\right] \tag{5}$$

where $m_{max}$is the maximum variation rate; $P_s$is the probability that the number of species in the habitat corresponds to S_1; $P_{max}$is the maximum value of P_s. The relationship of P_s to inflow rate λ or outflow rate μ is as follows.

$$P_s = \begin{cases} -(\lambda_S+\mu_S)P_S+\mu_{S+1}P_{S+1} & S=0 \\ -(\lambda_S+\mu_S)P_S+\lambda_{S-1}P_{S-1}+\mu_{S+1}P_{S+1} & 1\le S\le S_{max}-1 \\ -(\lambda_S+\mu_S)P_S+\lambda_{S-1}P_{S-1} & S=S_{max} \end{cases} \tag{6}$$

In BBO, if the variation probability of a certain habitat m_s is non-zero, a SIV is generated randomly according to the mutation operation to replace the existing SIV in the habitat, so as to increase the species diversity, improve the solution set of the habitat and obtain the optimal solution.

2.3 Principle of CPSO-BBO algorithm

CPSO-BBO can effectively prevent all individuals of the population from falling into local optimum by means of molecular grouping.That is, CPSO population is divided into several sub-populations at the beginning of each iteration, and CPSO optimization is conducted for each sub-population. A certain number of optimal individuals are generated in each sub-population to form “elite population”.Then, BBO is used to optimize the “elite population”. At the end of the operation, the “elite population” is put back into their subpopulations to continue the next iteration cycle.This algorithm avoids the defects of single algorithm in search ability and robustness and effectively coordinates the search ability between the two algorithms.

2.4 Typical function tests

To verify the optimization performance of this algorithm, use Schaffer function and Sphere function to calculate and compare the minimization problem of BBO algorithm and CPSO-BBO algorithm.The iterative optimization process of the two test functions is shown in figure 2.

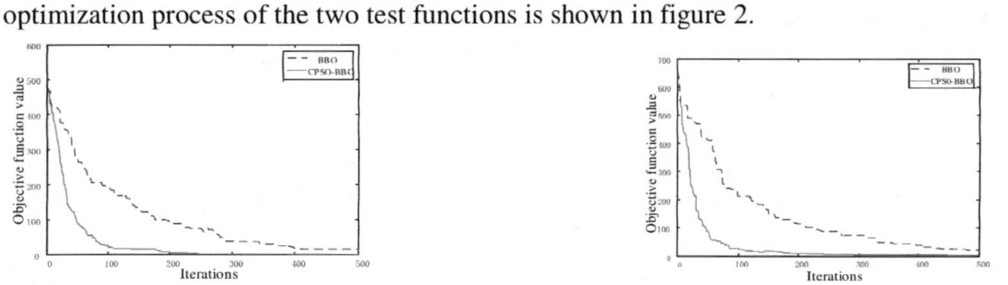

(a) Schaffer function training curve (b) Sphere function training curve

Figure 2. Function training curve

CPSO-BBO algorithm has a fast convergence rate, fewer iterations, is not easy to get into local optimization and can quickly enter into the optimal solution interval.

2.5 SVM parameter optimization of CPSO-BBO algorithm

In order to improve classification accuracy and recognition efficiency of SVM, CPSO-BBO algorithm is used to optimize RBF kernel function parameter σ and support vector machine parameter c. Moreover, CPSO-BBO algorithm is used to optimize SVM classification accuracy of training set as fitness function,

$$\psi = \frac{1}{m}\sum_{n=1}^{m}\left(\frac{x_m}{x_n}\times 100\%\right) \tag{7}$$

where x_m is the correct number of categories in the mth test set; and x_n is the number of samples in the step test set.

The steps for CPSO-BBO to optimize SVM model parameters are as follows.

Step1: Determine the suitability function, initialize the population and set initial parameters, including chaos search times T, maximum migration rate I_{max}, maximum migration rate E_{max}, maximum number of species that can be accommodated in the habitat S_{max}, and iteration times, etc.

Step2: Initialize a group of habitats with chaotic mapping. Set the initial fitness vector to be X_i and X_i include the σ and c parameters of SVM.

Step3: Divide the population into m sub-populations.

Step4: CPSO algorithm optimizes the divided population and calculates the fitness value to complete the mapping of each species. Take the one with high fitness value as the local optimal solution.

Step5: Use BBO algorithm to conduct global search for the local optimal solution obtained.

Step6: Get the global optimal solution.

Step7: Judge whether the termination condition has been reached. If so, output optimization results; Otherwise, continue repeating Step3~Step7 until the termination condition is satisfied.

The optimal penalty factor c and kernel parameter σ are obtained through the algorithm steps above, and then the diagnosis model of CPSO-BBO optimization SVM is constructed.

3. Experimental study on rotor fault diagnosis

3.1 Failure data acquisition

ZT-3 rotor test bed was used to simulate four different states of turbine rotor (normal state, action touch ground, imbalance, mismatched). Parameters of the test were set as follows: rotor speed was 3 000 r/min and sampling frequency was 5000 Hz, and the software platform is matlab 2014a.

3.2 Singular value denoising

In order to avoid the appearance of singular sample data, first of all, the vibration signal collected is normalized, so that the input sample set conforms to the standard normal distribution. Then, the phase space reconstruction of its time series is carried out by using c-c algorithm [10] to determine the optimal embedding dimension of phase space reconstruction m and best delay time τ . Finally, the singular value denoising of reconstructed phase space is carried out [11].

To evaluate the effect of noise reduction, vibration signal to select amount of improving SNR, ΔSNR (signal noise thewire, SNR) [12]. To test its noise reduction effect. The improvement of SNR is shown in table 1, and the noise reduction effect is obvious.

Table 1. Improvement of SNR before and after noise reduction

Type	Normal	Action touch ground	Mismatched	Unbalanced
SNR(dB)	19.1325	21.9834	16.1566	20.7293

3.3 Feature extraction

CEEMD can not only solve the modal aliasing problem of EMD decomposition process, but also effectively eliminate the influence of residual white noise. IMF components obtained after CEEMD decomposition of mismatched states are shown in figure 3.

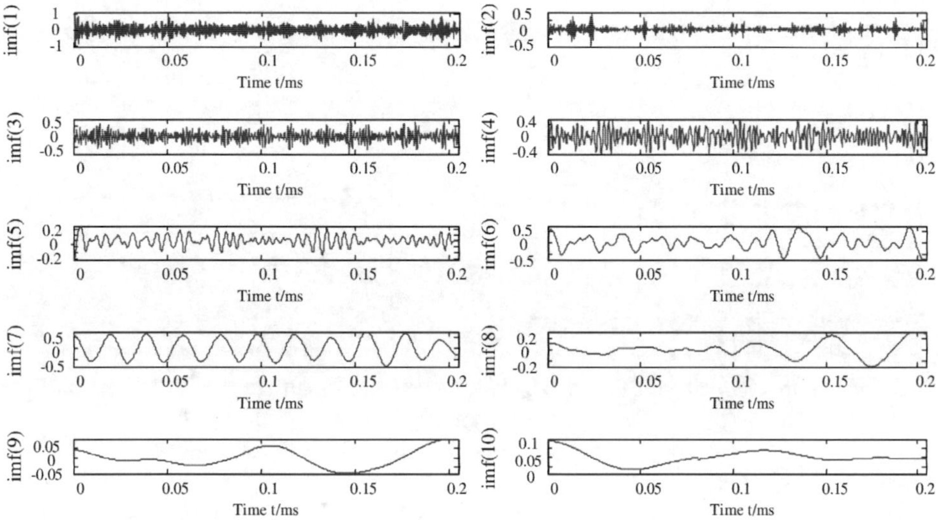

Figure 3 Misaligned state noise CEEMD decomposition results

It can be seen that CEEMD can obtain a clear waveform. According to the principle of variance contribution rate, fault feature sensitive components are selected to calculate their entropy values, and fault feature vectors are constructed as input components of SVM classifier.

To eliminate the influence of endpoint effect and false component, the first three IMF components were selected and their permutation entropy was calculated. To test the effect of arrangement entropy as the characteristic vector of four states, the boxplot of arrangement entropy was used to verify the degree of differentiation. This paper selects IMF_1 for example, the boxplot of entropy values of four states is shown in figure 4.

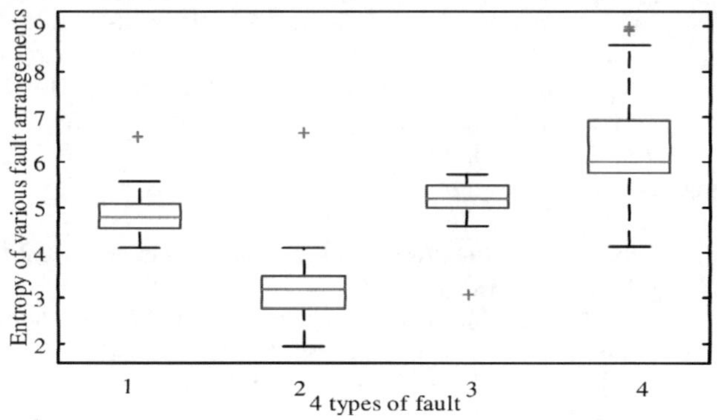

Figure 4 Permutation entropy boxplot of 4 states rotor

It can be seen that the selection of entropy of arrangement as the eigenvalue to construct the feature vector has an obvious distinction effect, so it can be used as the feature vector of pattern recognition.

3.4 Fault identification results and analysis

In order to verify the feasibility of CPSO-BBO algorithm to optimize SVM diagnosis model, the BBO algorithm were compared. Four simulated states were used as experimental data on ZT-3 test bench.

Figure 5 shows the classification diagram.

According to figure 5, CPSO-BBO optimized SVM diagnostic model after training, among the 60 test samples, there are 2 groups of sample identification errors. After testing 60 groups of test samples in the BBO-SVM diagnostic model, 6 groups of sample identification errors were found.

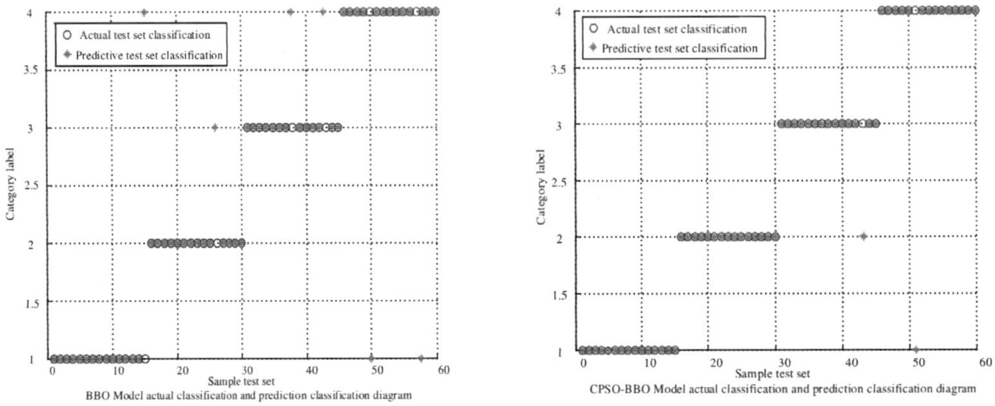

Figure 5 Comparison of CPSO-BBOand BBO algorithms to optimize the SVM test and prediction results

4. Conclusion

(1) Using chaotic particle swarm optimization biogeography algorithm to improve the convergence speed and optimization accuracy of the algorithm.

(2) CPSO-BBO algorithm can quickly and effectively determine the penalty factor and kernel parameters of SVM, thus improving the learning ability and generalization ability of SVM.

(3) Under the same test samples, the recognition accuracy of CPSO-BBO optimized SVM diagnostic model is 96.67%, 6.7% higher than that of the BBO-SVM diagnostic model, and the running time is 21.837s, 12.5 s shorter than that of BBO. The advantages of CPSO-BBO optimization SVM algorithm in turbine rotor fault diagnosis are fully verified.

Reference

[1] He Qing, Xie Fangfang, Li Hong, et al. Vibration feature extraction of steam turbine units based on manifold learning method [J]. Vibration, test and diagnosis, 2014, 34 (4) : 705-708.

[2] Jiang Jiuliang, Liu Wenyi, Hou Yujie, et al. Research on bearing fault diagnosis method based on inner product extension LMD and SVM [J]. Vibration and impact, 2016, 35(6): 104-108.

[3] Zhao Chongchong, Liao Mingfu, Yu Xiao. Application of Support Vecter Machine to Fault Diagnosis of Rotation Machinery[J]. Journal of Vibration, Measurement & Diagnosis. 2006, 26(01): 53-57.

[4] Feng Zonghui, Peng Dan, Yuan Rongli. Engine fault diagnosis based on PSO-SVM [J]. Computer measurement and control, 2014, 22(2): 355-360.

[5] Xiong Qing, Zhang Weihua. Fault diagnosis method of rolling bearing based on mf-dfa and PSO optimized LSSVM [J]. Vibration and impact, 2015, 34(11): 188-193.

[6] Sun Shuguang, Yu han, Du Taihang, et al. Fault vibration and acoustic diagnosis method of universal circuit breaker based on multi-feature fusion and improvement of QPSO-SVM [J]. Journal of electrical technology, 2017, 32 (19) : 107-117.

[7] ZHAO F Q, LI G Q, YANG C, et al. A human computer cooperative particle swarm optimization based immune algorithm for layout design[J]. Neuro computing, 2014, (132) : 68-78.

[8] MOJTAB S, BAMAKAN H, WANG H D, et al. An effective intrusion detection framework based on MCLP /SVM optimized by time-varying chaos particle swarm optimization [J]. Neurocomputing, 2016 (199) :

[9] QIN Q D, CHENG S, CHU X H, et al. The Solving nonconvex/non - smooth economic load dispatch problems via an enhanced particle swarm optimization [J]. Applied Soft Computing, 2017 (59) : 229-242.

[10] Xu Zili, Wang Yiyang, Zhou Jiliu. C-C average method for estimating the embedding delay time and delay time window of nonlinear time series [J]. Journal of SiChuan university: engineering science, 2007, 39(1): 151-155.

[11] Walton J, Fairley n. Noise Reduction in X - ray Photoelectron Spectromicroscopy by A Singular Value Decomposition Sorting Procedure [J]. Journal of Electron Spectroscopy and Related Phenomena, 2005,

[12] Hou Pingkui, Gong Yunfan, Yang Yuying, et al. Nonlinear noise reduction for underwater target radiated noise time series [J]. Acta acoustica sinica, 2001, 26(3): 207-211.

Analysis of Turbine Rotor Fault Diagnosis Method

Zhibiao Shi and Wenzhuo Dong*

School of Mechanical Engineering, Northeast Electric Power University, Jilin, China, 132012

*2274971760@qq.com

Abstract: The role of steam turbine in generating set is very important. During the operation, the rotor state is monitored by high-frequency vibration sensor. With the rapid development of computer, sensor and communication technology, modern machinery and equipment is developing towards the direction of electromechanical integration, as a result of the monitoring equipment group of large scale and the required point, single point more high sampling frequency and data collection time span is long, so the monitoring data of exponential growth, mechanical equipment health monitoring field enter the era of "big data". The traditional method of turbine rotor fault diagnosis has encountered some difficult technical problems and bottlenecks. This paper first introduces the fault mechanism of turbine rotor . Then, the fault diagnosis methods of turbine rotor proposed by many scholars this year are analyzed, the fault characteristics of turbine rotor are summarized and the challenges are discussed, and the direction of further research is prospected.

Key words: Turbine rotor; Failure mechanism; Fault diagnosis

1. Introduction

The turbine plays an important role in the generator set. During the operation, the rotor state is monitored by high-frequency vibration sensor. Because the rotor running environment is complex and the fault features are easy to be submerged in the noise, the fault diagnosis results are affected and the fault occurrence is difficult to predict. Therefore, it is very important to study the fault diagnosis of turbine rotor accurately and quickly[1].

Turbine rotor fault diagnosis is mainly divided into two parts, one is the selection of the detection signal, the other is the signal processing process. At present, the main signal used in turbine rotor fault diagnosis is vibration signal[2]. Signal processing includes two core stages. The first stage is to extract features from the original signal, and the second stage is to classify the extracted features.

2. Study on fault vibration mechanism of turbine rotor

2.1 mechanism of rotor imbalance fault

Due to the quality of the material distribution, installation error, machining error and rotor wear, corrosion and deposition in the operation process factors on the influence of the rotor of the rotating machinery parts, such as the rotor rotates center do not coincide with the location of the normal, causes the damage of eccentric mass of rotor or rotor, causing the rotor fault.

2.2 Mechanism of rotor misalignment fault

Misalignment faults exist in many forms. Usually, the rotor has many shafts connected with each other, so misalignment of shafting is one of the main faults. The rotating shafts are connected by coupling to transfer torque and motion. In the process of high-speed operation and load bearing for a long time, it is easy to cause deformation and bending of rotor, uneven thermal expansion of bearing seat, uneven settlement of foundation, etc., as well as installation errors, all of which will lead to misalignment faults.

2.3 Mechanism of rotor rubbing fault

Rotating machinery is a periodic rotary movement, in the movement when the axis of rotation occurs bending deformation or because the coupling is not medium cause the axis deviated from the normal range, it will cause the motion and static parts of the gap, and produce friction phenomenon.

3. Research status of turbine rotor fault diagnosis

This paper mainly explores the core stage of signal processing: pattern recognition, i.e. classification according to extracted features. Fault diagnosis based on pattern recognition develops rapidly. Artificial neural network (ANN) and support vector machine (SVM), as intelligent recognizers, have received the most extensive attention and research in the field of mechanical fault diagnosis. But they also have some disadvantages. In order to solve or reduce the inherent defects of SVM, Tipping M E proposed a new sparse probability model algorithm similar to SVM in 2000: relevance vector machine. Although RVM has better sparsity and generalization ability, and has achieved relatively ideal results in prediction probability, it also has its own shortcomings in the training process. Due to its high computational complexity and large overhead, RVM is only suitable for handling small sample problems. In order to solve this problem, Tipping proposed a fast marginal likelihood algorithm in 2003, which significantly improved the speed of model training[3]. In 2005, Thayananthan A popularized this model, which solved the training problem of multiple output regression and classification[4]. In 2006, Rasmussen C E from Cambridge, UK and Max Planck, Germany published gaussian process machine learning, which made gaussian radial basis (RBF) kernel function become a widely used RVM kernel function[5]. In 2007, with the development of multi-core RVM research, multi-core RVM began to enter the application field. In 2008, Damoulas T and Girolami M applied multi-core RVM method to protein folding recognition and homogeneous detection, and obtained higher classification accuracy than single-core RVM[6]. In 2010, PsorakisI et al. summarized the sparsity and accuracy of solutions of multi-classification RVM method, which extended the original RVM to the application of multi-classification data by introducing auxiliary variable Y[7]. In 2010, Wu Lianghai used particle swarm optimization (PSO) to optimize the parameters of RVM and train RVM to get the optimal parameters by using the method of PSO[8]. In 2010, Li Gang used genetic algorithm to optimize the core parameters of RVM[9]. In 2011, Liu Junshun put forward the study on fast classification algorithm of relevant vector machine based on clustering. In addition, Bit Reduction and SMOTE algorithm were adopted to pretreat the original training samples with clustering, and then multiple local classifiers were constructed for various data clusters, and the feasibility of the algorithm was verified by experiments[10]. In 2012, Zhao Tong deeply studied the optimization method of correlation vector machine. According to the low proportion of "correlation vector", he proposed a fast estimation and optimization algorithm of correlation vector machine, and proposed a solution to the problem of low detection accuracy caused by the use of a single kernel function[11]. In 2013, Liu Changyuan improved the "one-to-one" classifier, which is the most widely used and has the highest classification accuracy among multiple classification problems of related cameras, and greatly improved the classification time of the algorithm[12]. In 2015, Ouyang Ting proposed a fast multi-classification algorithm based on clustering related vector machines, which improved the computational speed of multi-classification problems[13].

With the continuous in-depth research by scholars at home and abroad, we will find that for the same classification task, when we use machine learning algorithm to do it, we should first identify the feature and label, then input this data into the algorithm for training, and finally save the model to predict the accuracy of classification.But this method exists the problem is that we need to determine beforehand good features, every feature is a dimension, characteristics of too little number, we may not be accurate classification, namely what we owe fitting, if the number of features is overmuch, may cause we are obsessed with a certain characteristics in the process of classification result in classification error, namely the fitting. As a result, we need to spend a lot of time and energy on feature engineering to achieve a good result of model training. However, the emergence of neural network makes it unnecessary for us to do a lot of feature engineering, such as designing the content of features or the number of features in advance. We can directly feed the data into it, let it train and "correct" itself, and then get a better effect.

In terms of fault diagnosis based on pattern recognition, deep learning, as a young intelligent pattern recognition algorithm, has been successfully applied in fault diagnosis and prediction, showing its advantages in many aspects. Compared with the traditional information processing technology, it saves money and reduces manpower. The processing speed of big data is also better than the traditional algorithm, which fully shows the huge computing power in the information age to automate the processing of data. With the help of big data from various aspects, deep learning has achieved success in various fields. With the structure and form of neural network, deep learning fully conducts feature mining of data. Meanwhile, with the advantage of massive data, the instability of neural network has been overcome to some extent. Applied to the fault diagnosis of mechanical and electrical equipment, it can not only effectively use the massive data collected by sensors, but also have a powerful feature mining ability.

Currently, the main deep learning algorithm is divided into convolution neural network and the neural network, and compared with the neural network cycles, convolution neural network has better adaptability and adjustability, so the convolution neural network under the researchers' study, there are already many improved methods, although ordinary one-dimensional convolutional neural network has a significant effect, but considering the factors under different working conditions of steam turbine rotor fault data accuracy is not high, is not sensitive to individual factors, in order to improve the practicability of diagnostic algorithm improve the convolutional neural network model, one of the most important principle is a complicate the convolutional neural network structure, improve its generalization ability.

4. Fault diagnosis based on convolutional neural network

Since 2006, HINTON et al[14]groundbreaking put forward the deep learning theory, it has attracted wide attention in academia and industry and opened up a new way for the utilization of big data. Convolution neural network (CNN) is the leader in the field of deep learning with its extraordinary feature learning and pattern recognition ability, and has achieved a series of breakthrough research results in the fields of speech recognition, image processing and target detection. In terms of traditional pattern recognition methods, fault diagnosis generally needs to rely on signal processing means to extract fault features and use pattern recognition algorithm for diagnosis. In practical application, these methods have their own limitations, and it is difficult to make full use of the increasing mass data of mechanical equipment. Therefore, CNN, as a "sharp tool" with strong feature learning and pattern recognition ability, has been widely concerned by field scholars and introduced into the field of mechanical equipment fault diagnosis.

4.1 Fault diagnosis principle of convolutional neural network

As one of the most important models of deep learning, CNN provides an end-to-end learning architecture. After years of development, CNN has derived multiple versions, and its capability of feature learning and modal recognition is also increasing day by day[15]. Usually, CNN first calculates its prediction error by means of supervised learning through feedforward operation (convolution operation, pooling operation, etc.), and then updates parameters through the back propagation algorithm based on gradient. The gradient feedbacks layer by layer from back to front until updating the first layer parameters of the network. Compared with the traditional neural network architecture, CNN introduces the concepts of weight sharing and receptive field, which greatly reduces the number of parameters to be learned and has stronger learning ability.

The thought method of diagnosis includes convolution layer, nonlinear activation layer, pooling layer, full connection layer, etc. Through the organic combination of these basic "components", the original feature space can be mapped to the feature domain with more representational significance, so as to realize the prediction of samples.Convolution layer and pooling layer play an important role in CNN diagnosis model, which is directly related to the advantages and disadvantages of feature extraction.

Convolution layer: carry out convolution operation layer by layer through the action of several convolution kernels, and obtain different feature graphs of each layer of network through the nonlinear activation layer. Each layer of network "combines" the convolution kernel output of the upper layer to extract the topological structure features hidden in the data layer by layer.

Pooling layer: it uses the overall statistical characteristics of the adjacent outputs of a certain position to replace the output of the network at this position. When the input makes a small amount of translation, the expression of the input is approximately unchanged, that is, translation invariance. The maximum pooling and average pooling are shown in equations 1 and 2.

$$p^{l(i,j)} = \frac{1}{W} \sum_{t=(j-1)W+1}^{jW} \alpha^{l(i,t)} \tag{1}$$

$$p^{l(i,j)} = \max_{(j-1)W+1 \le t \le jW} \{\alpha^{l(i,t)}\} \tag{2}$$

Where: $\alpha^{l(i,t)}$ is the activation value of tth neuron in l^{th} layer and i^{th} frame; W is the width of the pool area; $p^{l(i,j)}$ is the width of the pool area.

Pooling can be regarded as adding an infinitely strong priori: the function learned at this layer has invariance to a small amount of translation, and pooling can greatly improve the statistical efficiency of the network. On the other hand, because the pooling operation counts the surrounding data, the output can represent the input with less data, which means that the neural network has the ability to compress data and improve the computation rate of the neural network.

Through layer by layer convolution, pooling and other operations, CNN takes the minimum loss as the learning goal, and by extracting the translation, rotation and constant scaling feature representation of input data, it can effectively realize fault diagnosis and recognition with the help of top-level classifier.

4.2 Research status of fault diagnosis of convolutional neural network

According to the published literature, the application of CNN in fault diagnosis mainly has two ideas: 1) use CNN as a classifier; 2) use CNN as a method of feature extraction and pattern recognition.

CNN-based diagnostic methods are good at processing massive data, picking up features from them and distinguishing different patterns of information contained in different data. In addition, with the aid of CNN's unique advantages, the diagnosis model based on CNN characteristics can be adaptive extraction and pattern recognition for whole organic fusion, overcome traditional feature extraction method to extract the characteristics of shallow characterization of the problem of insufficient ability of fault signals, and to avoid the dependence on professional knowledge and experience, effectively reduce the complexity of the diagnostic process. It also has unique advantages in fault diagnosis. Therefore, CNN-based fault diagnosis technology is one of the development directions of intelligent diagnosis.

4.3 Challenges and prospects of fault diagnosis based on convolutional neural network

Due to the complexity and intelligence of modern mechanical equipment, it is more and more difficult to obtain accurate, complete and effective fault information, so it is urgent to use intelligent diagnosis method that integrates signal analysis, modeling and knowledge processing[16-17]. In a sense, how to obtain the characteristic distribution that can represent the fault state has become the focus of current mechanical fault diagnosis, which directly affects the effectiveness and reliability of fault monitoring and diagnosis. The CNN-based fault diagnosis method, by building a deep model and utilizing its unique advantages, can directly learn fault features from a large amount of data adaptively, so as to achieve intelligent diagnosis in the context of "big data", which is unified with the adaptive extraction of fault features and pattern recognition. After the development of recent years, the CNN-based fault diagnosis strategy has been published, but there are still some challenging problems in the application process that need to be further explored.

(1) Feature extraction of diagnostic model

Feature learning of the CNN diagnostic model is not known, and it is difficult to analyze the learned features in an abstract way. In order to better apply CNN for fault diagnosis, how to define the meaning of learned features of the CNN diagnostic model is still a problem to be solved.

(2) CNN model adaptive problem

Modern mechanical equipment produces a large amount of monitoring data, and its data types are diverse and inconsistent, which can reflect the state information of mechanical equipment from various aspects, such as vibration data and noise data. At the same time, signals generated by modern mechanical equipment are easily disturbed by the external environment and present complex and varied characteristics. It is difficult to ensure that the training data set used by the model is exactly the same as the distribution of the data set to be tested. Therefore, how to solve the adaptive problem of CNN model is also a direction worthy of research.

5. Conclusion

In this paper, the mechanism of turbine rotor fault is analyzed, the diagnosis method of traditional machine learning is summarized, the research idea and research status of the most advantageous CNN to realize mechanical fault diagnosis in deep learning are discussed, and the direction of further research is needed to realize fault diagnosis by CNN.

Reference

[1] He Qing, Xie Fangfang, Li Hong, et al. Vibration feature extraction of steam turbine units based on manifold learning method [J]. Vibration, test and diagnosis, 2014, 34 (4) : 705-708.
[2] Verma N K, Gupta V K, Sharma M, et al. Intelligent condition -based monitoring of rotating those using sparse auto - encoders [C]//Prognostics and Health Management (PHM), 2013 IEEE Conference on. IEEE, 2013:1-7.
[3] Faul A, Avenuse J t. Fast Marginal Likelihood Maximisation for failure Bayesian Models[J]. Proceedings of the Ninth International Workshop on Artificial Intelligence&Statistics, 2003,42 (22) : 3-6.
[4] Thayananthan a. temet-based pose estimation and tracking of 3D hand motion[D]. Cambridge: Degree thesis of Department of Engineering of University of Cambridge, 2005:25-30.
[5] Rasmussen C E, Williams C k. Gaussian Processes for Machine Learning[M]. Commonwealth of Massachusetts, MIT Press, 2006:211-244.
[6] Damoulas T, Girolami m. Probabilistic multi-class multi-kernel Learning: On Protein Fold Recognition And Remote Homology Detection[J]. Bioinformatics, 2008,24 (10) : 10-2390.
[7] Psorakis I, Damoulas T, Girolami M a. Multiclass relevance vector machines: sparsity and accuracy[J]. IEEE Neural Netw, 2010,21 (10) : 1588 -- 1598.
[8] Wu Lianghai. Network intrusion detection based on particle swarm optimization correlation vector machine [J]. Microelectronics and computer, 2010,27 (05) : 181-184.
[9] Li Gang, Wang Guilong, Xue Huifeng. Genetic algorithm optimization method for RVM kernel parameters [J]. Control engineering, 2010,17 (03) : 335-337.
[10] Liu junshun. Research on fast classification algorithm of correlation vector machine based on

IOP Conf. Series: Materials Science and Engineering **686** (2019) 012025 doi:10.1088/1757-899X/686/1/012025

clustering [D]. Master of science thesis of probability theory and mathematical statistics, south China university of technology, guangzhou, 2011:34-40.

[11] Zhao palm. Research on optimization methods of related vector machines [D]. Guangxi: master's thesis of computer application technology, guangxi university, 2012:24-27.

[12] Liu Changyuan. Research and application of multiple classification algorithm of correlation vector machine [D]. Harbin: engineering doctoral thesis of signal and information processing, Harbin engineering university, 2013:32-37.

[13] Ouyang Ting. Research on fast multi-classification algorithm of correlation vector machine based on clustering [D]. Master of science thesis of probability theory and mathematical statistics, south China university of technology, guangzhou, 2015:21-25.

[14] Hinton G E, Salakhutdinov R r. Reducing the dimensionality of data with neural networks[J]. Science, 2006,313 (5786) : 504-507.

[15] He K, Zhang X, Ren S, et al. Deep residual learning for image recognition[C]//IEEE Conference on Computer Vision and Pattern recognition[J]. IEEE Computer Society, 2016: 770-778.

[16] Li Wei. Current situation and development trend of intelligent fault diagnosis technology for complex systems [J]. Computer simulation, 2004,21 (10) : 4-7.

[17] Chen Jin. Application of signal processing in fault diagnosis of mechanical equipment [J]. Vibration and shock, 1999(4) : 83-86.

A Bi-Gaussian Acoustic Emission Model for Sliding Friction

Yuan Yin[1], Songtao Hu[2], Weifeng Huang[1,*], Xiangfeng Liu[1], Ying Liu[1] and Zixi Wang[1]

[1]State Key Laboratory of Tribology, Tsinghua University, Beijing 100084
[2]State Key Laboratory of Mechanical System and Vibration, Shanghai Jiao Tong University, Shanghai 200240

[*]Corresponding author's E-mail address: huangwf@tsinghua.edu.cn

Abstract. Acoustic emissions (AEs) generated by the sliding friction has been an effective media for analyzing the tribological interaction between surfaces. The existing AE model has revealed the relationship between the AEs generated by sliding friction and the surface parameters from a single-stratum Gaussian surface perspective, but researches have confirmed that two-process or worn surfaces show bi-Gaussian stratified characteristics. A bi-Gaussian AE model was established through analyzing the elastic deformation energy of the contacting asperities on the faces, which were represented using the bi-Gaussian stratified theory. Two numerically generated surfaces and four real worn surfaces were employed for analysis. The bi-Gaussian AE model and the existing Gaussian AE model, along with the deterministic AE model as the reference, were respectively applied to the surfaces to obtain the AE power variation with separation. The results show that the bi-Gaussian model has good accuracy for all the surfaces involved, exhibiting better stability than the Gaussian model, because the bi-Gaussian model better reflects the real contact status of the surfaces.

1. Introduction

Acoustic emission (AE) refers to the acoustic wave generated by the local energy release of materials. Besides the classical application of material damage detecting [1-3], it has been an effective approach for analyzing the tribological interaction between surfaces, exhibiting excellent sensitivity, information density and engineering convenience. Applications involved machining [4,5], bearings[6-9], and seals[10-15], etc.

In order to analyze the AEs generated by friction, it is imperative to model the correlation between the tribological behavior and the AEs. Boness et al [16] looked into the AE signals generated by friction, establishing an empirical relationship between the integrated RMS and the wear volume. Lingard et al [17] found a linear relationship between the AE count and the wear volume, together with a relationship between the AE count and the friction-pair condition. Although the abovementioned research has revealed the relationship between the friction AE and the tribological behavior, their conclusions were not based on a physical foundation. Fan et al [18] took a step forward by establishing an analytic relationship between the AE RMS and surface topography under the assumption that the energy of the AE arose from the elastic energy release of the asperity contact, modeling the AE RMS model under a slight sliding condition of Gaussian surfaces.

Fan's work assumed the surface analyzed to be Gaussian. However, the existing research has shown that the two-process [19-21] or worn [22-25] surfaces exhibit the bi-Gaussian stratified characteristic, which means that the surfaces consist of two Gaussian components with different roughness scales, other than the Gaussian characteristic assumption commonly used. For characterizing a bi-Gaussian surface, the segmented separation method was firstly proposed [19-23,26]. Hu et al [27] proposed a continuous separation method, which has better performance than the segmented separation. The bi-Gaussian characterization tools were employed for practical problems, including establishing elastic asperity contact model [28-29], analysing the cavitation in water-film lubrication [30], and describing the surface topography evolution [24, 31-35]. In regard of the AEs, Hu [36] compared the simulated AEs generated by the original surfaces and the reconstructed surfaces with bi-Gaussian stratified theory, showed that the bi-Gaussian characteristics do make difference.

In this study, the bi-Gaussian AE model is established based on the modeling of elastic energy release and the bi-Gaussian stratified theory. Two different forms were used in defining the statistics of the upper strata. Two numerically generated surfaces and four real worn surfaces are analyzed by the Gaussian AE model, the bi-Gaussian model and a deterministic model, respectively.

2. The Gaussian AE model

When under light load, AEs arise from the release of the elastic deformation energy of the contacting asperities during a sliding friction. The energy stored by the summit indexed $k(k = 1,2, \dots, N_s)$ can be expressed as [18]

$$U_{Ek} = \frac{2}{5} W_k \delta_k. \tag{1}$$

Here W_k refers to the squeezing force on summit k, and δ_k refers to the exceeding height of summit k beyond the matching ideal plane.

The time needed for the release of an individual summit contact τ_k can be expressed as [18]

$$\tau_k = \frac{a_k}{v} = \frac{\left(\frac{W_k R_{sk}}{E}\right)^{\frac{1}{3}}}{v}. \tag{2}$$

The equivalent radius $R_{sk}(k = 1,2, \dots, N_s)$ can be yielded from equivalent radius in x and y directions [29].

In the following deduction, we made a revision to the existing model, as the contact load and energy release rate of individual summit were not clarified by Fan [18]. According to the Hertz theory, the squeezing force of summit $k(k = 1,2, \dots, N_s)$ can be expressed as

$$W_k = E \delta_k^{\frac{3}{2}} R_{sk}^{\frac{1}{2}}. \tag{3}$$

The energy release rate of the individual summit k is yielded as below:

$$\dot{U}_{Ek} = \frac{U_{Ek}}{\tau_k} = \frac{2}{5} E \delta_k^2 v. \tag{4}$$

Under the assumption that the summit height obeys a Gaussian distribution, a statistical model can be established:

$$\dot{U}_E = \frac{2}{5} E A v \eta_s \int_h^\infty (z_s - h)^2 \varphi_{z_{sm}, \sigma_s}(z_s) dz_s. \tag{5}$$

Here $\varphi_{\mu,s}(*)$ is the PDF (probability density function) of a Gaussian distribution with mean value μ and height distribution RMS s:

$$\varphi_{\mu,s}(x) = \frac{1}{\sqrt{2\pi} s} \exp\left(-\frac{(x-\mu)^2}{2s^2}\right). \tag{6}$$

By employing a constant K that serves as the represent of the conversion rate from the elastic energy to the stress wave energy detected, the AE power, represented by the RMS of AE wave, can be expressed as

$$RMS_G = K\sqrt{\dot{U}_E} = K\sqrt{\frac{2}{5}EAv\eta_s\int_h^{\infty}(z_s-h)^2\varphi_{z_{sm},\sigma_s}(z_s)dz_s}. \tag{7}$$

This model will be mentioned as the Gaussian AE model.

3. The bi-Gaussian AE model

3.1. Representing surface topography

The probability material ratio (PMR) is an effective representation of surface height distribution. The PMR, which is obtained from the surface height distribution $z_{i,j}(i = 1,2,\dots,N_x, j = 1,2,\dots,N_y)$, is defined as:

$$MR(z) = \frac{\text{count}(z_{i,j} > z)}{N_x N_y} = 1 - CDF(z), \tag{8}$$

$$PMR(z) = \sqrt{2}\text{inverf}(2MR(z) - 1). \tag{9}$$

Here, $\text{count}(z_{i,j} > z)$ is the count of the nodes with height $z_{i,j}$ higher than a certain height z. $\text{inverf}(\cdot)$ is the inverse of error function $\text{erf}(\cdot)$.

For a Gaussian surface, there is

$$PMR(z) = \sqrt{2}\text{inverf}\left(1 - 2\Phi\left(\frac{z - z_m}{\sigma_u}\right)\right). \tag{10}$$

$\Phi(\cdot)$ refers to the CDF of the standard Gaussian distribution

$$\Phi(x) = 0.5 + 0.5\,\text{erf}\left(\frac{x}{\sqrt{2}}\right). \tag{11}$$

A bi-Gaussian surface is the mixture (obeying surface combination theory) of two Gaussian surfaces: z_u with mean height z_{mu} and height distribution RMS (root-mean-square) σ_u; z_l with mean height z_{ml} and height distribution RMS σ_l. According to the continuous separation method [27], the probability material ratio of the bi-Gaussian surface can be yielded as:

$$PMR(z) = \sqrt{2}\text{inverf}\left(2\left(1 - \Phi\left(\frac{z - z_{mu}}{\sigma_u}\right)\right)\left(1 - \Phi\left(\frac{z - z_{ml}}{\sigma_l}\right)\right) - 1\right). \tag{12}$$

The undetermined parameters z_{mu}, σ_u, z_{ml}, σ_l can be obtained by fitting the measured PMR with Eq. (12).

3.2. Establishing the bi-Gaussian AE model

For a bi-Gaussian stratified surface, if not under a heavy load, the actual contact occurs only on the summits within the upper component [28,29], indicating that the entire-surface summit parameters (i.e., σ_s and η_s) in Eq. (7) should be replaced by those of the upper component, as below:

$$RMS_{biG} = K\sqrt{\frac{2}{5}EAv\eta_{su}\int_h^{\infty}(z_s-h)^2\varphi_{z_{smu},\sigma_{su}}(z_s)dz_s}. \tag{13}$$

Note that it is the parameters of summit height distribution $z_{sk}(k = 1,2,\dots,N_s)$ instead of those of surface height distribution $z_{i,j}(i = 1,2,\dots,N_x, j = 1,2,\dots,N_y)$ that are used here, which is not directly obtained by the original continuous separation method above. Two adjustments are considered in order to define bi-Gaussian summit distribution parameters:

Definition A

Find all summits $z_{sk}(k = 1,2,\dots,N_s)$ on the entire surface, and then fit the summit PMR with Eq. (16) so that z_{smu}^A, σ_{su}^A, z_{sml}^A and σ_{sl}^A will be obtained[29].

$$MR_s(z_s) = \frac{\text{count}(z_{sk} > z_s)}{N_s}, \tag{14}$$

$$PMR_s(z_s) = \sqrt{2}\text{inverf}(2MR_s(z_s) - 1), \tag{15}$$

$$PMR_s^*(z_s) = \sqrt{2}inverf\left(2\left(1-\Phi\left(\frac{z_s - z_{smu}^A}{\sigma_{su}^A}\right)\right)\left(1-\Phi\left(\frac{z_s - z_{sml}^A}{\sigma_{sl}^A}\right)\right)-1\right). \tag{16}$$

The knee point concept from the segmented separation method is still remained, even though there is not an actual corner on the PMR curve yielded by continuous separation method [27]. The knee point for summit distribution is defined below:

$$r_{s,knee} = \frac{z_{smu}^A - z_{sml}^A}{\sigma_{su}^A - \sigma_{sl}^A}, \tag{17}$$

$$z_{s,knee} = \frac{z_{sml}^A \sigma_{su}^A - z_{smu}^A \sigma_{sl}^A}{\sigma_{su}^A - \sigma_{sl}^A}. \tag{18}$$

The $z_{sk}(k = 1,2,\dots,N_s)$ satisfying $z_{sk} > z_{s,knee}$ are considered to belong to the upper component, thus obtaining the summit density η_{su}^A, while the rest belonging to the lower component, obtaining η_{sl}^A.

Definition B

Divide the surface with the knee point of surface height distribution [27]

$$r_{knee} = \frac{z_{mu} - z_{ml}}{\sigma_u - \sigma_l}, \tag{19}$$

$$z_{knee} = \frac{z_{ml}\sigma_u - z_{mu}\sigma_l}{\sigma_u - \sigma_l}. \tag{20}$$

Then, find all summits $z_{sk}(k = 1,2,\dots,N_s)$ on the entire surface, and the z_{sk} satisfying $z_{sk} > z_{knee}$ are distinguished as the upper component to fit a Gaussian distribution to yield z_{smu}^B and σ_{su}^B, and obtain summit density η_{su}^B, while the rest belonging to the lower component and obtaining z_{sml}^B, σ_{sl}^B and η_{sl}^B.

4. Results and discussion

The analysis was proceeded on the following six surfaces (see Figure 1): (a) Simulated Gaussian surface; (b) Simulated bi-Gaussian surface; (c) SiC (a measured worn surface, and so were the following surfaces); (d) TC; (e) RiC; (f) MiC. Surfaces (c) to (f) were real worn surfaces of a mono-spring face seal, each of which had undergone a 24h running-in and a 100h service, and surface (b) was a simulated bi-Gaussian surface yielded by bi-Gaussian reconstruction[27]. Surface (a) was a simulated Gaussian surface reconstructed for reference.

The PMR curves of the target surfaces above were calculated in terms of the surface height and summit height, respectively, as shown in Figure 2. It can be seen that the PMR curves of the Gaussian surface (a) exhibit notable linear feature while those of the real worn surfaces do not. Instead, the PMR curves of real worn surfaces exhibit characteristic similar to that of the simulated bi-Gaussian surface, that is, consisting of two connecting sloping line (or being bilinear) approximately, revealing that the real worn surfaces do consist of two strata and should be characterized using bi-Gaussian stratified theory rather than a traditional entire-surface statistic. Then, the continuous separation method was applied to the surface height PMR and summit height PMR, respectively. The results are displayed in Figure 2, and listed in Table 1.

After obtaining surface statistics, the Gaussian model and bi-Gaussian model, along with a deterministic model as a reference, are applied to the surfaces above. The deterministic model can be simply described as summing up the $\dot{U}_{Ek}$ in Eq. (4). The dimensionless AE RMS, defined as

$$RMS^* = \frac{RMS}{K\sqrt{EAv}} \tag{21}$$

is yielded and illustrated in Figure 3.

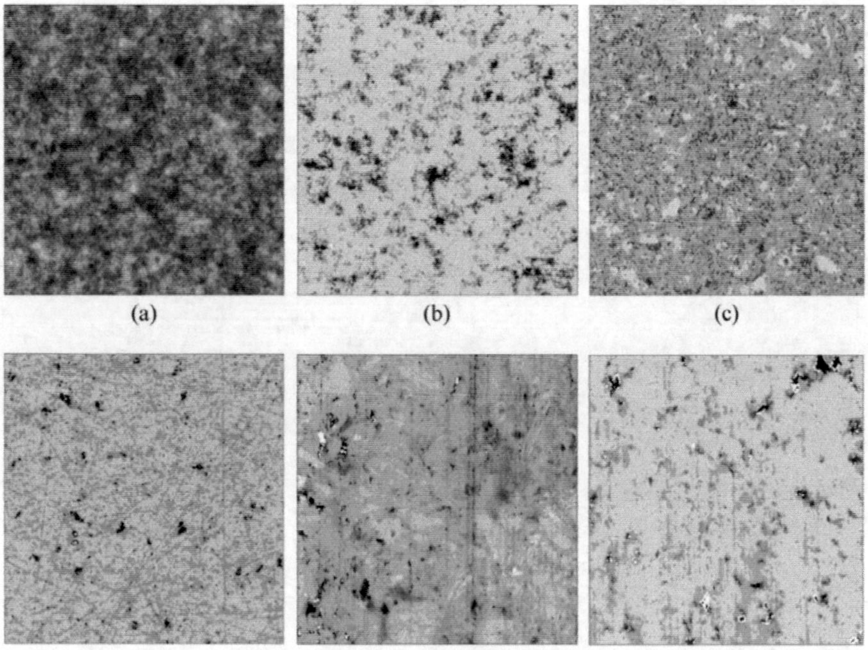

Figure 1. The analyzed surfaces

Table 1. Bi-Gaussian surface statistics

Surface data		(b)	(c)	(d)	(e)	(f)
Surface	$\sigma/\mu\mathrm{m}$	0.429	0.092	0.034	0.064	0.199
	$\sigma_u/\mu\mathrm{m}$	0.1009	0.0548	0.0115	0.0353	0.0488
	$z_{mu}/\mu\mathrm{m}$	0.1979	0.0419	0.0025	0.0090	0.0379
	$\sigma_l/\mu\mathrm{m}$	1.0570	0.1985	0.4246	0.3513	1.4473
	$z_{ml}/\mu\mathrm{m}$	0.7697	0.1421	0.8507	0.5364	2.2888
	r_{knee}	0.5981	0.6973	2.0534	1.6690	1.6095
	$z_{knee}/\mu\mathrm{m}$	0.1375	0.0037	-0.0211	-0.0499	-0.0406
Summit Entire-surface	$\sigma_s/\mu\mathrm{m}$	0.276	0.067	0.015	0.048	0.172
	$z_{sm}/\mu\mathrm{m}$	0.161	0.058	0.013	0.029	0.039
	$\eta_s/(\mu\mathrm{m})^{-2}$	0.491	0.110	0.136	0.107	0.100
Summit Definition A	$\sigma_{su}^A/\mu\mathrm{m}$	0.0986	0.0569	0.0080	0.0406	0.0524
	$z_{smu}^A/\mu\mathrm{m}$	0.2416	0.0780	0.0130	0.0338	0.0634
	$\eta_{su}^A/(\mu\mathrm{m})^{-2}$	0.3775	0.0959	0.1314	0.1043	0.0911
	$\sigma_{sl}^A/\mu\mathrm{m}$	1.0580	0.1617	0.2065	0.5061	1.6785
	$z_{sml}^A/\mu\mathrm{m}$	1.3558	0.2252	0.4775	1.0856	3.0428
	$\eta_{sl}^A/(\mu\mathrm{m})^{-2}$	0.1132	0.0142	0.0046	0.0027	0.0085
	$r_{s,knee}$	1.1612	1.4052	2.3398	2.2599	1.8322
	$z_{s,knee}/\mu\mathrm{m}$	0.1271	-0.0021	-0.0057	-0.0580	-0.0326
Summit Definition B	$\sigma_{su}^B/\mu\mathrm{m}$	0.1052	0.0603	0.0104	0.0412	0.0543
	$z_{smu}^B/\mu\mathrm{m}$	0.2194	0.0610	0.0114	0.0322	0.0589
	$\eta_{su}^B/(\mu\mathrm{m})^{-2}$	0.3671	0.0938	0.1347	0.1038	0.0917
	$\sigma_{sl}^B/\mu\mathrm{m}$	1.2482	0.1736	0.6372	0.8851	2.9599
	$z_{sml}^B/\mu\mathrm{m}$	1.0133	0.1412	0.2629	0.4338	1.6467
	$\eta_{sl}^B/(\mu\mathrm{m})^{-2}$	0.1237	0.0163	0.0013	0.0033	0.0079

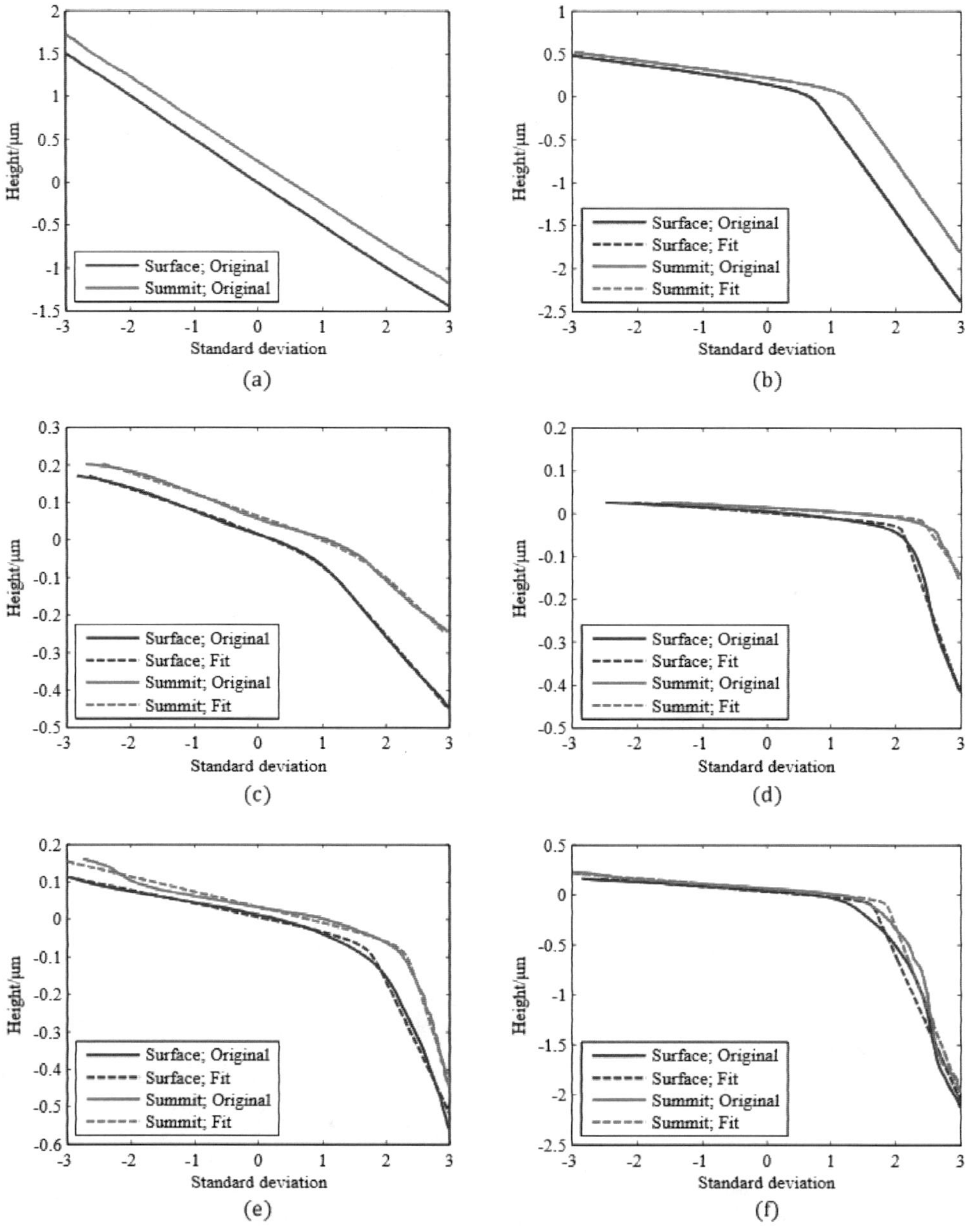

Figure 2. The PMR curves of the analyzed surfaces

For the Gaussian surface (a), the Gaussian model and the deterministic model agrees perfectly. However, for the simulated bi-Gaussian surface (b), the Gaussian model yields obviously greater AE power, while the result of the bi-Gaussian model, with either upper strata statistic definition A or B, shows perfect agreement with that of the deterministic model. This significant error of the Gaussian model is caused by taking the entire-surface statistics to analyze a friction process which only involves the higher summits, whose height distribution RMS is quite smaller than that of the entire surface. However, the bi-Gaussian model can correctly characterize the distribution of the summits involved.

For the real worn surfaces (c), (d), (e) and (f), analysis shows that:

(1) The deterministic model and the bi-Gaussian model result in similar results for either the surface (d) or (f), while the result of the Gaussian model shows obvious error.

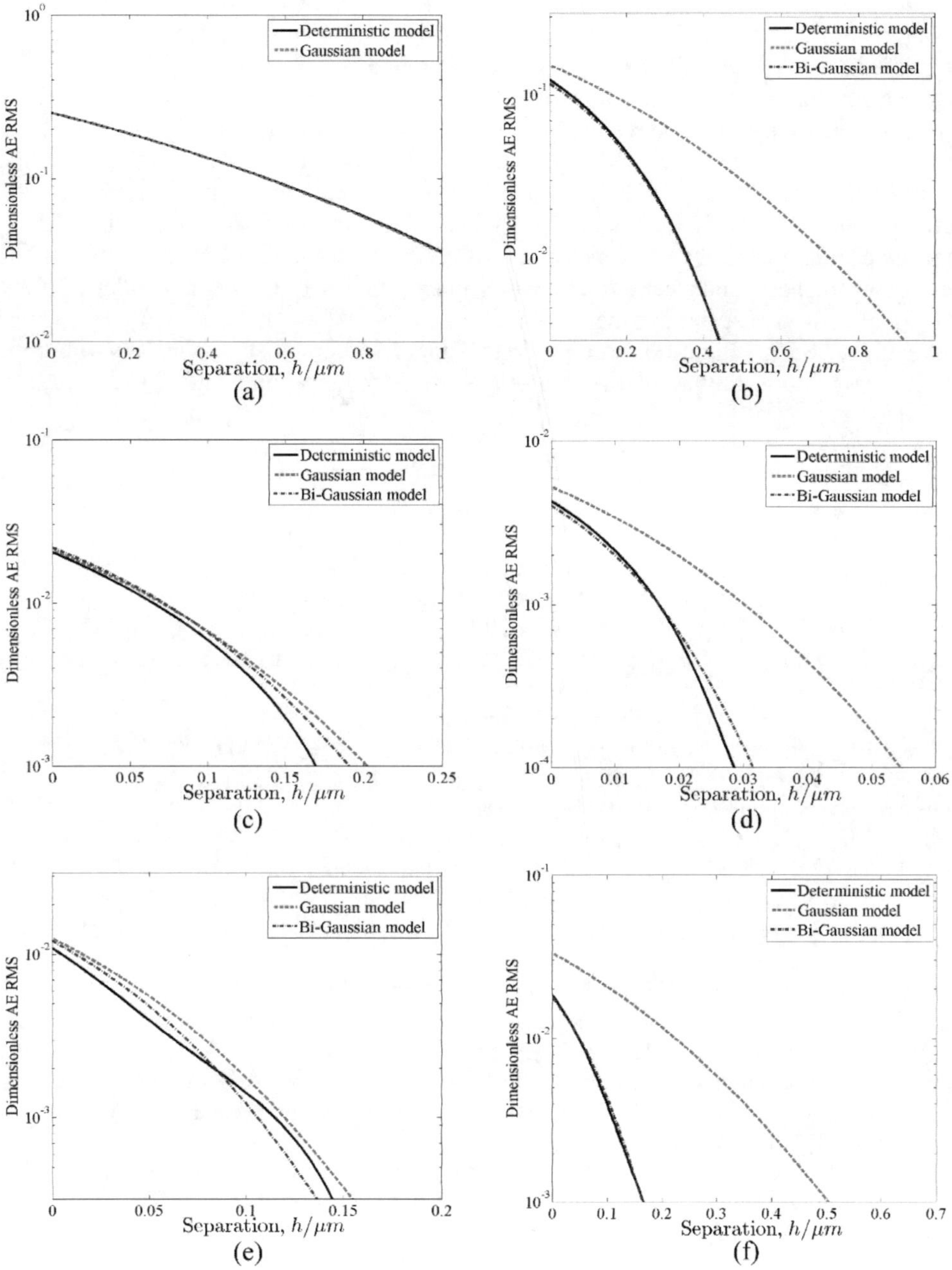

Figure 3. Dimensionless AE RMS

(2) The Gaussian model and the bi-Gaussian model both agree with the deterministic model for surfaces (c) and (e). The Gaussian model doesn't show an obvious error like what it shows for surfaces (b), (d) and (f), because that the two strata of either surface (c) or (e) do not differ significantly.

(3) For surface (e), both the Gaussian model and the bi-Gaussian model failed to capture the AE characteristics accurately, because of the relatively poor regularity of the highest summits.

As for the comparison of the two upper strata statistic definitions, the bi-Gaussian model performs well no matter using either A or B, but the definition A leads to slightly better performance. Therefore, the definition A is more recommended.

5. Conclusion

A bi-Gaussian model was established based on the existing Gaussian model and bi-Gaussian stratified theory. As a reference, a deterministic model is also employed.

The AE characteristics of several simulated surfaces and real worn surfaces are analyzed. The following can be concluded from the results:

(1) The bi-Gaussian model exhibits good accuracy for the bi-Gaussian surfaces, including the simulated bi-Gaussian surface and the real surfaces worn in experiments, thus exhibiting good stability.

(2) Although the existing Gaussian model shows small error for surfaces which don't possess strong stratified characteristic, the error, brought by the ignorance of stratified characteristic, increases with the enlarging disparity between strata, which means that a bi-Gaussian modeling for the AEs generated from bi-Gaussian surfaces is necessary.

(3) Between the two potential definitions for the upper strata statistics, the one which define the statistics by directly fitting the probability material ratio is more recommended to be used in the bi-Gaussian model.

References

[1] Roberts T M and Talebzadeh M 2003 *J. Constr. Steel. Res.* 59 pp 675-712
[2] Ben A I, Karra C, Mahi A, Guerjouma R and Haddar M 2014 *Appl. Acoust.* **86** pp 106-117
[3] Masmoudi S, Mahi A, Turki S and Guerjouma R 2014 *Appl. Acoust.* **86** pp 118-125
[4] Cho S S and Komvopoulos K 1997 *J Manuf. Sci. Eng.* **119** pp 238-246
[5] Beggan C, Woulfe M, Young P and Byrne G 1999 *Int. J. Adv. Manuf. Tech.* **15** pp 737-742
[6] Hawman M W and Galinaitis W S 1988 *IEEE Ultrason. Symp.* **2** pp 885-889
[7] Abdullah M A and Mba D 2006 *Mech. Syst. Signal. Pr.* **20** pp 1537-1571
[8] He Y, Zhang X and Friswell M 2009 *J. Vib. Acoust.* **131** 061012.
[9] Pandya D H, Upadhyay S H and Harsha S P 2013 *Expert Syst. Appl.* **40** pp 4137-4145
[10] Orcutt F K 1969 *Proc. 4th Int. Conf. Fluid Sealing* pp 205-217.
[11] Miettinen J and Siekkinen V 1995 *Wear* **181-183** pp 891-900
[12] Huang W, Lin Y, Gao Z, Fan W, Suo S and Wang Y 2013 *Tribol. Lett.* **49** pp 379-384
[13] Huang W, Lin Y, Liu Y, Liu X, Gao Z and Wang Y 2013 *Tribol. Lett.* **52** pp 253-259
[14] Braccesi C and Valigi M C 2014 *Tribol. Int.* **71** pp 125-131
[15] Towsyfyan H, Wei N, Gu F and Ball A. 2015 *Proc. 54th Annu. Brit. Conf. Non-Destruct. Test.*
[16] Boness R J and McBride S L 1991 *Wear* **149** pp 41-53
[17] Lingard S, Yu C W and Yau C F 1993 *Wear* **162-164** pp 597-604
[18] Fan Y, Gu F and Ball A 2010 *Wear* **268** pp 811-815
[19] Whitehouse D J 1985 *P. I. Mech. Eng.* B **199** pp 263-270
[20] Malburg M C, Raja J and Whitehouse D J 1993 *CIRP Ann-Manuf. Techn.* **42** pp 637-639
[21] Sannareddy H, Raja J and Chen K. 1998 *Int. J. Mach. Tool. Manuf.* **38** pp 529-536
[22] Thomas T R 1972 *Wear* **22** pp 84-90
[23] King T G, Watson W and Stout K J 1978 *Proc. 4th Leeds-Lyon Symp. Tribol.* pp 333-343.
[24] Hu S, Brunetiere N, Huang W, Liu X and Wang Y 2017 *Tribol. Int.* **112** pp 75-85
[25] Hu S, Huang W, Brunetiere N, Liu X and Wang Y 2017 *Friction* **5** pp 32-44
[26] 1998 *ISO 13565-3, Surface Texture: Profile Method*
[27] Hu S, Brunetiere N, Huang W, Liu X and Wang Y 2016 *Tribol. Int.* **102** pp 454-462
[28] Leefe S E 1998 *Tribol. Ser.* **34** pp 281-290
[29] Hu S, Brunetiere N, Huang W, Liu X and Wang Y 2017 *J. Tribol.* **139** 021403
[30] Hu S, Huang W, Brunetiere N, Song Z, Liu X and Wang Y 2016 *Tribol. Int.* **104** pp 328-341.
[31] Pawlus P and Grabon W 2008 *Precis. Eng.* **32** pp 342-347.
[32] Grabon W, Pawlus P and Sep J 2010 *Tribol. Int.* **43** pp 1882-1892
[33] Dzierwa A, Pawlus P and Zelasko W 2014 *P. I. Mech. Eng.* J **228** pp 1195-1210.
[34] Pawlus P, Michalczewski R, Lenart A and Dzierwa A 2014 *P. I. Mech. Eng.* J **228** pp 1374-1391.
[35] Hu S, Brunetiere N, Huang W, Liu X and Wang Y 2017 *Tribol. Int.* **114** pp 186-191
[36] Hu S, Huang W, Shi X, Peng Z, Liu X and Wang Y 2018 Tribol. Int. 119 pp 308-315

Research on Applicability of Fiber Bragg Grating Sensor for Ship Structure Monitoring System

Haoyun Tang[1,*], Huilong Ren[2] and Lianhui Jia[2]

[1] Navigation College, Dalian Maritime University, Dalian 116026, China;
[2] College of Shipbuilding Engineering, Harbin Engineering University, Harbin 150001, China

[*]Correspondence: tanghaoyun@dlmu.edu.cn

Abstract. A complete application research of Fiber Bragg Grating sensing technology in ship structure stress monitoring and evaluation system is studied in this paper. The working principles of Fiber Bragg Grating strain sensor in monitoring system are briefly introduced. And a kind of new differential FBG strain sensor for ship structure stress monitoring is applied and verified by a series of calibration and force measurement test systematically. Through finite element calculation and test towards ship structure, some defects of monitoring sensor used in bending deformation are found. Then, a three-dimensional strain sensor arrangement for stress monitoring of hull structure is designed and developed. The satisfactory test results are obtained by this sensor arrangement. Finally, the accuracy and reliability of hull structure monitoring system is improved further through this research results.

1. Introduction

Hull structural strength is one of the main factors affecting ship navigation safety. In order to ensure that ship has enough hull strength to resist the destructive effect caused by the environmental loads, various rules and maintenance plans have been taken in the process of ship design and construction[1-2]. However, for ship navigation in the complex environment, the external wave impact often shows some strong randomness. And these random factors cannot be predicted accurately in the ship design and construction process. Thus, the structural stress monitoring and evaluation system for ship is needed to ensure the safety of ship operation. The ship structure safety monitoring system is a kind of engineering intelligent system with sensors embedded in the hull structure, which endows similar life characteristics of structural health using self-diagnosis function[3]. Ship structural safety monitoring technology is also an interdisciplinary and comprehensive technology. It includes dynamics, material science, sensing technology, communication technology and database technology. Herein, the sensing technology is basic and critical. In the late 1970s, Lindemann[4] took some exploratory researches. Due to the limitation of computer technology and sensor technology at that time, the research did not obtain ideal results. With the development of related technology, the hull structure monitoring becomes an achievable goal gradually. Relevant ship companies, which are in the United States, Britain and France, have developed their own structural monitoring software[5-10]. And the monitoring sensor used in ship structure has gradually changed from strain gauge to optical fiber sensor[11-14]. Fiber optic sensors have incomparable advantages over strain gauges, so they are widely used in ships with complex structures. But the reliability and layout of fiber optic sensors to monitor various complex deformations in ship still have some difficulty. In this paper, the principle of

IOP Conf. Series: Materials Science and Engineering **686** (2019) 012027 doi:10.1088/1757-899X/686/1/012027

optical fiber sensor are introduced. And a new differential sensor monitoring method is presented. Through experiments, the error reason and distribution characteristics of sensor are analysed. In addition, three-directional strain sensor arrangement is given in paper to monitor the vertical bending deformation. Through the finite element calculation and experiment, the design of sensor frame is modified, which further improves the accuracy of optical fiber sensor on hull structure monitoring.

2. Fiber Bragg Grating Sensing Principle

Fiber Bragg Grating (FBG) is a kind of core made by fiber materials. Using its photosensitive properties, the core can make the optical waveguide change periodically or aperiodically. And it also has the function of wavelength selectivity. Because the grating fiber has the advantages of small size and low fusion loss, it has been widely used in the field of optical fiber communication and sensing. With the wide application of FBG, the types of FBG are varied. According to the distribution of refractive index along the grating axis, the ultraviolet-written FBG can be divided into uniform FBG and non-uniform FBG. Uniform fiber Bragg Grating is a kind of FBG whose amplitude and period of refractive index remain unchanged along the fiber axis. And, the amplitude and period of refractive index in Non-uniform Fiber Bragg Grating is varied. In hull structure monitoring, the designed Bragg Grating core belongs to uniform FBG. It only reflects the light about the central wavelength λ_B and transmits the light of other wavelengths. In fact, its basic optical properties are just like a reflective optical filter, and the optical equation is as follows.

$$\lambda_{cw} = 2r_e T_g \tag{1}$$

Where, r_e is the effective refractive index, and T_g is the grating period. Because the refractive index and grating period is sensitive to dynamic strain, the centre wavelength λ_{cw} will be offset when the environmental physical field like strain changes. Through measuring the offset of wavelength in fiber core, the dynamic waves of external field can be obtained, and the FBG sensing can be realized[15]. Herein, the reflection bandwidth is in the order of magnitude about 10^{-1} *nm*, and the reflectivity can reach 95%~100%.

2.1. *Effect Of Strain on Central Wavelength*

In ship structure monitoring, the sensors are mainly used to monitor the structural strain change at its location. Therefore, the establishment on relationship between the measured value of sensor and the structural strain is very important. In fiber sensors, the variation of central wavelength caused by strain is calculated based on Equation (2).

$$\Delta\lambda_{cw}^{\varepsilon} = 2\left\{\varepsilon T_g r_e - \varepsilon T_g \frac{r_e^3}{2}[(1-\nu)p_{12} - \nu p_{11}]\right\} \tag{2}$$

Herein, P_{12} and P_{11} are the elasto-optic coefficients. ν is the Poisson ratio of core material, and ε is strain variation. Then, the effective elasto-optic coefficient is defined as Equation (3).

$$p_e = \frac{r_e^2}{2}[p_{12} - \nu(p_{11} + p_{12})] \tag{3}$$

Finally, the relationship between the measured value of sensor and the structural strain is as follows.

$$\Delta\lambda_{cw}^{\varepsilon} = (1 - p_e)\varepsilon\lambda_{cw} \tag{4}$$

In fact, in addition to strain, the temperature change around the sensor will also cause the change of central wavelength. Similar method is used, and the relationship between the measured fluctuation of sensor and the temperature variation is also as follows.

$$\Delta\lambda_{cw}^{T} = (\alpha + \zeta)\Delta W_T \lambda_{cw} \tag{5}$$

Where α is the coefficient of thermal expansion. ζ is the thermo optic coefficient, and ΔW_T is temperature variation around the sensor. It is assumed that the effects of strain and temperature on the central wavelength of grating are independent at each other. Thus, it is derived from equation (4) and (5) that, when the strain and temperature change simultaneously, the relative change of central wavelength on the grating can be expressed as follows.

$$\frac{\Delta\lambda_{cw}}{\lambda_{cw}} = (1 - p_e)\varepsilon + (\alpha + \zeta)\Delta W_T \tag{6}$$

In summary, Equation (6) is the basic principle formula of Fiber Bragg Grating sensing technology.

3. Fiber bragg grating strain sensor

3.1. Sensor Structure and Working Principle

In section 2.1, it is found that both fluctuation of structure strain and temperature can affect the central wavelength of fiber bragg grating. Therefore, strain sensors using fiber bragg grating needs to consider the interference of temperature on structure strain measurement. A simple and common solution is to equip these strain sensors with some temperature compensation sensors. The strain sensor measurements is corrected by temperature-induced changes in the central wavelength of FBG measured by temperature sensors, then the influence of temperature on strain measurement can be eliminated. Because the different temperature changes at each compartment, a large number of temperature sensors are arranged at these monitoring positions. These temperature sensors greatly increase the burden of the monitoring system. And temperature data also occupy a large amount of system computing space[16]. In order to improve the efficiency of monitoring system, another advanced processing method is adopted using the FBG sensor, namely differential Fiber Bragg Grating strain sensor. The sensor uses two fiber bragg gratings encapsulated in the shield to eliminate the influence of temperature on the central wavelength by the differential method. Thus, the temperature compensation sensor is omitted. The structure of FBG sensor is shown in Figure 1(a).

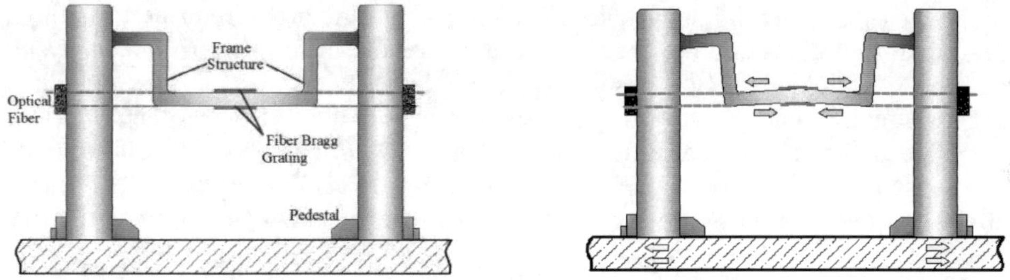

(a) (b)

Figure 1. Differential Fiber Bragg Grating sensor: (a) Structure of differential fiber bragg grating sensor; (b) Deformation of sensor under tension stress.

When the sensor on the hull structure suffers tension stress, the frame of Fiber Bragg Grating sensor produces mid-hog deformation. Because the two fiber gratings are attached to the upper and lower surfaces of frame beam, the upper grating is pulled and the lower grating is compressed, as shown in Figure 1(b). The central wavelength of upper and lower grating appears deviation by strain and temperature. Because the position of grating at upper and lower surfaces is close to each other, the temperature change is considered as the same value. Thus, when the differential method is used, the effect of temperature fluctuation is offset. And the influence caused by strain is enlarged due to the two deformations in opposite directions. Similarly, when the sensor on the hull structure is subjected to compressive stress, the frame structure produces mid-sag deformation. In the frame beam, the upper grating is compressed and the lower grating is pulled. This design improves the sensitivity of sensor towards strain, so that it can obtain larger range and higher accuracy.

3.2. Sensor calibration test

The linearity of monitoring sensor was taken in a calibration test. In order to simulate tensile deformation in sensors, the ends of the monitoring sensor frame were fixed on two supports of calibrator. Then, the sliding support of calibrator was driven to produce relative displacement and another support kept fixed. Strain was also calculated and compared with the measured value by differential FBG sensor.

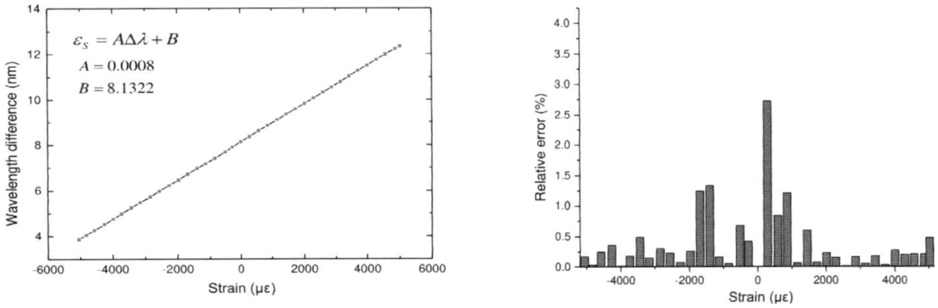

(a) (b)

Figure 2. Calibration test: (a) Relation between central wavelength difference and tensile deformation; (b) Distribution of relative error about differential FBG sensor.

In the calibration test, it is found that the central wavelength difference of FBG sensor is linear with the tensile deformation. And the strain measured formula is obtained by the least square method from the results of calibration experiment, as shown in Figure 2(a). The error between the theoretical strain and the measured value from the differential FBG sensor is also calculated. The maximum of absolute error is 7.795 $\mu\varepsilon$, which account for 2.72% of whole structural strain. Figure 2(b) shows the distribution of relative error about the differential FBG sensor. It observed that the relative error is remarkable when the structural strain approaches zero, and gradually reduces to both ends. In fact, when the structural strain is close to zero, the accuracy requirement of measurement system is higher. Thus, the error of differential FBG sensor is relatively large. For the actual ship structure monitoring, the structural stress, which is caused by waves and lead to the damage of hull structure, is often remarkable. The range of structural strain is more than 2000 $\mu\varepsilon$. Therefore, the differential FBG sensor always has small relative error in actual environmental structure response. The measurement by differential FBG sensors can meet the application requirements on hull structure monitoring.

3.3. One-directional sensor monitoring test

In practical engineering application, the differential FBG sensors need to be installed on column and crossbeam of hull structure, so the monitoring performance of sensors should be tested in detail through the one-directional sensor monitoring test.

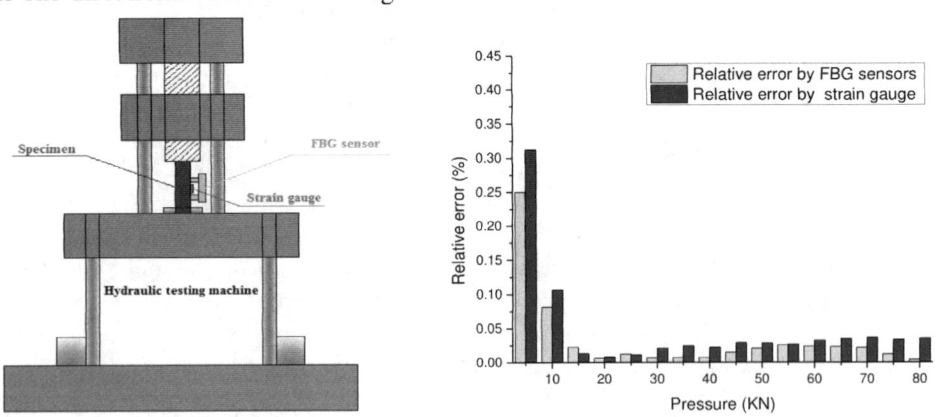

(a) (b)

Figure 3. One-directional sensor monitoring test: (a) Design of one-directional sensor monitoring test.; (b) Error analysis on the differential FBG sensor and strain gauge.

IOP Conf. Series: Materials Science and Engineering **686** (2019) 012027 doi:10.1088/1757-899X/686/1/012027

The hydraulic testing machine was used to simulate the structural response caused by wave load in the mechanical test, as shown in Figure 3. And the steel specimen is adopted to simulate the hull structure. During the mechanical test, the specimen was fixed on the testing machine firstly. Then, the FBG strain sensor was installed on the specimen, and the strain gauge was also pasted on the specimens under the frame of differential FBG sensor to observe the measurement difference between the FBG strain sensor and strain gauge. When the designed force using the hydraulic testing machine was applied on the specimen, the structural strain was monitored by two kinds of monitoring sensors. Through observation on the results of one-directional sensor monitoring test, it is found that the monitoring by two kinds of sensors have good linearity. But they have different errors compared with the theoretical value. Figure 3(b) show the relative error using FBG strain sensor and strain gauge. Obviously, the relative error of FBG sensor is smaller than that of strain gauge at the high and low pressure conditions. Especially in the violent strain, the one-directional sensor monitoring using the FBG sensors is more reliable.

3.4. Three-directional sensor monitoring test

Fiber Bragg Grating sensor can obtain high reliability data when measuring unidirectional tension and compression. In fact, the hull structure not only suffers tension and compression deformation, but also some bending deformation. In the stress monitoring of hull structures with complicated wave loads, the three-dimensional strain sensor arrangement is often used to obtain the composite stress[17]. And the corresponding strength theory is also applied to evaluate the hull structure strength. Thus, the three-directional sensor monitoring test was taken, as shown in Figure 4(a).

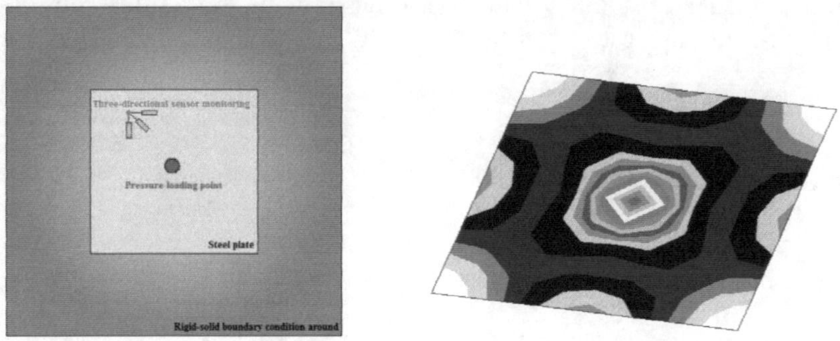

(a) (b)

Figure 4. Three-directional sensor monitoring analysis: (a) Design of the three-directional sensor monitoring test; (b) Finite element calculation on steel plate.

In order to compare the test results with the theoretical values, the finite element model about rigid steel plate was established, and the surface pressure was used to load this steel plate. Figure 4(b) show the stress distribution of steel plate under designed pressure. Obviously, the stress concentration occurs near the applied point of pressure. The stress decreases gradually as it expands around. Figure 5(a) shows the relationship between the measured and theoretical values of Mises stress with the different external load. It is found that the measured values deviate from the theoretical values gradually with the external pressure increasing. In fact, when the hull structure is suffering bending deformation, the bottom of differential FBG sensor frame will rotate inward. And the bracket of FBG sensor frame also incline inward θ under the influence of bending moment. Thus, the extrusion force at cross-beam increases, as shown in Figure 5(b). With the external load increasing, the inclination angle becomes more obvious, and the measured values deviate from the theoretical values more remarkable. It is suggested that some monitoring errors of differential FBG sensor rise with the large pressure loading. Through analysis on the reason caused monitoring error, the differential FBG sensor is revised. The height of sensor bracket is modified and reduced by a third in the improved planI. Also, the height of sensor bracket is reduce by half in the improved planII.The same sensor experiments are repeated. Figure 6(a) shows the relationship between the measured and theoretical values of the improved sensor with different external load. It is found that the coincidence between the measured and theoretical values of improved sensor is much higher with the sensor bracket height decreasing. Figure 6(b) also shows the distribution of measurement errors before and after this improvement. Obviously, with the implementation of improved plan, the relative error of differential FBG sensor decreases gradually. The relative error is reduced by 3.24% in the large pressure case. It is indicated that the improvement for differential FBG sensor can increase monitoring accuracy effectively. The max relative error of revised sensor is less than 2.50%, which also meets the requirements of structural monitoring.

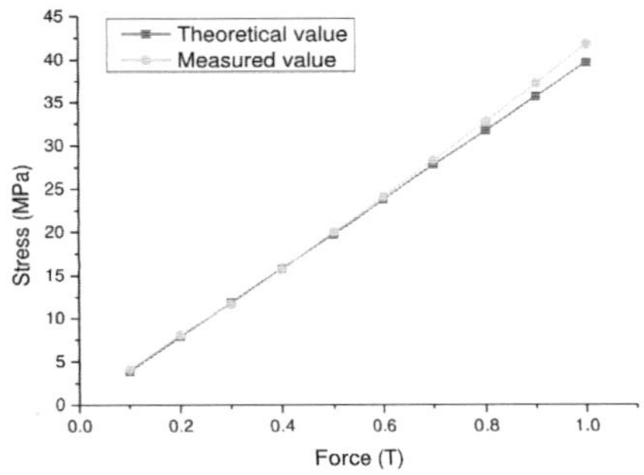

(a)

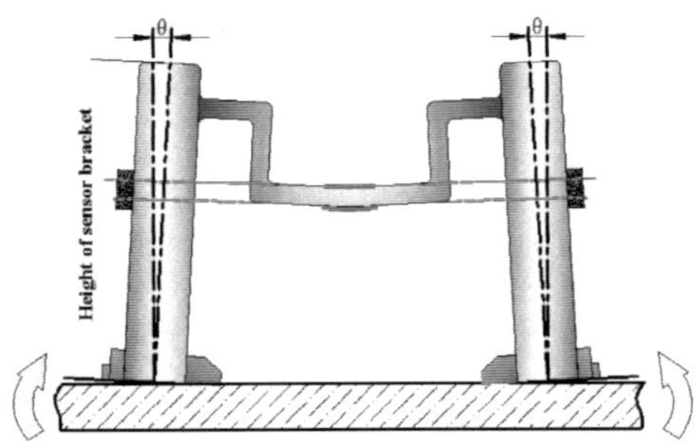

(b)

Figure 5. Results and analysis of three-directional sensor monitoring: (a) Results of three-directional sensor monitoring test; (b) Differential FBG sensor under bending deformation.

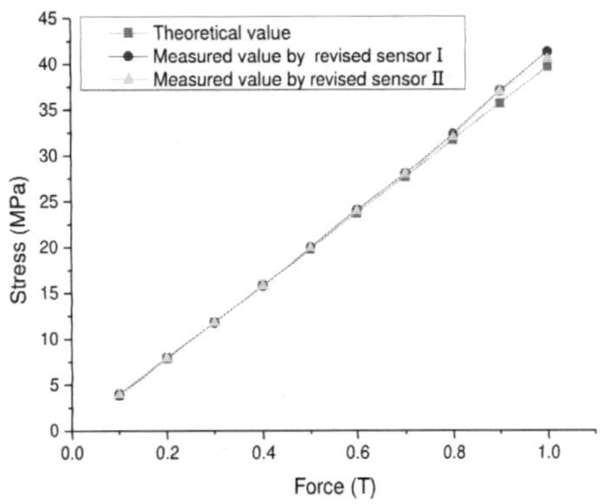

(a)

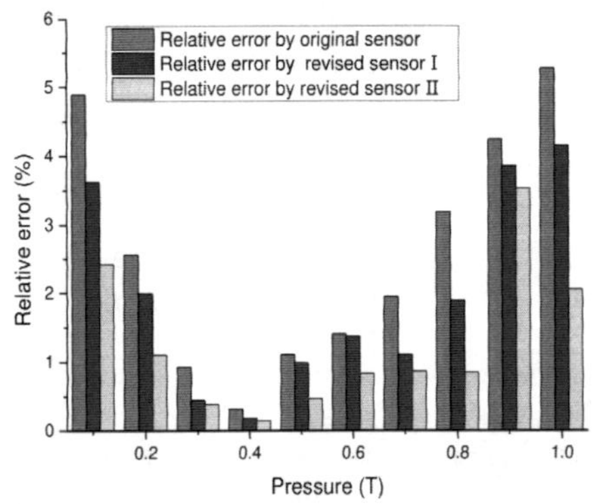

(b)

Figure 6. Results and error distribution of three-directional sensor monitoring after improvement: (a) Results of three-directional sensor monitoring revised test; (b) Error distribution on the improved differential FBG sensor.

4. Conclusion

The application of FBG sensing technology in stress monitoring towards hull structure is studied by a series of sensor tests. The test results show that the differential FBG strain sensor has high accuracy and good linearity on structural measurement, compared with the traditional strain gauge. And the defects of three-dimensional FBG sensor used in bending stress measurement are analyzed and improved. Finally, the improved three-dimensional fiber Bragg grating strain sensor is verified effectively and can be used to monitor the complex stress state of hull structure.

Acknowledgments

The project was supported by the Fundamental Research Funds for the Central Universities (3132019134). The authors express their gratitude for the fund support.

References

[1] Lloyd's Register of Shipping 2006: Rules for the Classification of Trimarans, *Lloyd's Register Rules and Regulations*.
[2] China Classification Society 2015: Guidelines for Fatigue Strength of Ship Structure, *China Classification Society Press*.
[3] San G.G 1986: Application of structural reliability theory ship longitudinal strength. *Shipbuilding*. **4**, pp102-123.
[4] Lindemann K., Odland J., Strengtheagen J. 1977: On the application of hull surveillance systems for increased safety and improved structural utilization in rough weather. *Transactions of SNAME*, **85**, pp131-166.
[5] Li G.X. 1994: Joint development of hull monitoring system in Europe. *Ships & Yachts*, **11**, pp40.
[6] Huston S.P. 2010: Structural health monitoring of a high speed naval vessel using ambient vibrations. *Georgia Institute of Technology*.
[7] Okasha N.M., Frangopol D.M., D Saydam, 2011: Reliability analysis and damage detection in high-speed naval craft based on structural health monitoring data. *Structural Health Monitoring*, **10**, pp361-379.
[8] Kokarakis J.E., Goye J. t, Jong G.D., Segretain J.F. 2009: Predictor-corrector system for health monitoring of the hull structure of bulk carriers. *Royal Institution of Naval Architects-International Conference on Design and Operation of Bulk Carriers*, pp145-159.
[9] Xu W.B, Yang C.L., Shi X.Y. 2006: Design and implementation of vessel structure safety monitoring system based on CAN bus. *Journal of Dalian Maritime University*, **32**, pp17-20.
[10] He J. 2007: Real time monitoring system for ship deck deformation. *Ship science and technology*, **29**, pp45-48.
[11] Hjelme D.R., Bjerka L., Neegard S. 1997: Application of Bragg grating sensors in the characterization of scaled marine vehicle models. *Applied Optic*, **36**, pp328-336.
[12] Wang G, Pran K, Sagvolden G, 2001: Ship hull structure monitoring using fiber optic sensors. *Smart Materials and Structures*, **10**, pp472-478.
[13] Karianne P., Geir B.H, Sagvolden G. 2002: Wavelength multiplexed fiber Bragg grating system for high-strain health monitoring applications. *Measurement Science and Technology*, **13**, pp471-476.
[14] Sagvolden G., Pran K., Vines L. 2002: Fiber optic system for ship hull monitoring. *The 15th optical fiber sensors conference technical digest*, **1**, pp435-438.
[15] Jia L.H. 2014: Research on Ship Structure Monitoring and Assessment System, Doctoral thesis, *Harbin Engineering University*.
[16] Zhao C.H., Chen L.W., Ma H.Z. and Luo T.F.: Digital signal processing, *Electronic Industry Press*, 2008.
[17] Ren H.L., Tang L.Y. Jia L.H., Wan Q. 2017: Optimal layout of hull structure monitoring points based on risk assessment. *Journal of Harbin Engineering University*, **38**, pp1-7.

IOP Conf. Series: Materials Science and Engineering **686** (2019) 012028 doi:10.1088/1757-899X/686/1/012028

Surfactant-Free Electrodeposition of Ag Dendrites as Photocatalyst for Methylene Blue Degradation

Lintang Dhanasmoro and Setia Budi*

Department of Chemistry, Faculty of Mathematics and Natural Sciences, Universitas Negeri Jakarta, Jl. Rawamangun Muka, Jakarta Timur 13220, Indonesia

*Corresponding author: setiabudi@unj.ac.id

Abstract. Ag dendrites have successfully fabricated through square-wave electrodeposition method. The electrochemical properties of Ag dendrites were examined via electrochemical impedance spectroscopy and cyclic voltammetry techniques in a solution containing 0.5 M KCl. It was found that Ag dendrites demonstrated equivalent series and charge transfer resistance of 91.3 Ω and 473.7 Ω, respectively. The highest specific capacitance value of Ag dendrites obtained 0.98 F/g at scan rate 10 mV/s. The recorded photocatalytic activities towards methylene blue (MB) degradation under UV light irradiation showed that photocatalytic performance of Ag dendrites reaches approximately 41.88%.

1. Introduction

The rapid development of the textile industry in recent decades has negative impacts on water quality due to untreated dye wastewater. Approximately 10-15% of dye wastewater discharged to the environment during the dyeing process takes place where the dye contained less than 50% potentially carcinogenic azo group [1]. Azo dye pollutants that released into the environment is continuously able to inhibit sunlight penetration into the water and disrupt the lives of aquatic plants [2]. One of azo dye which generally used as textile dyes and laboratory materials is methylene blue (MB). MB is difficult to degrade and can cause a serious health problem. Therefore some researchers have tried to find a decomposition technique of dyes, such as adsorption [3], biodegradation [4], and photocatalysis [2]. Using appropriate catalyst materials, photocatalysis is simpler and effectively applied to degrade organic pollutants. One of good material used as a photocatalyst is metallic nanoparticles. Among various metal, silver is one of the most potentials as a photocatalyst because silver has the highest visible light absorption [5]. Also, silver has high optical cross-sectional characteristics, showed superior electrocatalytic activity, stable, and has good reproducibility. One way to obtain a high catalytic activity could be done by optimizing the morphology of nanostructures. Dendritic structures can enhance photocatalytic activity due to their unique properties, such as low density, large surface area, and provides more active sites [6]. However, most of previously reported synthetic method of Ag dendrites used environmentally unfriendly organic additive such as surfactants [7] and templates [8]. Those additives not only make the synthetic procedure complicated but also introduce impurities and release toxic chemicals to the environment [9]. Herein, we report a simple approach for controllable synthesis of Ag dendrites by applying square wave electrodeposition. As-made Ag dendrites with well-defined branched morphologies exhibited enhanced catalytic activity, which is promising as photocatalyst for methylene blue degradation.

IOP Conf. Series: Materials Science and Engineering **686** (2019) 012028 doi:10.1088/1757-899X/686/1/012028

2. Experimental

2.1. Synthesis
Ag dendrites were prepared by electrodeposition method directly on ITO substrates PET-coated. The chemicals used were $AgNO_3$ (99.9% Ag, Merck) and HNO_3. In this work, the substrates were rinsed with ethanol and double distilled water before electrodeposition process.

All electrodeposition processes were performed in a three-electrode cell using a Pt wire as the counter electrode, ITO as the working electrode, and Ag/AgCl as the reference electrode. The upper potential limit, lower potential limit, and pulse duration were 1.75 V (Ag/AgCl), –0.3 V (Ag/AgCl) and 100 ms, respectively. The applied potentials were controlled with an eDAQ potentiostat model EA163 and Ecorder 401. After the electrodeposition, the resultant sample was rinsed using double-distilled water and dried.

2.2. Characterization
The morphology of the Ag dendrites was characterized by FEI Quanta 650 scanning electron microscope operated at 20 kV. To confirm Ag in deposits thus verified by energy dispersive X-ray Oxford Instruments EDX. Electrochemical impedance spectroscopy (EIS) measurement was performed using an electrochemical impedance analyzer (EDAQ; ERZ100) embedded to a potentiostat (EDAQ;163). A Platinum wire was used as the counter electrode, Ag/AgCl was used as a reference electrode and Ag dendrites films were used as the working electrode. All electrodes were soaked in a cell containing an electrolyte solution of 0.5 M potassium chloride (KCl).

2.3. Photocatalytic activity measurement
The photocatalytic activity of Ag dendrites was appraised by the degradation of MB. In this work, Ag dendrites film was prepared in 5 ppm MB solution. Then the solution was irradiated with UV and the absorbance of the solution was measured with GBC Cintra 2020 UV-VIS Spectrophotometer at a wavelength of 664 nm. Every 20-second reaction, approximately 3 mL of the solution measured its absorbance until 30 min irradiation. The same procedure was performed on the 1-9 ppm MB standard solution to obtain the calibration curve. Furthermore, the catalytic activity of MB degradation for ITO substrate was also measured. The results of absorbance measurements were then used in the linear regression equation of the MB solution so that MB concentrations were obtained. Thus, the photodegradation efficiency of MB is calculated as a degradation percentage.

3. Results and Discussion

3.1. Morphological analysis
In this work, Ag was synthesized by the electrodeposition method with 600 s deposition time. Figure 1a shows SEM images of Ag dendrites with the magnification of 10,000-fold magnification. Ag dendrites were formed showing the main trunk with secondary branches. The morphology of the crystal strongly depended on the distance of the formation conditions and thermodynamic equilibrium. According to the literature, dendrites are generally formed under non-equilibrium growth conditions [10]. The whole growth processes occur in a non-equilibrium state, as the fast nucleation and growth contribute to the formation of complicated structures. In this study, we used a lower negative potential compared than the standard reduction potential of Ag, which resulted in a non-equilibrium condition. Figure 1b shows EDX spectra of Ag dendrites with 600 s deposition time. From the spectra, it is indicated that Ag was presented in the deposit, while other elements come from the substrates.

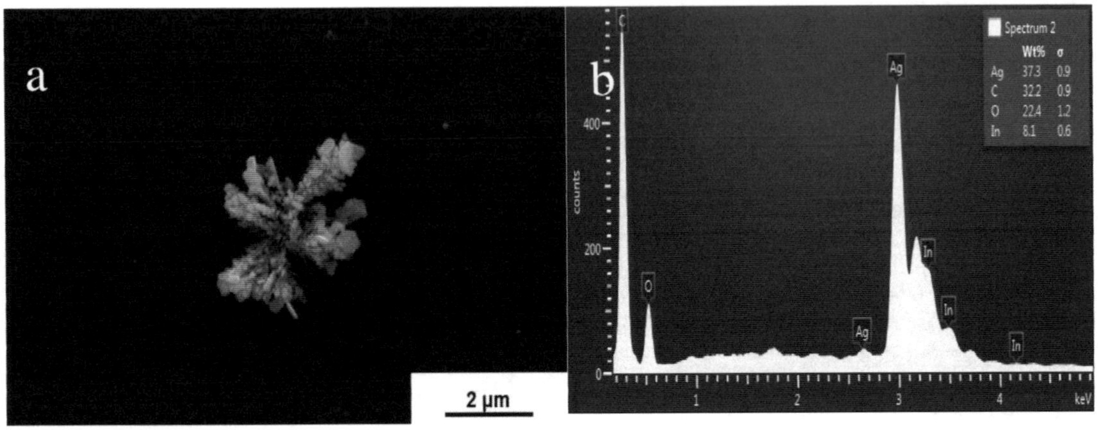

Figure 1. (a) SEM images and (b) EDX spectrum of Ag dendrites over deposition time of 600 s.

3.2 Electrochemical Properties

To understand the electrochemical properties of Ag dendrites on the ITO substrate, electrochemical impedance spectroscopy (EIS) was applied to investigate the electrical conductivity and ion transfer of the cells. Impedance testing is performed using EIS on the test cells in the frequency range between 1 Hz to 10 kHz in 0.5 M KCl solution as an electrolyte.

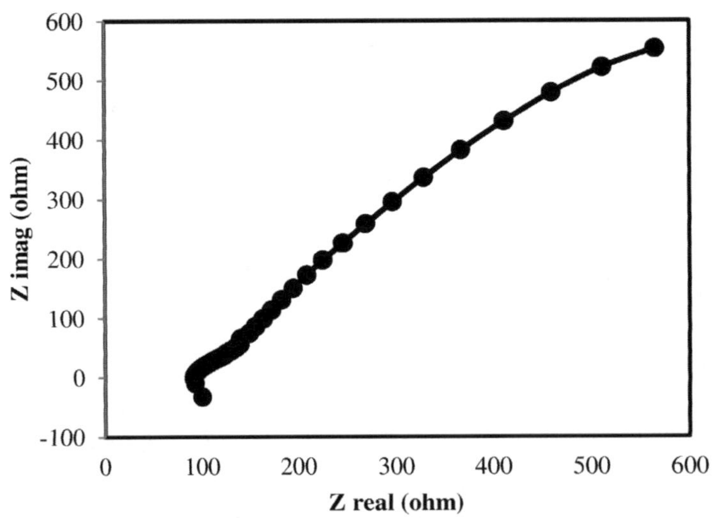

Figure 2. Nyquist plots of Ag dendrites at 600 s deposition time.

Figure 2 shows the Nyquist plot between real impedance and imaginary impedance in Ag dendrites of 600 s deposition time. The intercept value from the plot on the axis at high frequencies shows the equivalent series resistance (R_s) value, which includes inherent resistance to the material, electrolyte resistance, and contact resistance on the interface between electrodes and electrolytes [11]. The charge-transfer resistance R_{ct} or Faraday resistance is correlated with the intercalation and deintercalation of ions. The charge transfer resistance (R_{ct}), which results from the diffusion of electrons, can be calculated from the diameter of the semicircle in the high-frequency range. The result shows that R_s and R_{ct} value of Ag dendrites at 600 s deposition times were 91.3 Ω and 473.7 Ω. These results indicate good conductivity due to low internal electrode resistance.

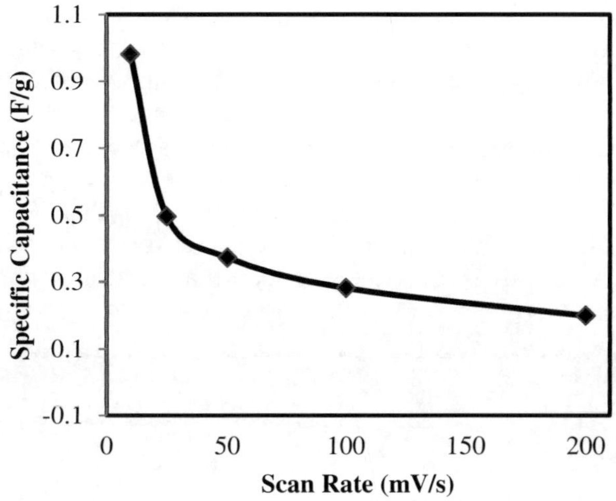

Figure 3. Specific capacitances of Ag dendrites at five scan rates.

The measurement of cyclic voltammetry of Ag dendrites was carried out at five different scan rates and each sample was measured 10 test cycles in the voltage range -0.9 V to -0.2 V. The calculated specific capacitances are plotted in Figure 3. Figure 3 displays a decreasing trend of specific capacitance value with the raise of scan rate. Slow scanning rate allows the electrolyte to penetrate the material pores more thoroughly and enlarge the contact with the electrode surface then producing a greater capacitance value [12]. Whereas at higher scanning speeds, positive ions in the electrolyte can only approach the outer surface of the electrode resulting in a smaller capacitance value [13].

3.3. Photocatalytic activity

The degradation of Methylene Blue (MB) was chosen to evaluate the photocatalytic performance of the as-prepared nanostructures. Methylene blue has a maximum absorption peak at 664 nm. Figure 4 reveals the absorption spectra of the MB solution under UV light until 30 minutes irradiation with Ag dendrites as photocatalyst. It can be seen that the value of intensity is decreases due to chromophore degradation. When illuminated by light, the electrons in the photocatalyst material will be activated to move from the valence band to the conduction band. Thus, it causes holes (positive charges) in the valence band, which then reacts with H_2O to form hydroxyl radicals which have strong oxidation so that it will degrade organic compounds. The electrons in the conduction band (negative charge) also reduce organic compounds [14].

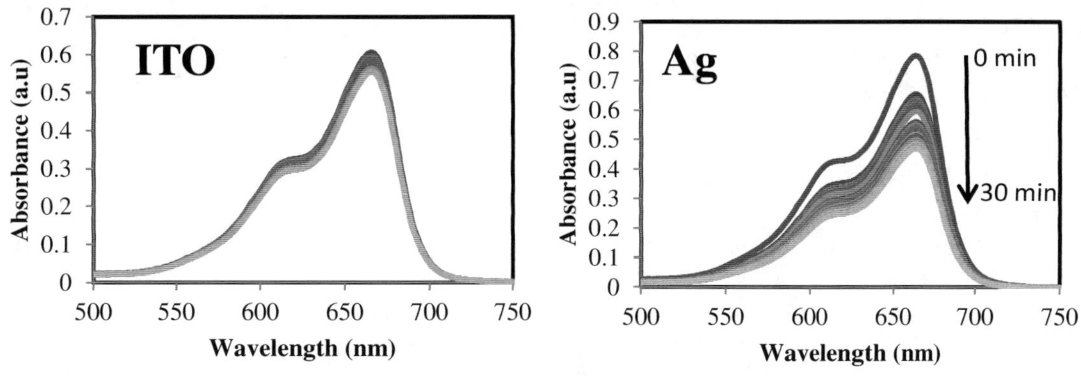

Figure 4. Absorbance spectrum of MB degradation under UV irradiation.

Figure 5 displays the degradation percentage for Ag dendrites 600 s deposition time was 41.88% after 30 minutes of UV irradiation. As a reference, the photocatalytic activity with only ITO is slightly degraded. There are two ways can be tried to enhance the photocatalytic efficiency of the as-prepared materials. First, the decrease in the recombination of the photo-generated electrons and holes allows them to take part in the photocatalytic reaction. Second, optimizing the morphology and structure of the products can get more reactive species. In our experiments, an enhanced photocatalytic activity could be attributed to dendritic morphology. This structure could provide more active site to adsorb reactive species and allow effective transport for the reactant molecules to get to the active sites due to reduced diffusion path of ions. From figure 5, it can be seen that Ag dendrites have good photocatalytic performance because of wider branch, lower charge transfer resistance, and higher specific capacitance value. Hence, it is clearly understood that dendrites morphology on silver has enhanced its photocatalytic activity.

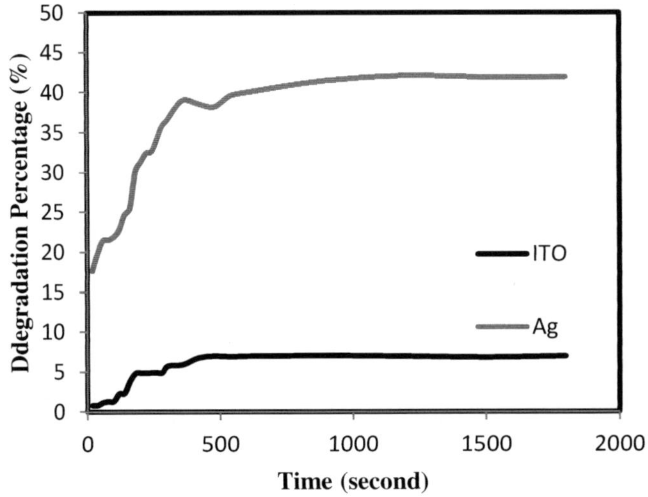

Figure 5. Photocatalytic degradation of MB under UV irradiation.

4. Conclusion

In this work, a novel Ag dendrites on ITO substrates has been achieved through square-wave electrodeposition method without any surfactant used. Such dendritic structures with broad branch give effective degradation of organic dye molecules. The as-prepared substrates show that Ag dendrites a good candidate for photocatalysis application towards methylene blue degradation with the degradation percentage of 41.88 % in just 30 minutes of UV irradiation.

Acknowledgment

The authors thank the Universitas Negeri Jakarta, which supported this work under research scheme Penelitian Unggulan Universitas 2019.

Reference

[1] Blackburn RS, Burkinshaw SM. A greener approach to cotton dyeings with excellent wash fastness. *Green Chem* 2002; 4: 47–52.
[2] Inamuddin. Xanthan gum/titanium dioxide nanocomposite for photocatalytic degradation of methyl orange dye. *Int J Biol Macromol* 2019; 121: 1046–1053.
[3] Tan IAW, Hameed BH, Ahmad AL. Equilibrium and kinetic studies on basic dye adsorption by oil palm fibre activated carbon. *Chem Eng J* 2007; 127: 111–119.
[4] Prado AGS, Bolzon LB, Pedroso CP, et al. Nb2O5 as efficient and recyclable photocatalyst for indigo carmine degradation. *Appl Catal B Environ* 2008; 82: 219–224.
[5] Lee M-K, Tu H-F. Au–ZnO and Pt–ZnO Films Prepared by Electrodeposition as Photocatalysts. *J Electrochem Soc* 2008; 155: D758.

[6] Shuanglong Lu KE, Danhua Ge, Jun Guo, Liang Wang, Hongjing Wang HG. One-Pot Synthesis of PtRu Nanodendrites as Efficient Catalyst for Methanol Oxidation Reaction. Epub ahead of print 2016. DOI: 10.1039/C6NR08895C.
[7] Agrawal VV, Kulkarni GU, Rao CNR. Surfactant-promoted formation of fractal and dendritic nanostructures of gold and silver at the organic – aqueous interface. 2008; 318: 501–506.
[8] Jiang G, Wang LI, Chen TAO, et al. Preparation and characterization of dendritic silver nanoparticles. 2005; 0: 1681–1683.
[9] Cai W, Pu K, Ma Q, et al. Insight into the fabrication and perspective of dendritic Ag nanostructures Insight into the fabrication and perspective of dendritic. 8080. Epub ahead of print 2017. DOI: 10.1080/17458080.2017.1335890.
[10] Fu L, Wang A, Zheng Y, et al. Electrodeposition of Ag dendrites/AgCl hybrid film as a novel photodetector. *Mater Lett* 2015; 142: 119–121.
[11] Rakhi RB, Chen W, Cha D, et al. Substrate Dependent Self-Organization of Mesoporous Cobalt Oxide Nanowires with Remarkable Pseudocapacitance.
[12] Mishra N, Shinde SM, Vishwakarma R. MWCNTs Synthesized From Waste Polypropylene Plastics and its Application in MWCNTs synthesized from waste polypropylene plastics and its application in super-capacitors. Epub ahead of print 2013. DOI: 10.1063/1.4810063.
[13] N. Sabari Arul, D. Mangalaraj, R. Ramachandran, A. Nirmala Grace and JIH. Fabrication of CeO2/Fe2O3 composite nanospindles for enhanced visible light driven photocatalyst and supercapacitor electrode. Epub ahead of print 2015. DOI: 10.1039/C5TA02630J.
[14] Putri NA, Fauzia V, Iwan S, et al. Mn-doping-induced photocatalytic activity enhancement of ZnO nanorods prepared on glass substrates. *Appl Surf Sci* 2018; 439: 285–297.

Experimental Study on the Stiffness and Shape Layout of Flapping Wing

Wenqing Yang, Jianhua Xu and Jianlin Xuan

School of Aeronautics, Northwestern Polytechnical University, Xi'an, 710072, China

Email: yangwenqing@nwpu.edu.cn

Abstract. Flapping-wing Micro Air Vehicle (FMAV) is a kind of bionic air vehicle, which looks like natural flying creatures and flies by flapping-wings only. Now, there is an upsurge of relative research about FMAV, and one focus research domain is to develop a flapping-wing with high aerodynamic performance. The wind-tunnel experiment is one of the main research manner, which can be used to test and control many parameters with high accuracy, like force, torque, flapping angle, frequency, power, etc. Based on the wind-tunnel system and inspired by birds, we mainly study the effects of stiffness and shape layout of the flapping wing, and get some conclusions. And this work may set up a good fundamental for further experiment about developing wing with high efficiency.

1. Introduction

The flapping-wing micro air vehicle (FMAV) is a type of aircraft inspired by the flight patterns of natural birds, bats or insects. It generates lift and thrust through the flapping wings to overcome the gravity and drag during flight. Ideally, FMAV can hover or even fly backward, such as dragonfly. It also has the characteristics of fast cruising speed and high aerodynamic efficiency. [1-4]

FMAV is flexible in operation, small in size, light in weight, low in cost, and easy to carry. In the near future, the use of FMAV will certainly become more and more extensive, in social public security work, and in special occasions (such as dangerous or inaccessible), FMAV can play an important role in the information acquisition and delivery work.

As a new type of micro air vehicle, researches on it is developing [5-9]. The mode of flapping wing flight is completely different from that of fixed-wing flight. It involves unsteady aerodynamics with low Reynolds number and many other fields that people have not fully understood. At present, wind tunnel experiment is one of the main research manners. [10,11]

The flapping-wing flight mode of natural creatures can be roughly divided into two modes: low-frequency flight modes like birds and bats, and high-frequency flight modes like insects such as bees and crickets. Future flapping wings are likely to move towards these two modes. In the first flight mode, the bird-like FMAV has a relatively large size and a low flapping frequency. It is suitable for outdoor flight over a long distance. In another flight mode, the insect-like FMAV has a smaller size and a higher flapping frequency. It is suitable for small-scale indoor flight at short range.

The size and flapping frequency in this experiment are similar to birds, so here we only study the flight of birds. By observing the flight of birds, it is known that the wings of birds are generally not straight when flying, but are divided into two sections, inside and outside, and have a certain forward or sweeping angle in the spanwise direction, as shown in Figure 1. There arises some questions: will this layout have some beneficial effects on the aerodynamics or other aspects of the wings? If so, can this bring some inspiration to our FMAV? It is for this purpose that we have designed the layout experiment of this paper.

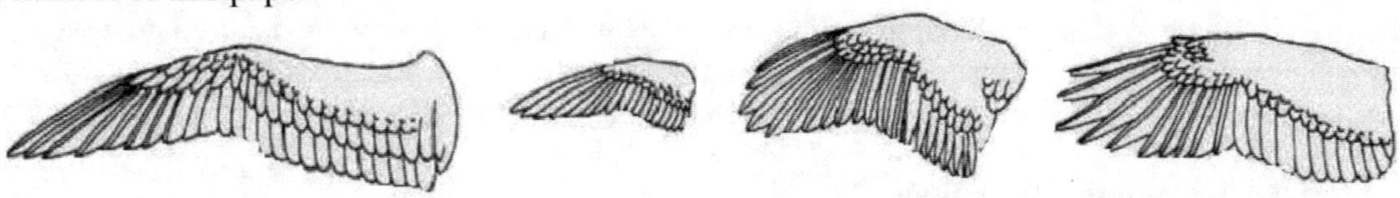

Albatross Swallow Pheasant Eagle

Figure 1. The wings of several birds

This paper is arranged as follows: In the section 2, the wind tunnel experimental system is shown, and the process of collecting and processing the experimental data. The steps of the experiment are introduced. In the section 3, the purpose of the experiment is described, and the experimental preparation, the parameters and fabrication of the wing. The experimental equipment configuration and experimental parameters, including wind speed, flutter frequency, angle of attack, and the processing of experimental data and the analysis of experimental results are also discussed. In the section 4, the experimental data are analyzed in detail, including the effects of the stiffness layout and shape layout of the wing on aerodynamics and power consumption.

2. Wind tunnel experiment system

2.1. Wind tunnel
The experiment was carried out at the low-turbulence wind tunnel in Northwestern Polytechnical University (NPU), shown in Figure 2(a). The wind tunnel is a DC low-speed wind tunnel, with the test section of 2m×1.05m×1.2m (length × width × height), which has a steady wind speed range from 5m/s to 25m/s, and the turbulence is 0.02%.

(a) Wind tunnel (b) Flapping mechanism

Figure 2. Low- turbulence wind tunnel in NPU and the flapping mechanism in experiment

2.2. Force/Torque sensor
The aerodynamic force generated by the flapping wing is relatively small, generally less than several hundred grams, and periodically changes with the flapping of the wing, which determines that the force/torque sensor should have high precision and high sensitivity. The Nano17 multi-axis force/torque sensor of ATI Company is suitable for our experiment.

2.3. Flapping mechanism
A flapping mechanism was specially designed and manufactured to achieve the flapping of the wing, shown in Figure 2(b). The mechanism consists of a four-bar mechanism, a brushless DC motor with a KV value of 6400/v, a diameter of 12 mm, a length of 30 mm, a working speed of 44800-52480 rpm, and a secondary gear reduction with a reduction ratio of I=60.2.

2.4. Measurement and control software

The experiment uses the LabView 8.2 to deal the measurement and control processes. The design of the whole software should include two modules. The measurement module is used to realize the acquisition and processing of experimental data, and the control module is used to realize the flapping control of the mechanism. The control module includes control of the power switch and the frequency.

3. Flapping wing layout experiment

In order to study the influence of different stiffness and shape layouts on aerodynamic forces, seven pairs of contrasting wings were specially fabricated. The main purpose of this paper is to study the influence of the layout. The seven wings are flat with no curvature. The basic shape of the wing is rectangular or parallelogram, and the initial installation angle of the wing is 0°. The wing is mainly composed of a skeleton and a skin, wherein the skeleton is made of carbon fibre rods with different diameters, and a membrane skin is laid on the surface, and the specific shape of the wing is shown in Figure 3. The skeleton of each wing consists of two main beams and six ribs. The wings have a span length of 220 mm and a chord length of 100 mm. The six ribs are evenly arranged along the span, with a spacing of 44 mm, and the ribs of each pair of wings are the same size.

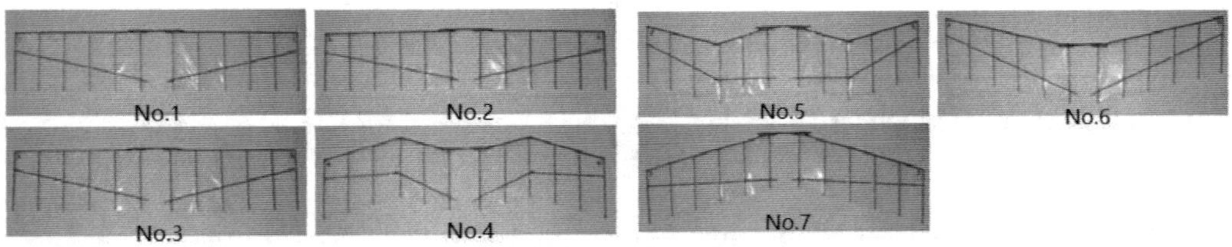

Figure 3. Experimental wings

The relevant parameters of the seven pairs of wings are shown in Table 1.

Table 1. Wing number and parameter arrangement

Grouping	Wing No.	Wing parameters			Remarks
First group (stiffness)		Front beam (mm)	Back beam (mm)	Rib (mm)	The wing is a flat wing
	1	2.0	1.5	1.0	
	2	2.0	1.5	1.5	
	3	1.5	1.5	1.0	
Second Group (shape)		Segment position	Sweep angle of inner segment	Sweep angle of outer segment	"-" means forward sweep "+" means backward sweeping
	4	40%	-15°	+15°	
	5	40%	+15°	-15°	
	6	----	-15°	-15°	
	7	----	+15°	+15°	

The first three pairs of wings (No.1-3) are the first group to explore the influence of the stiffness layout on the aerodynamic forces. From the previous analysis, the stiffness of the wing is mainly determined by the skeleton, so the stiffness of the wing can be changed by simply changing the diameter of the carbon rod. The advantage of this method is that it only changes the stiffness of the skeleton that makes the results more comparable.

The rear four wings (No.4-7) are the second group used to explore the effects of the wing's shape layout on aerodynamic forces. In order to ensure that their stiffness keeps the same, the carbon rod parameters of the skeleton are taken as the same as the wing No.1, and the relative positions of the spar and the rib are also kept as similar as the wing No.1. The wing No.4 and No. adopt two parts, and the segment positions are 40% from the wing root; the inner wing of the wing No.4 is swept forward, and the outer section is swept back, similar to the wing layout of the birds. The wing No. 5 is just the opposite. The inner section is swept back and the outer section is swept forward. The wing No.6 and No. 7 each have a certain forward or sweeping angle. The wing No. 6 is swept forward and the wing No. 7 is swept back.

The experimental wind speed is 6m/s, the flapping frequency is 8Hz, and the experimental angle of attack is taken from -3° to 36° with an interval of 3°.

4. Results and discussion

4.1. Effect of stiffness

The effects of the stiffness layout includes a total of three pairs of wings, No. 1, No. 2, and No. 3. The specific parameters of the three pairs of wings are shown in Table 1. It can be seen that the overall stiffness distribution of the three pairs of wings is approximately the same, but the stiffness is different. The beams of the wing No. 2 are the same as the wing No. 1. The ribs are thicker than the No. 1, so the No. 2 and the No. 1 have similar span stiffness, and the chord wise stiffness is larger than that of the No. 1. The rib of the No. 3 is the same as that of the No. 1, and the beams of the wing is thinner than that of the No. 1, so the chord stiffness of the wing No. 3 and the wing No. 1 is similar, and the span stiffness is smaller than that of the wing No. 1.

Generally speaking, for a rigid flat wing (which is very stiff and can be considered as barely causing structural deformation), when the angle of attack is zero, it cannot produce an effective average aerodynamic lift, but just some thrust. But for flexible wings, it can produce an effective average aerodynamic force including lift and thrust, which is mainly due to the flexible structural deformation. In the process of flexible wing flapping, the aerodynamic force acting on the wing causes the wing to be passively deformed, and this deformation in turn changes the aerodynamic force on the wing. This coupling phenomenon is beneficial in a certain extent. On the one hand it can generate additional effective average aerodynamic forces, on the other hand it can help the wing to unload and protect the structure.

Figures 4 shows the lift coefficient and thrust coefficient as a function of angle of attack for wings No. 1, No. 2 and No. 3.

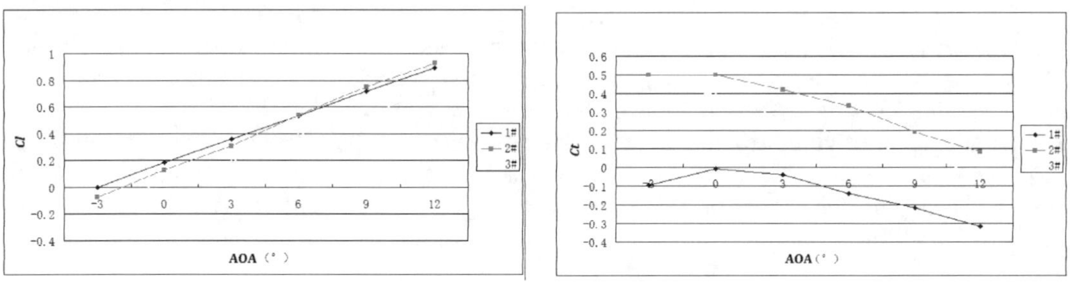

(a) Lift coefficient (b) Thrust coefficient

Figure 4. Aerodynamic coefficient of wings with different stiffness

It can be seen from the curves that the lift coefficient of the wing No. 2 is similar to that of the wing No. 1, and the thrust coefficient is larger than that of the wing No. 1; the lift coefficient of the wing No. 3 is smaller than that of the wing No. 1, but the thrust coefficient is bigger than the wing No. 1. Therefore, the chord wise stiffness has no significant influence on the lift force, and the impact of stiffness is greater on the lift force, and within a certain range, the lift stiffness is larger when the span wise stiffness is larger. For the thrust, the change in stiffness seems to be more sensitive, both the span wise stiffness and the chord wise stiffness have an obvious effect on it. The general trend is that the thrust is larger when the chord wise stiffness is larger.

4.2. Effect of shape layout

In the design of the FMAV, the shape of the wing directly determines the performance of the entire aircraft, which shows the importance of the shape layout of the wing. In fact, the shape selection of the flapping wing is crucial. A good shape can produce large aerodynamic force, but a bad wing shape may be counterproductive, making the aircraft performance reduced. As mentioned earlier, the wings of birds are always divided into two sections, then what is the advantage of this shape layout? Is it possible to generate a large effective aerodynamic force? To this end, we carried out a second set of experiments, which mainly consisted of four pairs of wings from No. 4 to No. 7. The specific parameters of the four pairs of wings are shown in Table 1. Generally speaking, the wing No. 4 is a bird-like layout. The inner section is swept forward and the outer section is swept backward; the wing No.5 is just opposite, the inner section is swept backward and the outer section is swept forward; the wing No.6 is swept forward; the 7th wing is swept backward.

According to the existing information, the inner and outer sections of the wing bear different "responsibility". The inner section mainly produces lift, and the outer section mainly generates thrust. The existing wing design method basically relies on this principle to achieve the requirements of large thrust and high lift. Figures 5 shows a comparison of the lift and thrust coefficients for the No. 1 and No. 4 to No. 7 wings, respectively.

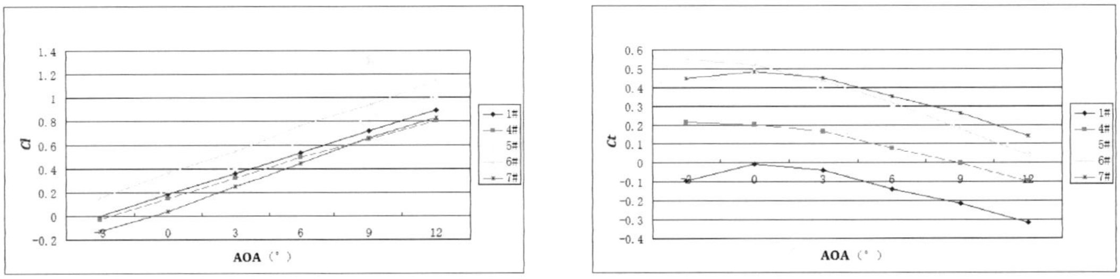

(a) Lift coefficient (b) Thrust coefficient

Figure 5. Aerodynamic coefficient of wings with different shapes

It can be seen from the curve in Figure 5(a) that the lift coefficient of the wing No. 6 is the largest, followed by the wing No. 5, and the wing No. 7 is the smallest. Therefore, in terms of lift generation, the forward swept wing is the best choice, the backward swept wing is the worst, and the wing No. 4, which imitates bird mostly, does not get the high lift we expected. But instead it is opposite to the shape of the wing No.4, the wing No.5 is capable of generating greater lift.

Referring again to Figure 5(b), the thrust coefficient curves of the respective wings generally show the following relationship: the wing No. 7 is the largest, the wing No. 5 and the wing No. 6 are equivalent, occupying the second position, and wing No. 1 is the smallest. That is, the swept wing produces the most thrust, and the imitation wing No.4 does not produce the large thrust as we would expect.

From the above analysis, the lift and thrust generated by the Wing No. 4 is not the biggest, that is, its aerodynamic characteristics are not good; while the No. 6 forward swept wing produces the largest lift and the thrust is relatively large. The comprehensive aerodynamic characteristics are the best.

5. Conclusions

The experiment explored the aerodynamic characteristics of the flapping wing. The effects of the stiffness layout and shape layout on the aerodynamic parameters are studied. The experimental results are helpful to understand the law of flapping flight in some extent.

In order to study the effects of stiffness layout and shape layout, seven pairs of wings were made and divided into two groups. The first group consists of three pairs of wings, each of which has different stiffness; the second group consists of four wings, each of which has a different shape. Through two sets of comparative experiments, the following can be concluded:

The aerodynamic force of the flapping wing has a great relationship with the stiffness, and the flexible deformation of the flapping wing enables it to produce an effective average lift and thrust. For lift, the chord stiffness has little effect on it, and the span stiffness has a great influence on it. Within a certain range, the lift increases as the span stiffness increases. For thrust, stiffness has a greater impact on it, and its magnitude is related to both the span and the chord stiffness. When the span stiffness is small and the chord stiffness is large, the thrust is large. The effect of the chord wise stiffness on the thrust is greater than the span stiffness.

By comparing the experimental data of the four wings No.4 to No.7, it can be found that the wing No.6 with forward swept shows best aerodynamic performance. Therefore, when the wing takes a certain forward sweep angle, it is advantageous for the aerodynamic characteristics and overall performance of the wing.

Acknowledgments
This work was supported in part by the National Natural Science Foundation of China under Grant 11872314.

References

[1] Shyy, W., Aono, H., Chimakurthi, S.K., Trizila, P., Kang, C. K., Cesnik, C.E.S., Liu, H., Recent progress in flapping wing aerodynamics and aeroelasticity, Progress in Aerospace Science, Vol.46, pp.284–327, 2010.
[2] Theodorsen, T., General Theory of Aerodynamic Instability and the Mechanism of Flutter, NACA Report, Vol. 496, 1935.
[3] Osoborne, M. F. M., Aerodynamics of Flapping Flight with Application to Insets, Journal of Experimental Biology, Vol.28, pp.221-245, 1950.
[4] Heathcote S, Wang Z, Gursul I. Effect of spanwise flexibility on flapping wing propulsion, Journal of Fluids and Structures. Vol. 24, Issue.2, pp. 183-199. 2008.
[5] Mueller, T. J., Stephen, B. M., Experimental Studies of Separation on a Two-Dimensional Airfoil at Low Reynolds Numbers, AIAA Journal, Vol. 20, Issue. 4, pp. 457-463, 1982.
[6] Dickinson, M. H., Lehmann, F. O., Sane, S. P., Wing rotation and the aerodynamic basis of insect flight, Science, Vol. 284, Issue. 284, pp. 1954-1960, 1999.
[7] Lauder, G. V., Flight of the robofly, Nature, Vol. 37, Issue. 4, pp. 341-347, 2001.
[8] Usherwood, J. R., Ellington, C. P., The aerodynamics of revolving wings I. Model hawkmoth wings, Journal of Experimental Biology, Vol. 205, Issue. 11, pp. 1547-1564, 2002.
[9] Srygley, R. B., Thomas, A. L. R., Unconventional lift-generating mechanisms in free-flying butterflies, Nature, Vol. 420, Issue. 6916, pp. 660-664, 2003.
[10] Wang, H., Zeng, L., Liu, H., et al. Measuring wing kinematics, flight trajectory and body attitude during forward flight and turning manoeuvres in dragonflies, Journal of Experimental Biology, Vol. 206, Issue. 4, pp. 745-757, 2003.
[11] Selig, M. S., Guglielmo, J. J., Broeren, A. P., Experiments on Airfoils at Low Reynolds Numbers, AIAA Paper, 1996-0062.

IOP Conf. Series: Materials Science and Engineering **686** (2019) 012030 doi:10.1088/1757-899X/686/1/012030

The Microstructural Evolution of a Cold Rolled AlSi9Mg Aluminum Alloy after Semi-Solid Isothermal Treatment Process

Yongfei Wang[1,*,a], Shengdun Zhao[1,b], Yi Guo[2], Shuangjiang Li[1] and Xiaoguang Fan[3]

[1]School of Mechanical Engineering, Xi'an Jiaotong University, Xi'an 710049, China
[2]School of Energy and Power Engineering, Xi'an Jiaotong University, Xi'an 710049, China
[3]State Key Laboratory of Solidifation Processing, Northwestern Polytechnica University, Xi'an 710049, China

[*,a]yongfeio@126.com; [b]sdzhao@xjtu.edu.cn

Abstract. The key factor in semi-solid metal forming process is the preparation of semi-solid billet with fine and spherical microstructure. The recrystallization and partial melting (RAP) process consists of cold deformation and semi-solid isothermal treatment (SSIT) is a promising method, because that many alloys are always supplied in the deformed state. In this work, a cold rolled AlSi9Mg aluminum alloy was used directly to fabricate semi-solid billets. During the RAP process, samples cut from the cold rolled AlSi9Mg aluminum alloy were treated with different isothermal holding temperatures and times, and the effects of isothermal treatment parameters on the microstructures of AlSi9Mg alloy semi-solid billets were investigated. The results show that, with increasing the isothemal holding temperature and time, the average grain size of solid grains is gradually increased, the spheroidization degrees of solid grains are sharply increased and then follows by a slight improvement. The high-quality semi-solid billets can be prepared with SSIT process at 570 °C for 5~10 min.

1. Introduction

The semi-solid metal process (SSMP) is an effective near-net shape technology, which is developed by Spencer and co-workers[1-3]. The SSMP requires the microstructure to consist of fine and spherical solid grains[4-6]. To prepare ideal semi-solid billets, many methods have been developed in industry or laboratory, such as, swirled enthalpy equilibration device[7], cooling slope [8], mechanical and electromagnetic stirring [9, 10], mechanical vibration [11], strain-induced melting activation (SIMA) [12]and recrystallization and partial melting (RAP) [13, 14]. Among these methods, RAP processes are simpler because many alloys are supplied in the deformed state.

The RAP process mainly consists of cold deformation and SSIT process. Firstly, the strain energy is stored in the defromed alloy by cold deformation process. Secondly, the deformed alloy is treated with SSIT process, and then the semi-solid billet can be prepared. Additionally, the cold deformation is usually performed by cold working below recrystallization temperature, which is a critical factor in RAP process. Currently, The researchers mostly uses compression [15], extrusion [6], equal channel angular extrusion (ECAE) [16], and accumulative plastic deformation (ADP) [17] to the cold deformation step. However, less researchers pay attention to the using of cold rolling process.

In this work, a cold rolled AlSi9Mg aluminum alloy is used directly in RAP process to fabricate semi-solid billets. The samples cut from cold rolled AlSi9Mg aluminum plate were heated with different isothermal holding temperatures and times, and the effects of isothermal treatment parameters on the microstructures of AlSi9Mg alloy semi-solid billets were investigated.

2. Experimental procedure

The cold rolled AlSi9Mg aluminum alloy with a 60% reduction in room temperature is used as starting material in this work, and its chemical composition is shown in Table 1. The semi-solid temperatures of this alloy are obtained by differential scanning calorimeter (DSC), which are 555 ~ 595 °C.

Table 1. Chemical composition of the AlSi9Mg aluminum alloy (wt.%)

Si	Mg	Mn	Zn	Fe	Cu	Al
9.32	0.24	0.3	0.2	0.2	0.05	Bal.

The schematic illustration of the present RAP process is shown in Figure 1. As shown in Figure1, firstly, the cold rolled alloys were machined into small samples with a dimension of 10 mm×10 mm×6 mm; secondly, the samples were carried out with SSIT process to fabricate semi-solid billet, and the alloy after SSIT process was quenched in water to retain the semi-solid microstructure. During the SSIT process, the samples were placed into a resistance furnace, the heating rate and temperature fluctuation of the furnace were controlled within 30 °C/min and ±1 °C, respectively.

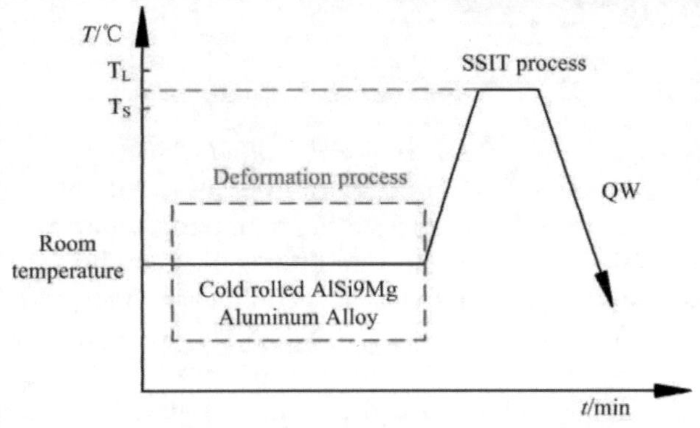

Figure 1. Schematic illustration of the present RAP process

The experimental parameters of SSIT process are shown in Table 2. The samples after SSIT process of as-received and cold rolled deformed AlSi9Mg alloys were ground and polished, and then etched with an aqueous solution of 5% HF. The microstructures and phases of samples were observed and examined by optical microscope and X-ray diffraction (XRD) device, respectively. The average grain size (*D*) and shape factor (*F*) of solid grains were measured by Image-Pro Plus 6.0 software, and they were calculated by Eqs. (1) and (2), respectively[18].

Table 2. The experimental parameters of SSIT process.

Sample	Isothermal holding temperature (°C)	Isothermal holding time (min)
As-received AlSi9Mg alloy	560, 565, 570	10
As-received AlSi9Mg alloy	570	5, 10 and 15

$$D = \frac{\sum_{N=1}^{N} \sqrt{4A/\pi}}{N} \quad (1)$$

$$F = \frac{\sum_{N=1}^{N} 4\pi A / P^2}{N} \quad (2)$$

Where *D*, *F*, *A*, *N* and *P* are average grain size, shape factor, area, the number and perimeter of grain, respectively.

3. Results and discussion

3.1. Microstructure of cold rolled AlSi9Mg alloy

Figure 2 shows the microstructure of cold rolled AlSi9Mg alloy in this work. As shown in Figure 2, the microstructure is typical of rolling alloy with fibrous grains and intermetallic particles aligned in the rolling direction. Moreover, there is no evidence of recrystallization phenomenon can be seen in Figure 2, possibly owing to the low deformation temperature of the cold rolled AlSi9Mg alloy [13].

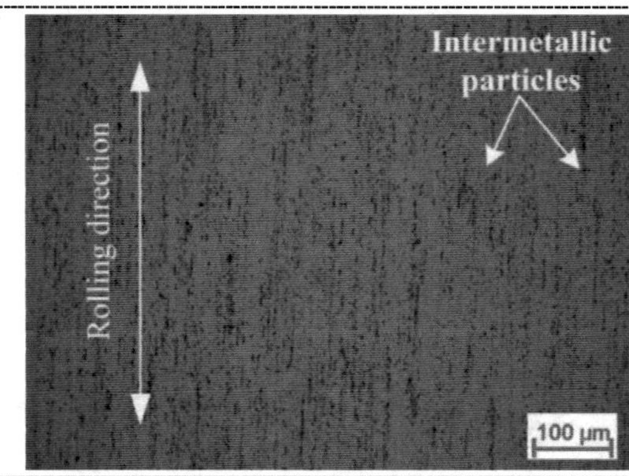

Figure 2. Microstructure of cold rolled AlSi9Mg alloy

3.2. Effect of isothermal holding temperature on the microstructure of cold rolled AlSi9Mg alloy

The microstructures of cold rolled AlSi9Mg alloy after SSIT process at different isothermal holding temperature for 10 min are shown in Figure 3~5. When the isothermal holding temperature is 560 °C (Figure 3), the microstructures mainly consist of solid grains and thin liquid films. Additionally, the sizes of solid grain are small, however, many solid grains (marked with A and B) are congregated, and the shapes of these solid grains are irregular. As shown in Figure 4, when the isothermal holding temperature increases to 565 °C, the liquid film becomes clear and thick, and the solid grains become separated and spheroidal. However, the grain size is increased and a small number of congregated solid grains (marked with A and B) can also be seen. As shown in Figure 5, when the isothermal holding temperature increases to 570 °C, the liquid film become more thicker, and the shape of solid grains become more spheroidal; however, the size of solid grains is obviously increased.

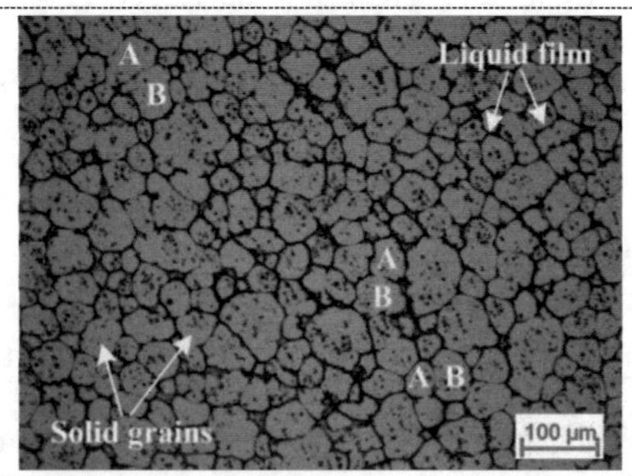

Figure 3. Microstructure of the cold rolled AlSi9Mg alloy after SSIT at 560 °C for 10 min.

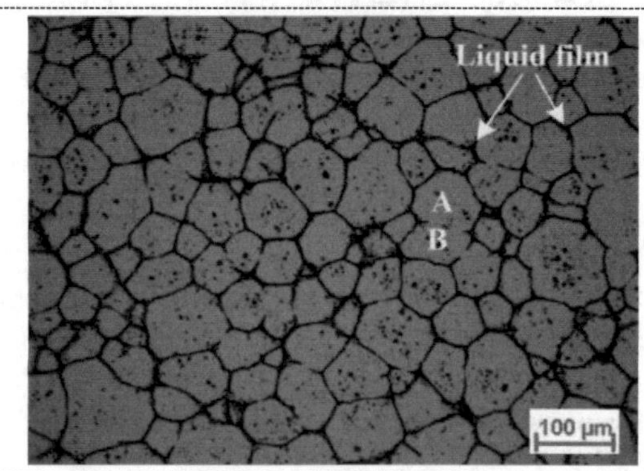

Figure 4. Microstructure of the cold rolled AlSi9Mg alloy after SSIT at 565 °C for 10 min.

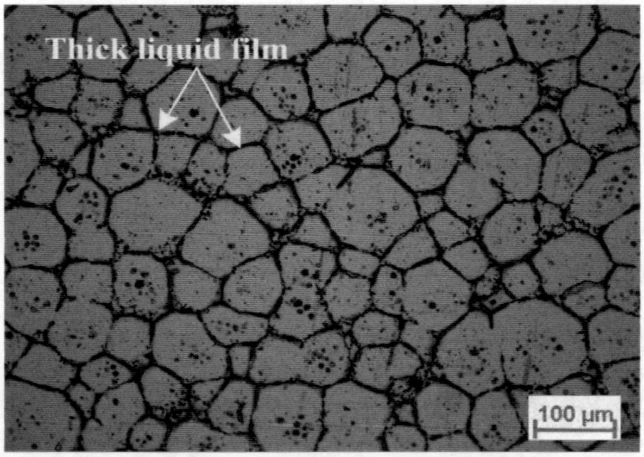

Figure 5. Microstructure of the cold rolled AlSi9Mg alloy after SSIT at 570 °C for 10 min.

3.3. Effect of isothermal holding time on the microstructure of cold rolled AlSi9Mg alloy

The microstructures of cold rolled ZL104 alloy after SSIT at 570 °C for different isothermal holding time are shown in Figure 6~8. When the isothermal holding time is 5 min, the fine and spheroidal solid grains shown in Figure 6 can be prepared, and some small liquid droplets can be seen inside the solid grains. As shown in Figure 7 and Figure 8, when the isothermal holding time is prolonged from 10 min to 15 min, the liquid film becomes more and more thicker, the solid grains is gradually bigger and bigger. However, there is no obvious improvement about the spheroidization degree of solid grains when the isothermal holding time increases from 10 min to 15 min, which can be attributed to the grain coalescence (such as solid grains C and D). Although some solid grains are relatively large and irregular resulting from grain coalescence (such as solid grains C and D), the liquid film becomes more and more thicker, and an obvious boundary penetration with liquid phase is observed as a result of high temperature (570 °C) and long time. Meanwhile, small solid grains (marked with the yellow crosses) and some large solid grains (marked with the yellow five-point stars) can be observed in Figure 6-8. Therefore, both the coalescence and Ostwald ripening may occur under these SSIT parameters, and Ostwald ripening plays an increasingly important role due to the more and more thicker liquid film. Moreover, the size of liquid droplets is increased while the number of liquid droplets is decreased.

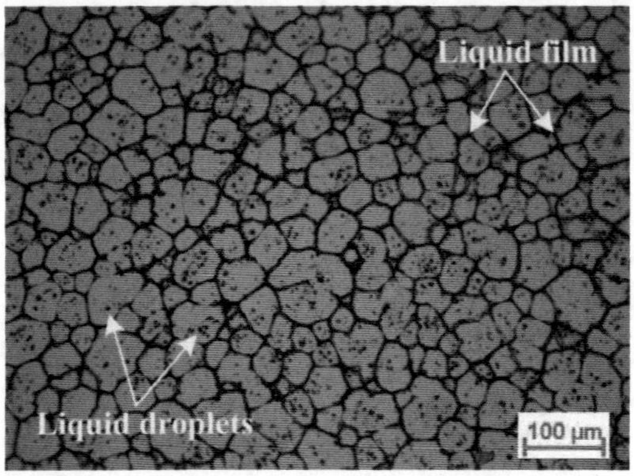

Figure 6. Microstructure of the cold rolled AlSi9Mg alloy after SSIT at 570 °C for 5 min.

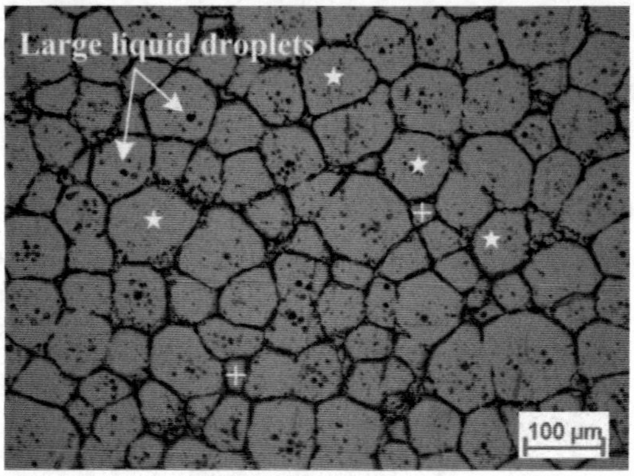

Figure 7. Microstructure of the cold rolled AlSi9Mg alloy after SSIT at 570 °C for 10 min.

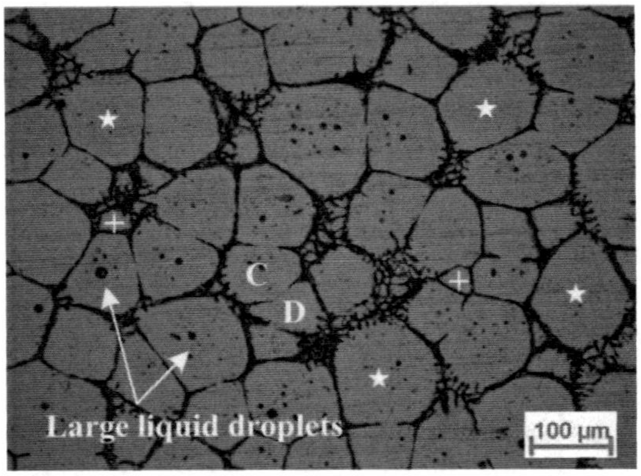

Figure 8. Microstructure of the cold rolled AlSi9Mg alloy after SSIT at 570 °C for 15 min.

3.4. Effect of isothermal treatment process parameters on the average grain size and shape factor

Figure 9 and Figure 10 show the quantitative analyses of microstructures from Figure 3~8. With the increase of isothermal holding temperature and time, the average grain size of solid grains are gradually increased. When the isothermal holding temperature increased from 560 to 565 °C and the isothermal holding time increased from 5 to 10min, the shape factor is sharply increased. When the isothermal holding temperature increased from 565 to 570 °C and the isothermal holding time increased from 10 to 15 min, the shape factor is slowly increased. Namely, the spheroidization degree of solid grains is sharply increased and then follows by a slight improvement. Additionally, it has been shown that the semi-solid billet can be suitable for SSMP when the grain size is less than 100 μm or the shape factor is more than 0.6[19]. Therefore, from the results of Figure 9 and Figure 10, the RAP process in this study is an effective method to produce ideal semi-solid billets. The optimal isothermal holding temperature and time are 570 °C and 5~10 min, respectively; the average grain size is 35.88~ 69.54 μm and the shape factor is 0.81~0.83.

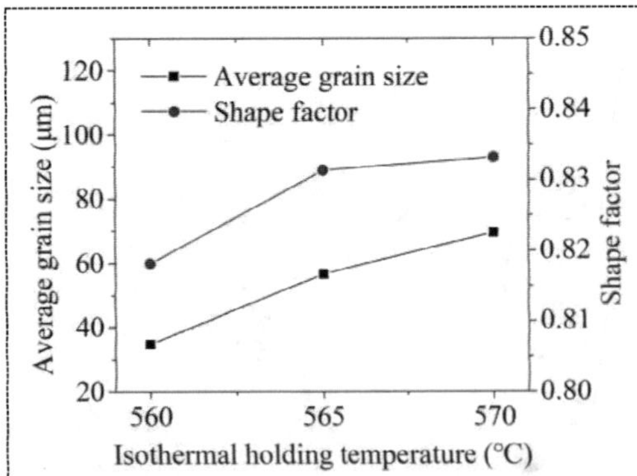

Figure 9. Variations of the average grain size and shape factor with the isothermal holding temperature.

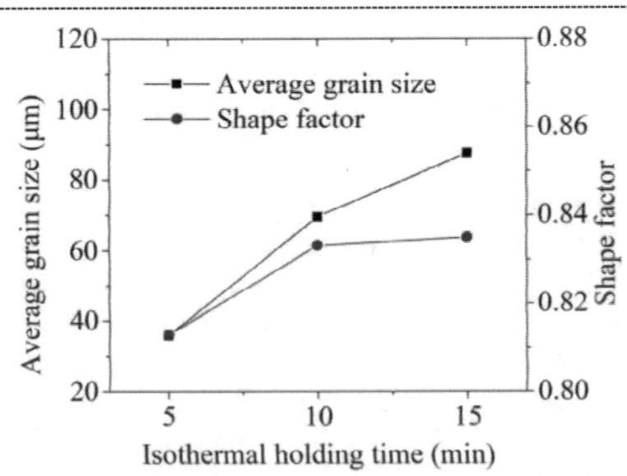

Figure 10. Variations of the average grain size and shape factor with the isothermal holding time.

4. Summary

In this work, a cold rolled AlSi9Mg aluminum alloy is used directly in RAP process to fabricate semi-solid billets, and the effects of isothermal treatment parameters on the microstructures of AlSi9Mg alloy semi-solid billets were investigated. The main conclusions are summarized as follows:

(1) With increasing the isothermal holding temperature and time, the average grain size of solid grains is gradually increased, the spheroidization degrees of solid grains are sharply increased and then follows by a slight improvement.

(2) Cold rolled AlSi9Mg aluminum alloy used directly in the RAP process is an effective method to produce ideal semi-solid billets. The optimal isothermal holding temperature and time are 570 °C and 5~10 min, respectively; the average grain size is 35.88~ 69.54 μm and the shape factor is 0.81~0.83.

Acknowledgments
This work is supported by the China Postdoctoral Science Foundation (Grant No.2018M643627), the fund of the state Key Laboratory of Solidification Processing in NPU (Grant No. SKLSP201921), the Fundamental Research Funds for the Central Universities (Grant No. XZY012019003, Grant No. XZD012019009, and XJH012019010).

References

[1] Spencer D. B., R. Mehrabian and M. C. Flemings. 1972 Metall. Mater. Trans. B, 3 (3), 1925.
[2] Wang Y, Zhao S and Zhang C. 2017 Mate. Trans. 58 (02), 176.
[3] Wang Y, Zhao S and Zhang C. 2017 J. Mater. Eng. Perform. 26 (9), 4354.
[4] Wang Y, Zhao S and Zhang C. 2017 Rare Metal Mat. Eng. 46 (12), 3875.
[5] Xu Y, Hu L, Jia J and Xu B. 2016 Mater. Charact. 118, 309.
[6] Jiang J, Wang Y, Xiao G and Nie X. 2016 J. Mater. Process. Tech. 238, 361.
[7] Ragab K A, Mohamed B, Amine B and Hatem M. 2016 Key Eng. Mater. 710, 315.
[8] Van T N., Zuhailawati H., Seman A. A., Huy T. D. and Dhindaw B. K.. 2015 J. Mater. Eng. Perform. 24 (5), 2108.
[9] Jang C. H., Jin C. K., Bolouri A. and Kang C. G. 2014. P I Mech Eng C-J Mec. 228 (3), 399.
[10] Chen Z, Lu L, Rong F, Jiang Y and Zhou R. 2015 Solid State Phenom. 217-218.
[11] Murakami Y, Kenji M and Masayuki K. 2016 Mater. Trans. 57, 168.
[12] Binesh B. and Aghaie-Khafri M. 2015 Mater. Charact. 106, 390.
[13] Wang Y, Zhao S, Zhao X and Zhao Y. 2017 J. Mech. Sci. Tech. 31 (8), 3917.
[14] Wang Y, Zhao S, Fan Shu and Zhao Y. 2017 Rare Metal Mat. Eng. 46 (10), 2900.
[15] Nayyeri M. J. and Dehghani K. 2014 J. Mater. Eng. Perform. 23 (9), 3077.
[16] Jiang J., Wang Y., Liu J., Qu J, Du Z and Luo S. 2013 T. Nonferr. Metal. Soc 23 (3), 576.
[17] Chen Q., Zhao, Z. D. Chen G. and Wang B. 2015 J. Alloy Compd. 632, 190.
[18] Wang Y, Zhao S and Zhang C.2016. Solid State Phenom. 256, 257.
[19] Mohammadi H., Ketabchi M. and Kalaki A. 2011 J. Mater. Eng. Perform. 20 (7), 1256.

IOP Conf. Series: Materials Science and Engineering **686** (2019) 012031 doi:10.1088/1757-899X/686/1/012031

A Shifting Strategy That Reduces the Number of Shifts for Urban Conditions

SHENG Jia-wei[1,*], GU Yi-hui[2] and TIAN Jin-yue[1]

[1]Jiangsu University, School of Automotive and Traffic Engineering, Zhenjiang 212013, China

[2]Shanghai automobile gear works, Shanghai, 200000, China

[*]Correspondence E-mail: 280260847@qq.com.

Abstract. Nowadays, urban conditions have become the main operating conditions for vehicles. Urban traffic congestion leads to a higher rate of vehicle idle speed and frequent shifts. This paper proposes a new shifting strategy in order to effectively improve the problem of frequent shifting, which is called flexible shift. In view of the fact that the electric vehicle cancels the clutch, shifting is completed directly by the control of the motor during the shifting process. Reducing the number of shifts can effectively reduce wear and increase the life of the transmission, which can be realized by the flexible shift. This paper uses MATLAB to simulate the shifting times of this shift strategy under FTP conditions. The simulation results show that electric vehicles with flexible shifting control strategy can significantly reduce the number of shifts, improve comfort, reduce hardware loss, and do not reduce economic efficiency.

1. Introduction

The shift schedule of the automatic transmission is essentially a control process that replaces the driver's judgment and operation with a previously established control strategy [1]. According to the number of inputs in the control process, it can be divided into single parameters, dual parameters and multi-parameter shift rules [2,3,4]. According to the goal of control, it can be divided into dynamic shift law and economical shift law[5].The general engine is expected to have better economy when it is in medium or low load, and it is desirable to have better dynamic performance in medium and high loads. Therefore, in order to balance power and economy, a combination shift control strategy is usually adopted, that is, a threshold is set [6]. When the accelerator opening is smaller than the threshold, an economical shift control strategy is adopted. When it is greater than the threshold, the dynamic shift control strategy is adopted.

However, it is often necessary to follow the vehicle due to congestion in urban roads. At this time, the vehicle speed is unstable and the accelerator pedal opening degree changes greatly. The transmission will often shift between adjacent gears, which has a great influence on noise, thermal model and vibration, resulting in poor driving performance. In addition, because electric vehicle motors can quickly adjust the torque and speed [7], it is possible to choose to cancel the clutch and complete the shift process directly through the control of the motor during the shift under the premise of improving the transmission efficiency and simplifying the transmission mechanism. However, this will result in the input shaft of the transmission being splined to the output of the motor when synchronizing and engaging the gear, causing wear to the synchronizer. Therefore, reducing the number of shifts can effectively reduce wear and increase the life of the transmission. So, this paper hopes to optimize the vehicle's shifting frequency under complex conditions by trying new control methods.

2. Traditional shift law

The central controller looks up the corresponding gear position on the shifting curve map according to the accelerator pedal opening value and the speed to determine whether a shift is needed during the running of the vehicle. Each of the upshift and downshift curves in this control strategy is fixed. The gear position is determined according to the correspondence between the vehicle speed and the accelerator opening degree. Generally, between the adjacent gears, the speed corresponding to the upshift curve of the accelerator opening is set to be higher than the speed corresponding to the downshift curve. This is to prevent frequent shifts. However, this control strategy will still cause the vehicle to enter a cyclic shift in a complex condition, especially in urban condition.

When a vehicle is driving in the urban area, it often encounters traffic jams. When the road is slightly congested, the driver needs to follow the vehicle. The driver reacts to factors such as the distance between the two vehicles and the speed of the vehicle. The vehicle operating conditions are shown in figure 1.

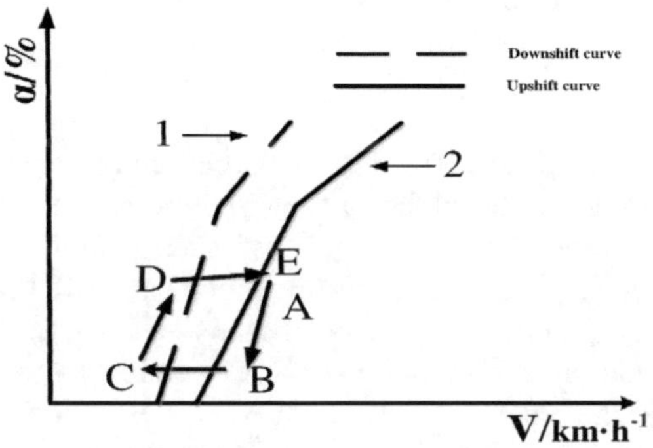

Figure 1 The diagram of transmission's running condition in traffic jam

The speed at this time may be between the downshift curve and the upshift curve. Due to the large weight of the vehicle, there will be a certain degree of lag from the change of accelerator opening to the change of speed. However, the traditional control strategy will determine the target gear based on the vehicle speed and the accelerator pedal signal value, which can easily cause frequent shifting problems. Assuming that the vehicle passes through a crowded road, the driver sees that the speed of the vehicle in front slows down and releases the accelerator pedal. At this moment, due to the inertia of the vehicle, the speed changes little in a short time. The control system judges the state of the vehicle from the A area to the B area based on the vehicle speed and accelerator pedal opening signal. Afterwards, the speed drops and enters the C area, and the transmission shifted down. Then the preceding vehicle accelerates and the inter-vehicle distance increases. In order to follow the vehicle, the driver depresses the accelerator pedal. At this time, the accelerator pedal opening increases. Because the inertia of the vehicle is greater, the vehicle speed rises more slowly. The control system judges that the vehicle enters the D area according to the corresponding relationship between the accelerator opening and the vehicle speed. Then the vehicle speed increases to enter the E area, and the transmission is upshifted. In the course of a deceleration and acceleration of the preceding vehicle, the transmission needs to complete the downshift and upshift process. When passing through a congested road section, such conditions are always encountered, which will greatly increase the wear of the internal friction parts of the transmission and result in a poor driving experience.

It can be seen that simply using the open-loop look-up table to select the target gear will cause the vehicle to shift frequently under complex conditions. When the vehicle passes through a crowded road, the accelerator pedal opening value is from large to small and then from small to large. The vehicle decelerates and then accelerates, that is, the acceleration goes from negative to positive. The changes of accelerator pedal opening, speed and acceleration are shown in figure 2.

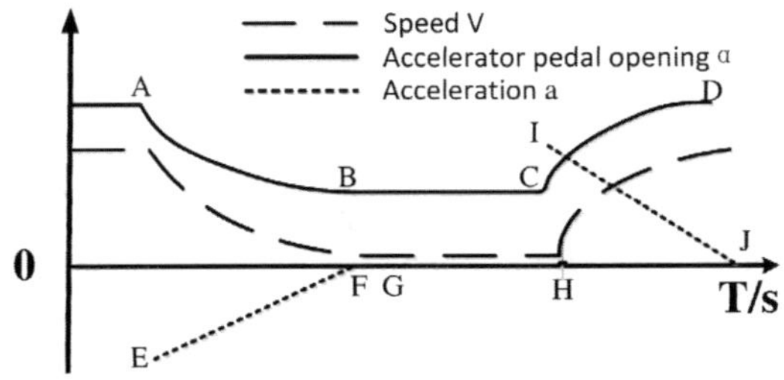

Figure 2 The curve of speed, acceleration, accelerator pedal opening of vehicle in traffic jam

Through analysis, we can see that there are three main reasons for frequent shifts under complex conditions. The first point is that the driver's control of the accelerator pedal has some lag with respect to the road conditions [8]. The second point is that due to the inertia of the vehicle, there is a certain lag in the change of the opening of the accelerator pedal to the change of the vehicle speed. The third point is that the simple open-loop look-up table shifts have drawbacks [9].

3. Flexible shifting

In view of the phenomenon of frequent traffic shifts in urban congestion, two optimization methods are proposed. The first is to start from the selection time and increase the limit of the target gear time limit for the transmission shift. Before the control system issues a shift instruction, determine whether the length of the selected target range exceeds the threshold value. If the threshold is exceeded, an instruction to perform a shift is issued. If the target gear is selected for a short time and does not exceed the threshold, the original gear is maintained. The second method is based on the change of acceleration opening value and the change of acceleration to determine the driving condition of the vehicle, and setting the threshold based on the average speed within a certain period of time. When the speed is faster, the downshift is delayed. When the speed is slow, the upshift is delayed. So it can achieve the purpose of reducing the number of shifts. The advantage of the first method is that it can be optimized with a threshold value, which has the advantage of saving time and labor. However, because only the time threshold is considered, it has a certain defect. First, the time threshold needs to be established. When the time threshold is set too large, it will affect the power and economy, and it will give the driver a bad feeling of shift delay. When the setting is too small, the effect of optimizing the number of shifts for the vehicle under complex conditions is limited. Secondly, the most direct feeling to the driver is the speed [10]. This method does not take into account the speed factor in the optimization process, so it is difficult to link the driving experience with the calibration of the threshold. If it is simply to optimize with a threshold, the freedom considered is less and the effect of optimization is limited [11].

Through the above analysis, this paper chooses to study the accelerator pedal opening value and acceleration in the case of complex conditions and correct the input speed to change the shift timing so as to achieve dynamic shifting and reduce the number of shifts. This is the proposed flexible shift.

In the design of control strategies for flexible shifts, the impact of vehicle speed needs to be considered. On the one hand, different drivers have different perceptions of speed. On the other hand, at different speeds, it is necessary to consider the effects of flexible shifts on dynamics and economy. In the shift map, there are many conditions for entering the cycle shift, and the corresponding speed range is also large. In order to ensure that the control strategy can better achieve the economical and dynamic requirements, it is necessary to set the threshold of the vehicle speed. So a satisfactory speed as input is needed. However, frequent shifts are caused by a certain lag from the driver's intention to the vehicle speed, so the current instantaneous speed is not suitable for the driver's ideal speed. The vehicle has undergone one or more changes in the speed before entering the cycle shift. When the

IOP Conf. Series: Materials Science and Engineering **686** (2019) 012031 doi:10.1088/1757-899X/686/1/012031

speed is lower than the driver's ideal speed, the driver depresses the accelerator pedal and the vehicle accelerates. When the speed is higher than the driver's ideal speed, the driver releases the accelerator pedal and the vehicle slows down. It can be seen that these changing speeds oscillate above and below the ideal speed of the driver, so the average of multiple cycle speeds before entering the cycle shift is taken as the input of the driver's ideal speed.

When the vehicle speed is lower than the threshold, the system enters the delay upshift. At this time, the vehicle is driving on the crowded road section. The driver's demand for power has reduced and TCU choose the delayed upshift control strategy. Once the vehicle accelerates and the transmission reaches the second gear, TCU will choose a delayed downshift control strategy. When the vehicle speed is faster and TCU enters the delayed shift strategy, the driver wants higher dynamics. At this time, there is a higher requirement for the output speed of the transmission. The low gear may not meet the requirements even when the motor rotates at the maximum speed. Even when the requirements are met, the motor efficiency will be lower because of the higher motor speed. Therefore, the delay downshift control strategy is entered at this time. Once the vehicle decelerates and the transmission enters the first gear, the TCU will choose to enter a delayed upshift control strategy.

When the vehicle is driving on a congested road, TCU will choose the flexible shift. When the vehicle has passed through the congested road, the control system can make a judgment and choose the normal shift control strategy. This requires a judgment mechanism. When the vehicle is working on a good road condition, the accelerator opening value should be constant or change linearly, so it is judged based on the change of the opening degree of the accelerator pedal whether it is necessary to exit the flexible shift. The optimization flow chart is shown in figure 3.

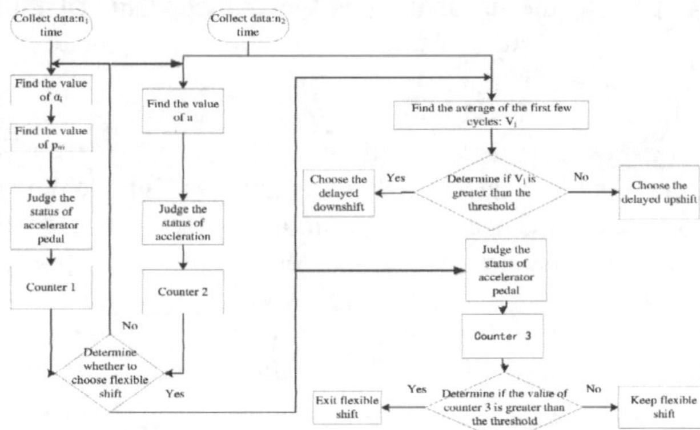

Figure 3 The optimization flow chart of flexible shift

The input parameters in the optimization method are accelerator opening value, vehicle speed and acceleration. The optimization steps are:

(1) Accelerator pedal opening, acceleration, speed sampling

Accelerator pedal sampling: Assuming that the potentiometer sampling time of the accelerator pedal is Δt_1, sampling is performed n_1 times, and the accelerator opening degree value of each sampling is $\propto_{ij}$. The pedal opening degree of the n_1 acquisition is taken as the average value of the pedal opening degree value $\propto_i$ at the current moment. The calculation is as shown in Eq.1.

$$\propto_i = \frac{\Sigma \propto_{ij}}{n_1} x \tag{1}$$

Where, $\propto_i$: the average value of accelerator pedal opening. $\propto_{ij}$: accelerator opening value for each sample.

Speed sampling: Assuming that the output shaft speed sensor sampling time is Δt_2, sampling n_2 times. The sampling rotation speed is ω_t each time. Convert the n_2 sampling rotation speed into speed and take the average as the input of the current speed V_i .The calculation is as shown in Eq.2.

$$V_i = 0.3377 \cdot \frac{\sum_0^{n_2} \omega_t \cdot r}{n_2 \cdot i_g \cdot i_0} \tag{2}$$

Where, V_i : current driver's ideal speed, km/h; ω_t :the rotation speed of transmission input shaft., RPM.

Acceleration sampling: Convert the rotation speed of n_2 samples into acceleration and take the average as input a_i of current acceleration. The calculation is as shown in Eq.3.

$$a_i = \frac{\sum_0^{n_2} \frac{d\omega_t}{dt} \cdot 2\pi \cdot r}{n_2 \cdot i_g \cdot i_0} \tag{3}$$

Where, a_i: average acceleration after sampling n_2 times.

(2) Judgment of driver's status and analysis of the vehicle's status

Select appropriate $\sigma_1$and σ_2 as threshold parameters. Assume that the previous sampled accelerator opening is $\propto_{i-1}$.Take $p_{\propto i}$ as the ratio of the current opening value to the previous opening value. When $p_{\propto i}$ is less than σ_2 and greater than σ_1, it is marked as state 1, otherwise it is marked as state 2. If it is the state 2, the counter 1 is incremented. When the state 1 occurs, the counter 1 does not change.

There are three states for setting the accelerator opening: large opening, middle opening, and small opening. State 1 includes the middle opening, which is less changeable than the original accelerator pedal opening. The state 2 includes a large opening and a small opening which change more greatly than the original accelerator opening. σ_1 is the critical value of small opening and middle opening. Therefore, the value of σ_1 must ensure that when $p_{\propto i}$ is less than σ_1, the accelerator pedal opening is at a smaller opening compared to the original. σ_2 is the critical value of middle opening and large opening. Therefore, the value of σ_2 must ensure that when $p_{\propto i}$ is greater than σ_2 , the accelerator pedal opening is at a larger opening compared to the original.

When the input current acceleration is $\propto_i$ which is less than σ_4 and greater than σ_3, it is marked as state 1, otherwise it is marked as state 2. If it is the state 2, the counter 2 is incremented. When the state 1 is present, the counter 2 is not changed. σ_3 and σ_4 are also suitable threshold parameters.

There are three states for setting the acceleration: rapid deceleration, steady, and rapid acceleration. State 1 includes the steady, relatively small change from the original speed. State 2 includes rapid acceleration and rapid deceleration, and this is a relatively large change from the original vehicle speed. σ_3 is the critical value of rapid deceleration and steady, so the value of σ_3 must ensure that if $\propto_i$ is smaller than σ_3, the vehicle has a larger deceleration trend. σ_4 is the critical value for rapid acceleration and steady, so the value of σ_4 must ensure that the vehicle has greater acceleration when $\propto_i$ is greater than σ_4.

(3) Whether to choose flexible shift

Counter 1 is reset to zero when counter 1 is in state 1 for k1 consecutive times. Counter 2 is reset to zero when counter 2 is in state 1 for k3 consecutive times. This phenomenon shows that the speed does not change too much and road conditions are good. When the counter 1 accumulates to k2 and the counter 2 accumulates to k5, it proves that the vehicle is driving on a congested road. TCU will decide to choose the flexible shift strategy.

(4) Flexible shift control

The control system calculates V_i as the ideal speed for the driver by averaging the speed of the previous cycles. The ideal vehicle speed is compared with the threshold value. When the ideal vehicle speed is high, it is hoped that the transmission gear will remain in the second gear, and the control strategy of delay downshift will be entered. Through the vehicle speed correction, the downshift curve is shifted to the left, allowing the transmission to shift at a lower speed. The revised vehicle speed calculation is shown in Eq.4.

$$V_{in} = (1+\propto \cdot \mu_1) \cdot V \tag{4}$$

Where, $\propto$: accelerator pedal opening. μ_1: compensation coefficient. The value of μ_1 is determined by the amount of leftward shift of the downshift curve.

Once the transmission is downshifted from second gear to first gear, TCU will choose the delayed upshift.

When the ideal vehicle speed is low, it is hoped that the transmission gear will remain in the first gear, and the control strategy of delay upshift will be chosen. Through the vehicle speed correction, the upshift curve is shifted to the right, allowing the transmission to shift at a higher speed. The revised vehicle speed calculation is shown in Eq.5.

$$V_{in} = (1+\propto\cdot \mu_2)\cdot V \tag{5}$$

Where,μ_2: compensation coefficient. The value of μ_2 is determined by the amount of rightward shift of the downshift curve.

Once the transmission is upshifted from first gear to second gear, TCU will choose the delayed downshift.

(5) Exit the flexible shift strategy

The calculator 3 starts counting after entering the flexible shift. Taking the pedal opening degree $\propto_i$ as an input, it is assumed that the accelerator pedal opening degree $\propto$ is maintained at the state 1. The state of the accelerator opening value is analysed by the next sampling data. If it remains in state 1, counter 3 is incremented by one, otherwise the counter is set to zero. When the counter is greater than k_6, the control system defaults that the vehicle has exited the complex condition and TCU exits the flexible shift.

4. Simulation results and analysis

Flexible shifting is mainly aimed at operating conditions in which the accelerator pedal opening degree constantly changes and the driver wants the vehicle speed to change continuously. In the FTP-75(federal test procedure) [12] operating condition, the range of vehicle speed change is large and the working conditions are more complicated. Therefore, this operating condition is selected as a simulation condition for flexible shifting, as is shown in figure 4.

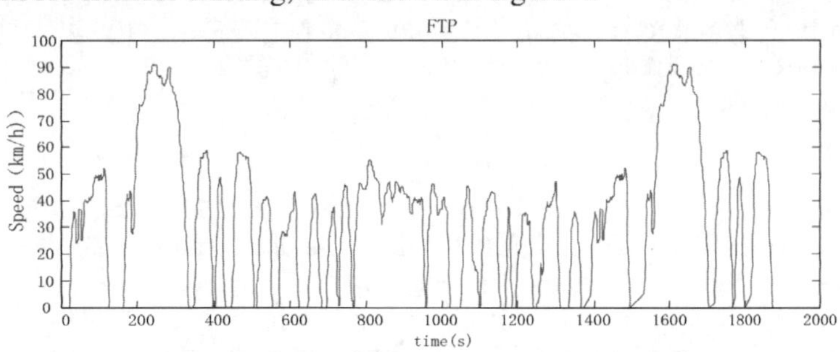

Figure 4 The speed-time curve of FTP

As can be seen from the figure, the change of vehicle speed with time is complex and close to the actual traffic conditions of the city, so it is reasonable to select this condition to test the effect of flexible shifting.

When shifting using the traditional look-up table, the shift curve during the entire FTP cycle is shown in figure 5. With flexible shifting, the shift curve for the entire FTP cycle is shown in figure 6.

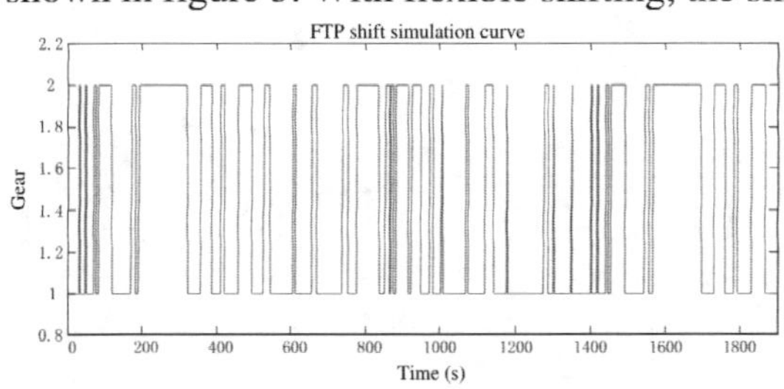

Figure 5 The shifting curve with the application of traditional look-up table

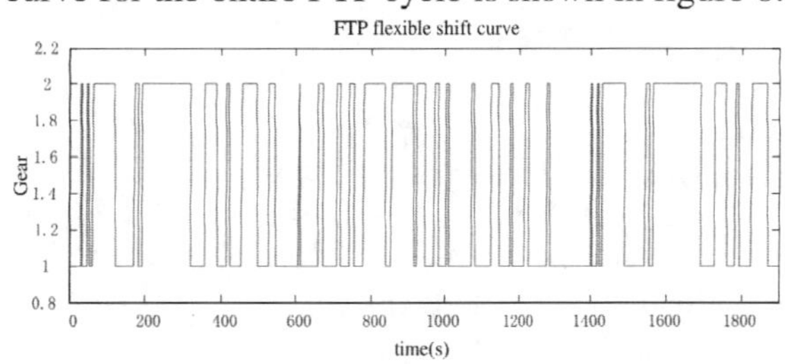

Figure 6 The shifting curve with the application of flexible shift

According to figure 5 and figure 6, it can be seen that when choosing the traditional look-up table shift strategy, the number of shifts is 74 times. With the flexible shifting control strategy, the number of shifts is 62 on the same road segment, and 12 cycles can be reduced in one cycle. Especially in the period of 600s~1000s, which is less than 7 minutes, when using the traditional look-up table shift strategy, the number of shifts is 21 times, and the number of shifts when using flexible shifts is 16 times, which can reduce 5 times of shifts. This proves that choosing flexible shifting under complex conditions can effectively reduce the number of vehicle shifts, increase the service life of the gearbox and increase comfort.

When using the traditional table look-up mode to shift gears, the battery SOC curve during the entire FTP cycle is shown in figure 7. With flexible shifting, the battery SOC curve during the entire FTP cycle is shown in figure 8.

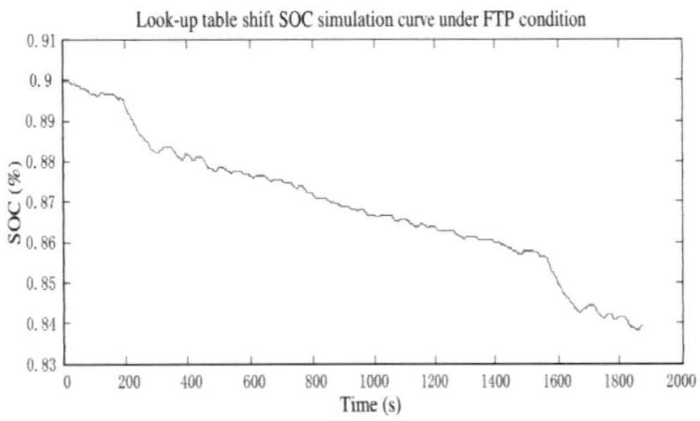

Figure 7 The SOC change curve with the application of traditional look-up table

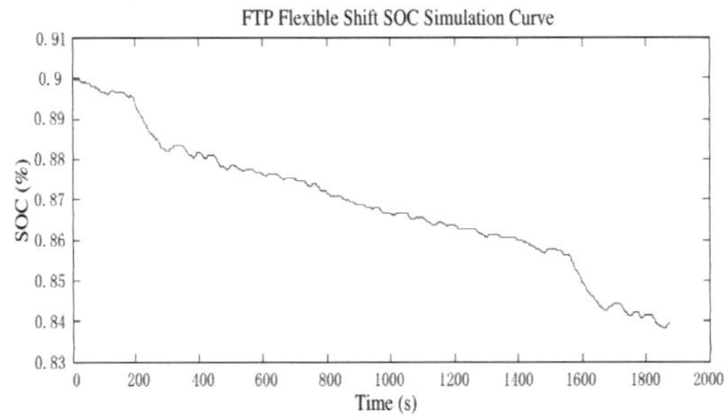

Figure 8 The SOC change curve with the application of flexible shift

As can be seen from the figure, after the flexible shift is adopted, the number of shifts of the transmission is significantly reduced, and the SOC curve is not significantly different. This shows that the flexible shift strategy can reduce the number of shifts and improve comfort while ensuring economic efficiency.

5. Conclusion

For traffic congestion in the city, this paper proposes a flexible shift strategy to solve the problem of frequent shifts. By using the vehicle speed, acceleration, and accelerator pedal opening as the basis for the system to judge whether to switch the flexible shift, a complete entry and exit flexible shifting standard and a control principle in the shifting process are proposed. FTP-75 is selected as the simulation condition for flexible shifting. The simulation model established by Matlab proves that flexible shifting significantly improves the number of shifts of the vehicle and comfort in FTP condition. At the same time, economical efficiency is not reduced. So this strategy has a good application prospect in future urban traffic.

Acknowledgments

This research was financially supported by National Key Research and Development Project (2016YFD0700402).

References

[1] Kim S., Oh J J. and Choi S B. 2018 Driveline torque estimations for a ground vehicle with dual-clutch transmission. *IEEE Transactions on Vehicular Technology,* **67(3)**, 1977-1989.

[2] Cairano S D, Doering J, Kolmanovsky I V, et al. 2014 Model Predictive Control of Engine Speed During Vehicle Deceleration, *IEEE Transactions on Control Systems Technology*, **vol.22, no.6**,pp.2205-2217.

[3] Li J. and Wei H. 2014 Coordinated control of downshift powertrain of combined clutch transmissions for electric vehicles, *Mathematical Problems in Engineering*, **no** 963725.

[4] Zhao Z , Chen H and Wang Q. 2014 Sliding mode variable structure control and real-time optimization of dry dual clutch transmission during the vehicle's launch, *Mathematical Problems in Engineering*, **no** 494731.
[5] Zhenkun D , Peng D and Wei G. 2017 Adaptive strategy for the automatic transmission shifting improvement, *Control & Decision Conference. IEEE.*
[6] Guo L, Gao B, Liu Q, Tang J and Hong C. On-line optimal control of the gearshift command for multi-speed electric vehicles, *IEEE/ASME Transactions on Mechatronics*, PP(99), 1-1.
[7] Alizadeh H V, Helwa M K and Boulet B. 2014 Constrained control of the synchromesh operating state in an electric vehicle's clutchless automated manual transmission, *Control Applications. IEEE.*
[8] Chen H Y W. , and Donmez B . 2016 What drives technology-based distractions? a structural equation model on social-psychological factors of technology-based driver distraction engagement. *Accident Analysis & Prevention*, 91, 166-174.
[9] You Y, Sun D Y , Qin D T. 2018 Shift strategy of a new continuously variable transmission based wheel loader, *Mechanism and Machine Theory*, **vol.**130, pp. 313-329.
[10] Chen S H, Pan, J. S, Lu, K, and Xu H. 2015 Driving behavior analysis of multiple information fusion based on adaboost, *Advances in Intelligent Systems and Computing*, **vol.**329, pp.277-285.
[11] Lei Y L, Liu K. Zhang Y X,et al. 2015 Adaptive Gearshift Strategy Based on Generalized Load Recognition for Automatic Transmission Vehicles, *Mathematical Problems in Engineering*, **no**.614989.
[12] Yang Zhengjun. 2014 Feasibility Study of WLTP for Road Light Vehicle Test Cycle in China. *Tsinghua University*.

AUTHOR INDEX

Institute of Physics
Dirac House, Temple Back
Bristol BS1 6BE UK

ISSN: 1757-8981
ISBN 978-1-7138-0938-8

9 781713 809388

CURRAN ASSOCIATES INC.
proceedings.com